Stigmatic Optics (Second Edition)

Online at: https://doi.org/10.1088/978-0-7503-6423-2

IOP Series in Emerging Technologies in Optics and Photonics

Series Editor

R Barry Johnson, a Senior Research Professor at Alabama A&M University, has been involved for over 50 years in lens design, optical systems design, electro-optical systems engineering, and photonics. He has been a faculty member at three academic institutions engaged in optics education and research, has been employed by a number of companies, and has provided consulting services.

Dr Johnson is an IOP Fellow, an SPIE Fellow and Life Member, an OSA Fellow, and was the 1987 President of SPIE. He serves on the editorial board of *Infrared Physics & Technology* and *Advances in Optical Technologies*. Dr Johnson has been awarded many patents, has published numerous papers and several books and book chapters, and was awarded the 2012 OSA/SPIE Joseph W Goodman Book Writing Award for Lens Design Fundamentals (second edition). He is a perennial co-chair of the annual SPIE Current Developments in Lens Design and Optical Engineering Conference.

Foreword

Until the 1960s the field of optics was primarily concentrated in the classical areas of photography, cameras, binoculars, telescopes, spectrometers, colorimeters, radiometers, etc. In the late 1960s optics began to blossom with the advent of new types of infrared detector, liquid crystal display (LCDs), light emitting diode (LEDs), charge coupled device (CCDs), laser, holography, and fiber optics along with new optical materials, advances in optical and mechanical fabrication, new optical design programs, and many more technologies. With the development of the LED, LCD, CCD, and other electro-optical devices, the term 'photonics' came into vogue in the 1980s to describe the science of using light in the development of new technologies and the operation of a myriad of applications. Today optics and photonics are truly pervasive throughout society and new technologies are continuing to emerge. The objective of this series is to provide students, researchers, and those who enjoy self-education with a wide-ranging collection of books, each of which focuses on a topic relevant to the technologies and applications of optics and photonics. These books will provide knowledge to prepare the reader to be better able to participate in these exciting areas now and in the future. The title of this series is *Emerging Technologies in Optics and Photonics*, in which 'emerging' is taken to mean 'coming into existence', 'coming into maturity', and 'coming into prominence'. IOP Publishing and I hope that you will find this series of significant value to you and your career.

A full list of titles published in this series can be found here: https://iopscience.iop.org/bookListInfo/emerging-technologies-in-optics-and-photonics.

Stigmatic Optics (Second Edition)

Rafael G González-Acuña
Huawei Technologies, Tampere Camera Lab, Korkeakoulunkatu 7, 33720 Tampere, Finland

Héctor A Chaparro-Romo
Independent researcher Av. Eugenio Garza Sada 2501 Sur, Mexico

IOP Publishing, Bristol, UK

ISBN 978-0-7503-6423-2 (ebook)
ISBN 978-0-7503-6426-3 (print)
ISBN 978-0-7503-6425-6 (myPrint)
ISBN 978-0-7503-6424-9 (mobi)

DOI 10.1088/978-0-7503-6423-2

Version: 20240701

IOP ebooks

British Library Cataloguing-in-Publication Data: A catalogue record for this book is available from the British Library.

Published by IOP Publishing, wholly owned by The Institute of Physics, London

IOP Publishing, No.2 The Distillery, Glassfields, Avon Street, Bristol, BS2 0GR, UK

US Office: IOP Publishing, Inc., 190 North Independence Mall West, Suite 601, Philadelphia, PA 19106, USA

To God

Contents

Preface

This treatise focuses on a particular concept of geometric optics, stigmatism. Stigmatism refers to the image property of an optical system that focuses a single point source in object space at a single point in image space. Two of these points are called a stigmatic pair of the optical system.

The treatise starts from the foundations of stigmatism: Maxwell's equations, the eikonal equation, the ray equation, the Fermat principle and Snell's law. Then we study the most important stigmatic optical systems without any paraxial or third-order approximation or without any optimization process. These systems are the conical mirrors, the Cartesian ovals and the stigmatic lenses.

Conical mirrors are studied step by step with clear examples.

In the case of the Cartesian ovals, two paradigms are studied. In the first, the Cartesian ovals are obtained by means of a polynomial series and the second by means of a general equation of the Cartesian oval.

For stigmatic lenses, the case is studied when the two refractive surfaces are Cartesian ovals. Then the general equation for stigmatic lenses is obtained.

Finally, the similarities of optical systems and their nature are studied.

It is recommended to read this treatise in order.

The Authors

Acknowledgments

Acknowledgments of Rafael G González-Acuña

Almighty God, creator of the Universe, this book aims to honor your glory. Thank you Lord for giving me the intelligence, the desire, the faith and the means to complete it. Lord, please help me to be a faithful servant of your will. Help me to deserve the promises of your son Jesus Christ, give me a sign, and show me the way ....

I want to thank my family.

I want to thank my mother Carmen Leticia Acuña Medellín, my father Rogelio González Cantú and my brothers Rolando and Rogelio.

Héctor A Chaparro-Romo, comrade, once again! We did it comrade! Time rewards!

To Professor R. Barry Johnson for your support, patience and fruitful discussions.

To Ashley Gasque and Robert Trevelyan for all the support!

I would also like to thank Yoshio Catillejos, Israel Meléndez, Gustavo Medina, Daniel Lomas, Roberto Martinez, César López, Roberto Vera, Ileana Paulette Zambrano, Miguel Rojas, Joel Guerra, Homero Pérez, Mauricio Arroyo, Adad Yepiz, Erick Patiño, Alberto Silva, Michelle C. Rocha, Adrian Lozano, Luis Garza, Eero Salmenin, Mina Yasmina, Guillermo Tamargo Martin Tamargo, Onna Niklas-Salminen, Dr. Benjamin Perez-Garcia, Dr. Maximino Avendaño, Dr. Genaro Zavala, Dr. Carlos Hinojosa, Dr. Francisco Cuevas, Dr. Rafael Torres, Dr. Blas Manuel Rodríguez Lara, Professor Reinhard Klette, Professor Alois Herkommer, Professor Russell Chipman, Professor Simon Thibault and Stephen Wolfram.

To Professor Julio C Gutiérrez-Vega and Dr Bernardino Barrientos García, my advisors during the PhD and master's degree, respectively.

I would also like to thank several institutions: Institute of Physics, Conacyt, Instituto Tecnológico y de Estudios Superiores de Monterrey, Wolfram Research, Centro de Investigaciones en Optica A.C, Auckland University of Technology, Institut für Technische Optik at Universität Stuttgart, Universidad Abierta y a Distancia de México, Universidad Yachay Tech and Oxford Immune Algorithmics.

Thank you very much!

Acknowledgments of Héctor A Chaparro-Romo

I want to thank the great spirit, this work is a sample of its existence.

I thank the IOP team for their support in continuing to spread the rigorous theory of optical design and showing that it is always possible to find better results.

I thank my family, my parents Maria and José, and my brother Eleazar, for their unconditional support and wise advice.

I thank all the readers who seek to find an understanding of this theory in this book; their time and interest are a motivation to continue being happy for the effort they now have in their hands.

I thank all my friends, new and old, because without them life would be very different.

Finally, I want to thank all those women who have shared with me part of their path, their experience and their feelings, always present.

About the authors

Rafael G González-Acuña

Rafael G González-Acuña studied industrial physics engineering at the Tecnológico de Monterrey and studied a master's degree in optomechatronics at Centro de investigaciones en Óptica, A.C. He is currently studying his PhD at the Tecnológico de Monterrey. His doctoral thesis focuses on the design of free spherical aberration lenses. He is co-author of the solution to the problem of designing bi-aspheric singlet lenses free of spherical aberration. He is co-author of the book *Analytical Lens Design.*

Héctor A Chaparro-Romo

Héctor A Chaparro-Romo obtained his bachelor's in electronic engineering at the Universidad Autónoma Metropolitana, and is currently studying for a degree in Economics at the Universidad Nacional Autónoma de México. He is co-author of the solution to the problem of spherical aberration in lens design, he is also co-author of several peer-reviewed scientific articles and a book on analytical design of optical systems. Héctor is an independent and self-employed researcher in his home office, where he fully focuses his capabilities in the complex field of computer networks and the Internet. As a pioneer, his main goal is to develop http://www.biaspheric.com as a reference portal for all those who want to learn deeply about the theory of optical design working from the rigorous analytical paradigm. He is co-author of the book *Analytical Lens Design.*

IOP Publishing

Stigmatic Optics (Second Edition)

Rafael G González-Acuña and Héctor A Chaparro-Romo

Chapter 1

The Maxwell equations

In this chapter, we give a brief review of the Maxwell equations for electromagnetic theory. After a concise explication, we obtain the step by step electromagnetic wave equation. Maxwell equations are the fundamental basis for optical theory, and therefore to the stigmatic optics discipline.

1.1 Introduction

In this chapter, we are going to review Maxwell's equations. The main goal is to get the wave equation. From the wave equation, we can get the eikonal equation. From the eikonal equation, we can derive the concept of ray and set the bases of geometrical optics.

The general idea of this book is to take Maxwell's equations as axioms and their implications as theorems. In this language, the wave equation would be a theorem, a direct consequence of Maxwell's equations. The equation of the eikonal, under one approximation, is an implication of the wave equation and from it we develop the theory of stigmatism.

The purest and most exquisite branch of geometric optics is stigmatic optics, the branch to which this book owes its name.

1.2 Lorentz force

Let's start with the definition of the electric field. An electric field can be described as a vector field in which a point electric charge of value q suffers the effects of an electrical force $\vec{\mathbf{F}}$ given by the following equation:

$$\vec{\mathbf{F}} = q\vec{\mathbf{E}}, \tag{1.1}$$

where $\vec{\mathbf{E}}$ is the electric field. Electric fields can originate from both electrical charges and variable magnetic fields.

doi:10.1088/978-0-7503-6423-2ch1

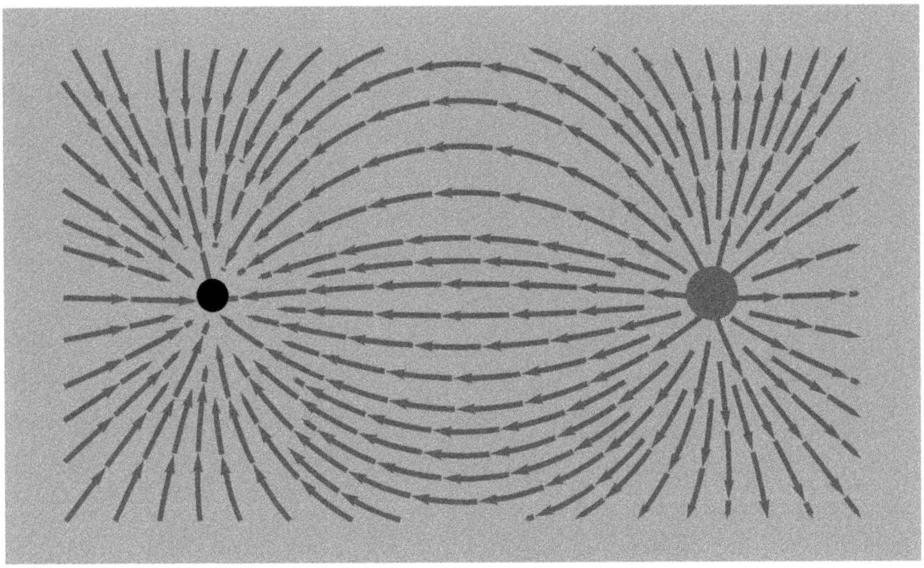

Figure 1.1. Charges with a different sign are attracted.

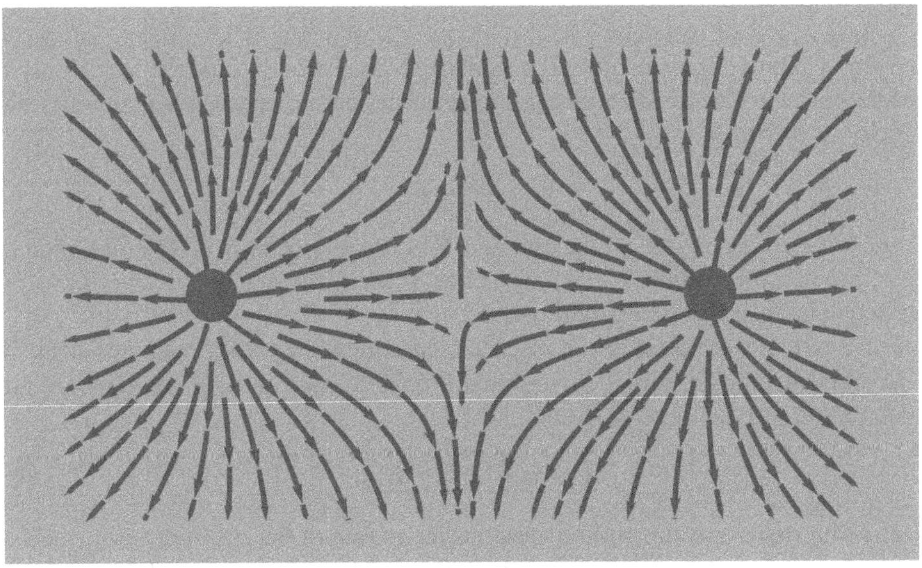

Figure 1.2. Positive charges repel each other.

Electric fields can be positive or negative. Positive if they are generated by positive charges, and negative if they are generated by negative charges. Charges with a different sign are attracted and with similar charge repel each other. Since the electric field is a vector space it can be represented as vectors, thus it is usually represented as vector lines. The lines emerge from positive charges and end in negative charges, as can be seen in figures 1.1–1.3.

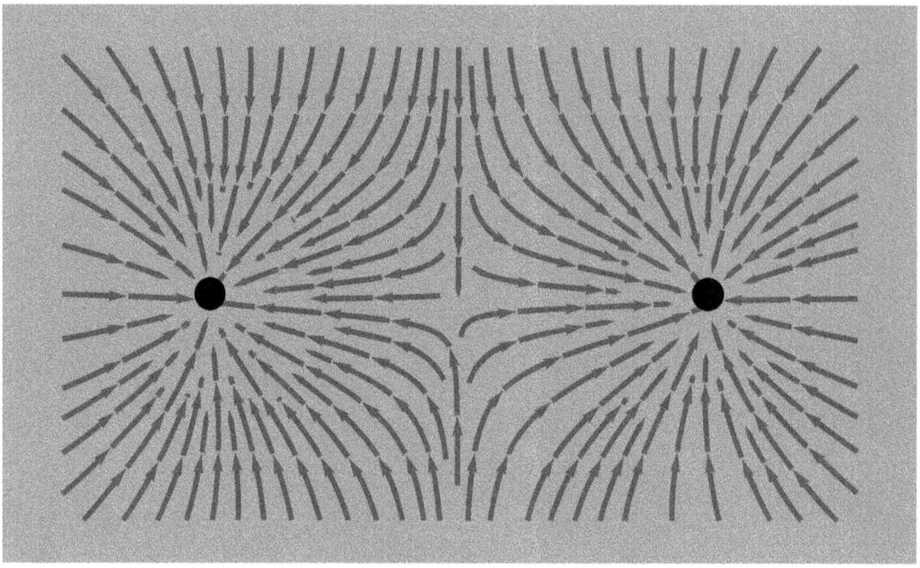

Figure 1.3. Negative charges repel each other.

A magnetic field is a vector field that specifies the magnetic influence of electric charges in relative movement and magnetized materials. A charge that is moving parallel to a current of other charges experiences a force perpendicular to its own velocity described by

$$\vec{\mathbf{F}} = q\vec{\mathbf{v}} \times \vec{\mathbf{B}}, \tag{1.2}$$

where $\vec{\mathbf{v}}$ is the velocity of the charge and $\vec{\mathbf{B}}$ is the magnetic field. Please notice the cross product in equation (1.2) describes that if the charge is moving along the magnetic field $\vec{\mathbf{B}}$ its force will be zero.

For a particle subjected to an electric field combined with a magnetic field, the total electromagnetic force or Lorentz force on that particle is given by the combination of equations (1.1) and (1.2),

$$\boxed{\vec{\mathbf{F}} = q(\vec{\mathbf{E}} + \vec{\mathbf{v}} \times \vec{\mathbf{B}}).} \tag{1.3}$$

The Maxwell equations entirely describe the nature of the electromagnetic fields $\vec{\mathbf{E}}$ and $\vec{\mathbf{B}}$. In the following sections, we are going to describe them briefly.

1.3 Electric flux

An initial concept needed to enter Maxwell's equations entirely is electric flux. The electric flux, or electrostatic flux, is a scalar quantity that expresses a measure of the electric field that passes through a defined surface, or expressed in another way, is the measure of the number of electric field lines that penetrate a surface.

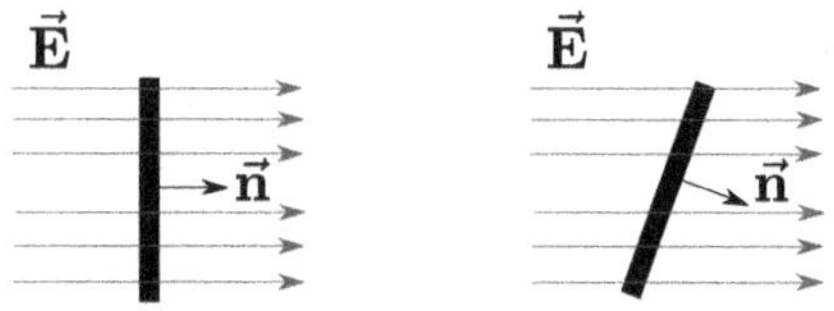

Figure 1.4. Flux of an electric field through a surface. On the right, the normal vector $\vec{\mathbf{n}}$ of the surface is parallel to the electric field $\vec{\mathbf{E}}$. On the left, there is inclination on the surface. Thus, there is an angle between $\vec{\mathbf{n}}$ and $\vec{\mathbf{E}}$.

The portion of electric flux $d\Phi_{\vec{\mathbf{E}}}$ through an infinitesimal area da is given by

$$d\Phi_{\vec{\mathbf{E}}} = \vec{\mathbf{E}} \cdot \vec{\mathbf{n}}\, da. \tag{1.4}$$

The electric field $\vec{\mathbf{E}}$ is multiplied by the component of the area perpendicular to the field. $\vec{\mathbf{n}}$ is the normal unit vector of the infinitesimal area da.

The electric flux through a surface S is therefore expressed by the surface integral,

$$\Phi_{\vec{\mathbf{E}}} = \int_S \vec{\mathbf{E}} \cdot \vec{\mathbf{n}}\, da, \tag{1.5}$$

where $\vec{\mathbf{E}}$ is the electric field and $\vec{\mathbf{n}}\, da$ is the differential surface vector that corresponds to each infinitesimal element of the entire surface S (see figure 1.4).

1.4 The Gauss law

We start with Gauss's law. Although there are many ways to express this law and notation differs, the integral form of the Gauss law is customarily given by the following expression

$$\boxed{\oint_S \vec{\mathbf{E}} \cdot \vec{\mathbf{n}}\, da = \frac{q_{\text{in}}}{\varepsilon_0},} \tag{1.6}$$

where $\vec{\mathbf{n}}$ is the normal unit vector of the closed surface S, q_{in} is the charge inside the closed surface S and ε_0 is a constant called the permittivity of free space. In the International System of Units, where force is in newtons (N), distance in meters (m) and charge in coulombs (C),

$$\varepsilon_0 = 8.85 \times 10^{-12} \frac{\mathrm{C}^2}{\mathrm{N} \cdot \mathrm{m}^2}. \tag{1.7}$$

First, let's pay attention to the left side of equation (1.6). The left side of this equation is the mathematical representation of the electric flux—the number of electric field lines—crossing into a closed surface S. On the right side, the total amount of charge contained within that surface divided by a constant is called the permittivity of free space. Therefore, what Gauss's law tells us is that an electric charge produces an electric field, and the flux of that field passing through any closed surface is proportional to the total charge inside the closed surface.

Let's assume that you have a closed surface S, where the shape and size of S are arbitrary. If there is no charge inside S, then the electric flux is zero. If there is a positive charge inside S, then the electric flux through the surface is positive. But, if you added an equal amount of negative charge, thus the total amount of charge inside S is zero, then the electric flux again is zero.

There is another way to express Gauss's law using the divergence theorem. The form is the following expression

$$\boxed{\nabla \cdot \vec{\mathbf{E}} = \frac{\rho}{\varepsilon_0},} \tag{1.8}$$

where ρ is the density of charge inside and ∇ is the nabla operator,

$$\boxed{\nabla = \frac{\partial}{\partial x}\vec{\mathbf{i}} + \frac{\partial}{\partial y}\vec{\mathbf{j}} + \frac{\partial}{\partial z}\vec{\mathbf{k}}}, \tag{1.9}$$

$\nabla \cdot \vec{\mathbf{E}}$ is the divergence of the field $\vec{\mathbf{E}}$. The divergence of the vector field is a scalar computation that indicates the tendency of the field to flow away from a point. Hence, Gauss's law tells us that the divergence of the field $\vec{\mathbf{E}}$ is the density of charge divided by the permittivity of free space.

Here, we limit ourselves to present this form of the Gauss law because the derivation of the divergence theorem is beyond the scope of this book. For a more detailed analysis of the divergence theorem, the reader is invited to read the references presented in the bibliography of this chapter.

1.5 The Gauss law for magnetism

The Gauss law for magnetism has the same structure as the Gauss law of the previous section, with the condition that there are no magnetic charges.

So, we start with the definition of magnetic flux through a surface S,

$$\Phi_{\vec{\mathbf{B}}} = \int_S \vec{\mathbf{B}} \cdot \vec{\mathbf{n}}\, da, \tag{1.10}$$

the magnetic field, $\vec{\mathbf{B}}$, multiplied by the component of the area perpendicular to the field, where $\vec{\mathbf{n}}$ is the unit normal vector of infinitesimal area da.

Therefore, over a closed surface S, Gauss's law of magnetism is given by

$$\boxed{\oint_S \vec{\mathbf{B}} \cdot \vec{\mathbf{n}}\, da = 0}. \tag{1.11}$$

As we mentioned in the introduction, there are no magnetic charges. What Gauss's law of magnetism tells us is that the total magnetic flux passing through any closed surface is zero.

Taking the knowledge acquired in the last section, we can obtain the vector form of Gauss's law of magnetism, hence,

$$\boxed{\nabla \cdot \vec{\mathbf{B}} = 0.} \tag{1.12}$$

This happens as expected because there are no magnetic charges. Therefore, the density of magnetic charge is zero.

1.6 Faraday's law

Faraday's electromagnetic induction law establishes that the electromotive force induced in a closed circuit is directly proportional to the speed with which the magnetic flux passing through any surface with the circuit as edge changes in time. Thus,

$$\boxed{\oint_c \vec{\mathbf{E}} \cdot d\vec{\mathbf{l}} = -\frac{d}{dt}\int_S \vec{\mathbf{B}} \cdot \vec{\mathbf{n}}\, da,} \tag{1.13}$$

where $\vec{\mathbf{E}}$ is the electric field, $d\vec{\mathbf{l}}$ is the infinitesimal element of the length of the circuit represented by contour C, $\vec{\mathbf{B}}$ is the magnetic field and S is an arbitrary surface, for which the edge is C. The right-hand rule gives the directions of contour C and $\vec{\mathbf{n}}da$.

The electromotive force or induced voltage (represented by emf) is any cause capable of maintaining a potential difference between two points in an open circuit or of producing an electric current in a closed circuit.

According to Stokes' theorem, the differential form of Faraday's law is generally written as

$$\boxed{\vec{\nabla} \times \vec{\mathbf{E}} = -\frac{\partial \vec{\mathbf{B}}}{\partial t},} \tag{1.14}$$

where $\vec{\nabla} \times \vec{\mathbf{E}}$ is the curl of the electric field $\vec{\mathbf{E}}$. The curl operates on a vector field and provides a vector result that designates the tendency of the field to circulate around a point and the direction of the axis of greatest circulation.

What the differential form of Faraday's law tells us is that a circulating electric field is produced by a magnetic field that changes with time.

1.7 Ampère's law

Ampère's law, also called the Ampère–Maxwell law, is generally written in its integral form as

$$\boxed{\oint_c \vec{\mathbf{B}} \cdot d\vec{\mathbf{l}} = \mu_0\left(\mathbf{I}_{\text{enc}} + \varepsilon_0\frac{d}{dt}\int_S \vec{\mathbf{E}} \cdot \vec{\mathbf{n}}\, da\right).} \tag{1.15}$$

The left side of equation (1.15) tells us about the circulation of the magnetic field around a closed path C. On the right side, we have two elements that produce the magnetic field. The first one is a steady current given by $\mathbf{I}_{\text{enc}}$. The other one is the change in time of the electric flux through a surface bounded by C.

Please notice in equation (1.15) that the factor μ_0 is a constant called the magnetic permeability of free space. In the International System of Units, where force is in newtons (N) and current in amperes (A),

$$\mu_0 = 1.256\,637\,0614 \times 10^{-6}\frac{\text{N}}{\text{A}^2}. \tag{1.16}$$

What equation (1.15) tells us is that an electric current or a changing electric flux through a surface produces a circulating magnetic field around any path that bounds that surface.

Now, due to Stokes' theorem, we can express the Ampère–Maxwell law as its differential form,

$$\boxed{\nabla \times \vec{\mathbf{B}} = \mu_0\left(\vec{\mathbf{J}} + \varepsilon_0 \frac{\partial \vec{\mathbf{E}}}{\partial t}\right).} \tag{1.17}$$

The left side of equation (1.15) is the circulating magnetic field. On the right side are the sources of the circulating magnetic field. Notice that in the first term on the right side of equation (1.15), $\vec{\mathbf{J}}$ is the current density vector. The second term, on the right side of the mentioned equation, is the rate of change of the electric field with time.

Therefore, what the Ampère–Maxwell law in its differential form tells us is that a circulating magnetic field is produced by an electric current and by an electric field that changes with time.

1.8 The wave equation

We have only briefly reviewed the Maxwell equations, but enough to obtain the wave equation. So we recall the set of Maxwell equations, equations (1.8), (1.12) (1.14) and (1.17), as equation (1.18),

$$\begin{cases} \nabla \cdot \vec{\mathbf{E}} = \dfrac{\rho}{\varepsilon_0}, \\ \nabla \cdot \vec{\mathbf{B}} = 0, \\ \nabla \times \vec{\mathbf{E}} = -\dfrac{\partial \vec{\mathbf{B}}}{\partial t}, \\ \nabla \times \vec{\mathbf{B}} = \mu_0\left(\vec{\mathbf{J}} + \varepsilon_0 \dfrac{\partial \vec{\mathbf{E}}}{\partial t}\right). \end{cases} \tag{1.18}$$

If we apply the curl on Faraday's law, equation (1.14), we get

$$\nabla \times (\nabla \times \vec{\mathbf{E}}) = \nabla \times \left(-\frac{\partial \vec{\mathbf{B}}}{\partial t}\right) = -\frac{\partial \nabla \times \vec{\mathbf{B}}}{\partial t}. \tag{1.19}$$

Now, we use the vector calculus identity expressed in equation (1.20),

$$\nabla \times (\nabla \times \vec{\mathbf{A}}) = \nabla(\nabla \cdot \vec{\mathbf{A}}) - \nabla^2 \vec{\mathbf{A}}. \tag{1.20}$$

Using the identity of equation (1.20), in equation (1.19),

$$\nabla \times (\nabla \times \vec{\mathbf{E}}) = \nabla(\nabla \cdot \vec{\mathbf{E}}) - \nabla^2\vec{\mathbf{E}} = -\frac{\partial \nabla \times \vec{\mathbf{B}}}{\partial t}. \tag{1.21}$$

The last term of equation (1.21) can be replaced using Ampère's law, equation (1.17),

$$\nabla \times \vec{\mathbf{B}} = \mu_0\left(\vec{\mathbf{J}} + \varepsilon_0\frac{\partial \vec{\mathbf{E}}}{\partial t}\right), \tag{1.17}$$

thus, replacing equation (1.17) in equation (1.21),

$$\nabla \times (\nabla \times \vec{\mathbf{E}}) = \nabla(\nabla \cdot \vec{\mathbf{E}}) - \nabla^2\vec{\mathbf{E}} = -\frac{\partial\left[\mu_0\left(\vec{\mathbf{J}} + \varepsilon_0\frac{\partial \vec{\mathbf{E}}}{\partial t}\right)\right]}{\partial t}. \tag{1.22}$$

Now, with Gauss's law we can reformulate equation (1.22); thus, let's recall Gauss's law,

$$\nabla \cdot \vec{\mathbf{E}} = \frac{\rho}{\varepsilon_0}, \tag{1.9}$$

replacing equation (1.8) in equation (1.22),

$$\nabla\left(\frac{\rho}{\varepsilon_0}\right) - \nabla^2\vec{\mathbf{E}} = -\mu_0\frac{\partial \vec{\mathbf{J}}}{\partial t} - \mu_0\varepsilon_0\frac{\partial^2 \vec{\mathbf{E}}}{\partial t^2}. \tag{1.23}$$

If we are working in a charge- and current-free region, $\rho = 0$ and $\vec{\mathbf{J}} = 0$, then equation (1.23) becomes

$$\boxed{\nabla^2\vec{\mathbf{E}} = \mu_0\varepsilon_0\frac{\partial^2 \vec{\mathbf{E}}}{\partial t^2}.} \tag{1.24}$$

Equation (1.24) is called the wave equation. The wave equation is an important second-order linear partial differential equation that describes the propagation of a variety of waves, such as sound waves, light waves and waves in water. It is important in various fields such as acoustics, electromagnetism, quantum mechanics and fluid dynamics. The form represented in equation (1.24) is for an electric field $\vec{\mathbf{E}}$. But if we apply $\vec{\nabla}\times$ to Ampère's law, and then apply a similar procedure, we can get

$$\boxed{\nabla^2\vec{\mathbf{B}} = \mu_0\varepsilon_0\frac{\partial^2 \vec{\mathbf{B}}}{\partial t^2}.} \tag{1.25}$$

Equation (1.25) is the wave equation for $\vec{\mathbf{B}}$. The process to get equation (1.25) is left as an exercise for the reader.

1.9 The speed and propagation of light

The speed that a wave propagates is presented in the standard form of the wave equation given by the following equation

$$\nabla^2\vec{\mathbf{A}} = \frac{1}{v^2}\frac{\partial^2\vec{\mathbf{A}}}{\partial t^2}, \tag{1.26}$$

$\vec{\mathbf{A}}$ is the wave and v_2 is the speed of the wave. Therefore, we can take the speed of the waves of equation (1.24) and (1.25) as

$$\frac{1}{v^2} = \mu_0\varepsilon_0, \tag{1.27}$$

and evaluating the values of the constants we get the speed of the electromagnetic waves in the vacuum, which is given as

$$v = 2,9979 \times 10^8\, m/s. \tag{1.28}$$

Normally, the above quantity is called the speed of light and is denoted by the letter c, thus we get

$$\boxed{c \equiv 2,9979 \times 10^8\, m/s.} \tag{1.29}$$

Throughout the book we are going to use the notation presented in equation (1.29).

1.10 Refraction index

When light is not traveling in a vacuum, its speed is modified by the refraction index of the medium in which it is traveling.

The refraction index of a homogeneous medium is constant, and it is a dimensionless number that describes how fast light travels through the medium. The refraction index is defined by

$$\boxed{n = \frac{c}{v},} \tag{1.30}$$

where c is the speed of light in vacuum and v is the phase velocity of light in the medium.

1.11 Electromagnetic waves

It is time to find the solution of the wave equation of the electric field. We start by recalling equation (1.24),

$$\nabla^2\vec{\mathbf{E}} = \mu_0\varepsilon_0\frac{\partial^2\vec{\mathbf{E}}}{\partial t^2}. \tag{1.24}$$

The method that we are going to apply to solve it is called variable separation. The variable separation method refers to a procedure to find a particular complete solution for specific problems involving partial differential equations such as a series, whose terms are the product of functions that have *separate variables*. It is one of the most productive methods in mathematical physics to find solutions to physical problems using partial differential equations.

Therefore, we are going to take the electric field as a function by the multiplication of a part that only depends on the position vector $\vec{\mathbf{r}}$ and another function that solely depends on scalar time t.

$$\vec{\mathbf{E}}(\vec{\mathbf{r}}, t) = \vec{\mathbf{R}}(\vec{\mathbf{r}})T(t), \tag{1.31}$$

replacing equation (1.31) in the wave equation, equation (1.24),

$$\nabla^2[\vec{\mathbf{R}}(\vec{\mathbf{r}})T(t)] - \mu_0\varepsilon_0\frac{\partial^2[\vec{\mathbf{R}}(\vec{\mathbf{r}})T(t)]}{\partial t^2} = 0. \tag{1.32}$$

Let us pause for a moment and set a simple notation to clear the procedure,

$$\vec{\mathbf{R}}(\vec{\mathbf{r}}) = \vec{\mathbf{R}}, \quad T(t) = T. \tag{1.33}$$

Notice that ∇ only affects $\vec{\mathbf{R}}(\vec{\mathbf{r}})$ and $\frac{\partial}{\partial t}$ only affects $T(t)$, therefore equation (1.31) becomes

$$\frac{\nabla^2(\vec{\mathbf{R}})}{\vec{\mathbf{R}}} - \mu_0\varepsilon_0\frac{1}{T}\frac{\partial^2 T}{\partial t^2} = 0. \tag{1.34}$$

Since both terms are being differentiated by different independent variables, they must be equal to the same constant,

$$\frac{\nabla^2(\vec{\mathbf{R}})}{\vec{\mathbf{R}}} = \mu_0\varepsilon_0\frac{1}{T}\frac{\partial^2 T}{\partial t^2} = -k^2. \tag{1.35}$$

This leads us to a time-dependent differential equation given by

$$\frac{\partial^2 T}{\partial t^2} + \frac{k^2 T}{\mu_0\varepsilon_0} = 0, \tag{1.36}$$

where,

$$\frac{1}{c^2} = \mu_0\varepsilon_0. \tag{1.37}$$

The solution of equation (1.36) is given by

$$T(t) = e^{-ikct}. \tag{1.38}$$

Therefore, we can write the electric field as

$$\vec{\mathbf{E}}(\vec{\mathbf{r}}, t) = \vec{\mathbf{R}}(\vec{\mathbf{r}})e^{-ikct}, \tag{1.39}$$

where the spatial part of equation (1.39) is given by

$$\boxed{\nabla^2\vec{\mathbf{R}}(\vec{\mathbf{r}}) + k^2\vec{\mathbf{R}}(\vec{\mathbf{r}}) = 0.} \tag{1.40}$$

Equation (1.40) is the Helmholtz equation. The Helmholtz equation is a partial differential equation which is the special part of the wave equation.

Where k is the wave number defined as the number of radians per unit distance, we more often use

$$k = \frac{2\pi}{\lambda} \tag{1.41}$$

and λ is the wavelength. Thus, from the definition of the refraction index,

$$n = \frac{c}{v} \Rightarrow w = kv', \tag{1.42}$$

where w is the angular frequency. When light enters a medium, the wavelength is modified as

$$\lambda = \frac{\lambda_c}{n}{}', \tag{1.43}$$

where λ_c is the wavelength inside the medium. The wave number is also modified as

$$k = k_c n', \tag{1.44}$$

where k_c is the wave number inside the medium.

Recapitulating, we are working with a three-dimensional vector field such that

$$\vec{\mathbf{E}}(\vec{\mathbf{r}}, t) = \vec{\mathbf{E}}(x, y, z, t) = E_x(x, y, z, t)\hat{\mathbf{e}}_x + E_y(x, y, z, t)\hat{\mathbf{e}}_y + E_z(x, y, z, t)\hat{\mathbf{e}}_z. \tag{1.45}$$

Therefore, solving the Helmholtz equation is not the simplest task. It depends on the dimension of the wave, if it is in one, two or three dimensions and in the coordinate system. The coordinate system affects ∇. ∇ has different expressions for different coordinate systems. In the next subsection, we explore some particular solutions of the Helmholtz equation.

1.11.1 One-dimensional way

We start with the one-dimensional case. We assume that there is electric field in the z direction:

$$\vec{\mathbf{E}}(\vec{\mathbf{r}}, t) = \vec{\mathbf{E}}_z(z)e^{-ikct}. \tag{1.46}$$

Replacing equation (1.46) in the Helmholtz equation,

$$\frac{\partial^2 \vec{\mathbf{E}}_z}{\partial z^2} + k^2\vec{\mathbf{E}}_z = 0. \tag{1.47}$$

The solution of the equation is given by

$$\vec{\mathbf{E}}_z = Ae^{ikz} + Be^{-ikz}. \tag{1.48}$$

Therefore, $\vec{\mathbf{E}}(\vec{\mathbf{r}}, t)$ is given by

$$\vec{\mathbf{E}}(\vec{\mathbf{r}}, t) = (Ae^{ikz} + Be^{-ikz})e^{-ikvt} = Ae^{i(kz-kvt)} + Be^{i(kz+kvt)}; \tag{1.49}$$

simplifying,

$$\vec{\mathbf{E}}(\vec{\mathbf{r}}, t) = A\cos(kz - kvt) + B\cos(kz + kvt). \tag{1.50}$$

The minus sign of the first cosine means that the wave is traveling to the right of positive z. The plus sign of the second cosine implies that the wave is moving to the right of negative z.

Also, notice that the time is being multiplied by the angular frequency,

$$w = kv. \tag{1.51}$$

Notice that v is the speed inside a medium and c is the speed in vacuum. Therefore, if we pick the wave that is traveling to positive z, $\vec{\mathbf{E}}(\vec{\mathbf{r}}, t)$ is given by

$$\vec{\mathbf{E}}(\vec{\mathbf{r}}, t) = E_0\cos(kz - wt), \tag{1.52}$$

where we set $A \to E_0$. The last expression is the equation of the plane wave.

1.11.2 Spherical coordinates

Now, let's pay attention to Helmholtz equation spherical coordinates. First, let's recall the Helmholtz equation,

$$\nabla^2\vec{\mathbf{E}}(\vec{\mathbf{r}}) + k^2\vec{\mathbf{E}}(\vec{\mathbf{r}}) = 0. \tag{1.53}$$

Hence, in spherical coordinates, the Helmholtz equation is expressed as

$$\frac{1}{r^2}\frac{\partial}{\partial r}\left(r^2\frac{\partial\vec{\mathbf{E}}(\vec{\mathbf{r}})}{\partial r}\right) + k^2\vec{\mathbf{E}}(\vec{\mathbf{r}}) = 0. \tag{1.54}$$

To solve it let's assume that $\vec{\mathbf{E}}(\vec{\mathbf{r}})$ has the following form,

$$\vec{\mathbf{E}}(\vec{\mathbf{r}}) = \frac{E'(r)}{r}\hat{\mathbf{e}}_r, \tag{1.55}$$

where $E'(r)$ is a function of r. Thus, replacing equation (1.55) in equation (1.54),

$$\frac{1}{r^2}\frac{\partial}{\partial r}\left[r^2\left(-\frac{E'}{r^2} + r\frac{\partial E'}{\partial r}\right)\right] + k^2\frac{E'}{r} = 0, \tag{1.56}$$

expanding,

$$\frac{1}{r^2}\left(-\frac{\partial E'}{\partial r} + r\frac{\partial^2 E'}{\partial r^2} + \frac{\partial E'}{\partial r}\right) + k^2\frac{E'}{r} = 0, \tag{1.57}$$

simplifying,

$$\frac{1}{r}\frac{\partial^2 E'}{\partial r^2} + k^2\frac{E'}{r} = 0. \tag{1.58}$$

Notice that it is the same equation that we solved in the previous section. Therefore, we can conclude that the solution of the wave equation in spherical coordinates has the following form

$$\vec{\mathbf{E}}(\vec{\mathbf{r}}, t) = \frac{E_0}{r}\cos(kz - wt). \tag{1.59}$$

Notice that the amplitude of the wave decreases as $r \to \infty$.

As an exercise for the reader, please study the Helmholtz equation in cylindrical coordinates. The Helmholtz equation in cylindrical coordinates is the following expression

$$\frac{1}{r}\frac{\partial}{\partial r}\left[r\frac{\partial \vec{\mathbf{E}}(\vec{\mathbf{r}})}{\partial r}\right] + k^2\vec{\mathbf{E}}(\vec{\mathbf{r}}) = 0. \tag{1.60}$$

1.12 End notes

In this chapter, we briefly studied Maxwell's equations, from which we found the wave equation. From the latter, we obtained some particular solutions and their spatial part—the Helmholtz equation.

The Helmholtz equation will be of great help to us because through it we will find the eikonal equation and, in turn, the ray equation. These last equations lay the foundations of geometric optics. Geometric optics is the playing field of stigmatism, which will be presented in depth in chapter 5 and the following chapters; stigmatic systems will be studied in detail.

References

Arfken G B and Weber H J 1999 *Mathematical Methods for Physicists* (New York: Academic)

Boas M L 2006 *Mathematical Methods in the Physical Sciences* (New York: Wiley)

Born M and Wolf E 2013 *Principles of Optics: Electromagnetic Theory of Propagation, Interference and Diffraction of Light* (Amsterdam: Elsevier)

Buchdahl H A 1993 *An Introduction to Hamiltonian Optics* (North Chelmsford, MA: Courier Corporation)

Campbell L 1882 *The Life of James Clerk Maxwell* (London: Macmillan)

Fleisch D 2008 *A Student's Guide to Maxwell's equations* (Cambridge: Cambridge University Press)

Goodman J W 2005 *Introduction to Fourier Optics* (Greenwood Village, CO: Roberts and Company Publishers)

Griffiths D J 2005 *Introduction to Electrodynamics* (Cambridge: Cambridge University Press)

Hecht E 1974 *Schaum's Outline of Optics* (New York: McGraw-Hill)

Hecht E 2012 *Optics* (India: Pearson Education)

Jackson J D 1999 *Classical Electrodynamics* (Hoboken, NJ: Wiley)

Lakshminarayanan V, Ghatak A and Thyagarajan K 2002 *Lagrangian Optics* (Berlin: Springer)

Lax M, Louisell W H and McKnight W B 1975 From Maxwell to paraxial wave optics *Phys. Rev.* **A11** 1365

Luneburg R K 1964 *Mathematical Theory of Optics* (Berkeley, CA: University of California Press)

Mahon B 2004 *The Man Who Changed Everything: The Life of James Clerk Maxwell* (New York: Wiley)

Maxwell J C 1990 *The Scientific Letters and Papers of James Clerk Maxwell: 1846–1862* **vol 1** (Cambridge: Cambridge University Press)

Perko L 2013 *Differential equations and Dynamical Systems* **vol 7** (Berlin: Springer)

Ronchi V and Barocas V 1970 *The Nature of Light: An Historical Survey* (Cambridge, MA: Harvard University Press) p 12+ 288

Zill D G 2016 *Differential equations with Boundary-Value Problems* (Toronto: Nelson Education)

IOP Publishing

Stigmatic Optics (Second Edition)

Rafael G González-Acuña and Héctor A Chaparro-Romo

Chapter 2

The eikonal equation

In this chapter, we study the fundamental equation of geometric optics, the eikonal equation. The eikonal equation is derived from the wave equation under the circumstances studied in this chapter. Also in this chapter, we study the direct implications of the eikonal, such as the ray equation.

2.1 From the wave equation, through the Helmholtz equation, to end with the eikonal equation

In chapter 1 of this book we had three objectives. The first objective was to make a summary of Maxwell's equations. The second was to see that when we do not have charges or currents, we can derive the wave equation from Maxwell's equations, and finally we wanted to study some particular solutions to the wave equation. To do this we had to obtain, as an intermediate step, the Helmholtz equation, which is the spatial part of the wave equation.

In this chapter, we ask ourselves: if Maxwell's equations are our axioms, they are true, and we do not doubt that they are; then, if the wave equation and Helmholtz's equation imply that Maxwell's equations are true, they are our theorems; what else is true? What more effects do the wave equation and the Helmholtz equation have and in which circumstances? The goal of this chapter is to obtain from the wave equation and through the Helmholtz equation, the equation of the eikonal equation. The eikonal equation is an implication of the equations already mentioned in very particular but useful circumstances that define the geometric optics branch of optics.

The eikonal equation is a partial differential equation with non-linearity found in wave propagation. It is an approximated version of the wave equation.

We will study the spatial part of the wave equation, the Helmholtz equation, and given certain approximations. We will arrive at the eikonal equation.

Later, we will see that from the eikonal equation, we can formulate the equation of the ray and with it the concept of ray—a fundamental concept in stigmatic optics.

doi:10.1088/978-0-7503-6423-2ch2

It is imperative to understand these steps because they are the pillars of geometric optics and stigmatic optics.

We can see the eikonal equation as the intermediary of two worlds, two paradigms, wave optics and ray optics, commonly called geometric optics.

Geometrical optics, also called ray optics, is one of the oldest sciences. It is the study of light through geometry. Euclid proposed the first premise used in geometrical optics in his book *Optics*. The premise is that light propagates in straight lines, making a connection between the paths of light and the geometry proposed in his masterpiece, *Elements*. Euclid's *Elements* focuses on the nature of straight lines. *Elements* is a mathematical and geometric treatise that consists of thirteen books. In the books lies the foundation of Euclidean geometry.

In geometric optics, it is common to describe the propagation of light in terms of rays. The ray is a useful abstraction to approximate the paths along which light propagates under certain circumstances. Euclid's premise on the propagation of light in straight lines is only valid when light travels around a homogeneous medium. The refraction index of a homogeneous medium is constant, and it is a dimensionless number that describes how fast light travels through the medium. All these properties of light described by geometrical optics are inherent in the eikonal equation.

Geometric optics does not deal with specific optical effects, such as diffraction and interference. Diffraction is the term that comes from the Latin 'diffractus', which means broken. The etymology refers to the phenomenon by which a wave can contour an obstacle in its propagation, moving away from the behavior of rectilinear rays.

However, geometric optics deals with reflection and refraction, which are the two main phenomena studied in geometrical optics. The reflection of light is the phenomenon of returning the rays of light that fall on the surface of an object; commonly, these objects are mirrors. Refraction is the redirection of a light ray when it enters a medium where its speed is different. Refraction occurs when the refractive index of the input medium is different from the refractive index of the output medium.

Ignoring diffraction and interference is useful in practice when the wavelength is small compared to the size of the optical elements in which the light interacts. This paradigm is particularly useful for describing geometric aspects of images including mirrors, lenses, axicons and other optical devices. Geometrical optics is the leading theory used in optical design. The art of designing cameras, microscopes, telescopes and optical systems in general is called optical design.

Geometric optics is a very challenging science, where most of the results are particular cases obtained using sophisticated optimization algorithms, for example, the design of a specific camera or telescope. The difficulty of geometrical optics occurs because the equations that model the light rays and their interactions with different media can be quite long, complicated and, in many cases, non-linear. Non-linearity has led optical design to resemble art rather than a scientific discipline. The non-linearity presented in geometric optics is also due to the eikonal equation since, as we mentioned before, it is a partial differential equation with non-linearity found in wave propagation.

2.2 The eikonal equation

We start our journey with a wave such that no term-dependent term t is presented. This wave stands for the following equation

$$\vec{\mathbf{E}}(\vec{\mathbf{r}}) = \vec{\mathbf{A}}(\vec{\mathbf{r}})e^{ik_c g(\vec{\mathbf{r}})}, \tag{2.1}$$

where $\vec{\mathbf{A}}(\vec{\mathbf{r}})$ is the amplitude of the wave and it is multiplied by an exponential part, something that should not be surprising according to the results obtained in the previous chapter. This exponential part has, as an exponent, the product of the complex number i, the wave number k_c and a function $g(\vec{\mathbf{r}})$ that depends on the vector r, how the waveform of equation (2.1), $g(\vec{\mathbf{r}})$, is proportional to wavefront.

Hence, k_c is given by the following expression

$$k_c = \frac{2\pi}{\lambda_c}, \tag{2.2}$$

where λ_c is the wavelength inside a homogeneous medium.

Until now, we have not discussed much about $\vec{\mathbf{A}}(\vec{\mathbf{r}})$ and $g(\vec{\mathbf{r}})$; the idea is to study how $\vec{\mathbf{A}}(\vec{\mathbf{r}})$ and $g(\vec{\mathbf{r}})$ should be so that equation (2.1) will be a solution to the Helmholtz equation. So, do not be troubled by what $\vec{\mathbf{A}}(\vec{\mathbf{r}})$ and $g(\vec{\mathbf{r}})$ are, we will only assume that they are functions that have continuous derivatives with respect to x, y and z.

The first step in studying the nature of amplitude $\vec{\mathbf{A}}(\vec{\mathbf{r}})$ and the function to the wavefront proportional $g(\vec{\mathbf{r}})$ is to recall the Helmholtz equation

$$\vec{\nabla}^2\vec{\mathbf{E}}(\vec{\mathbf{r}}) + k^2\vec{\mathbf{E}}(\vec{\mathbf{r}}) = 0. \tag{2.3}$$

In it, we will insert equation (2.1) and note that the procedure is simple but long so we will do it in steps. First, we will focus purely on the Laplacian that is present in the Helmholtz equation. We begin with the derivative of $\vec{\mathbf{E}}(\vec{\mathbf{r}})$, with respect to x, we have

$$\frac{\partial\vec{\mathbf{E}}(\vec{\mathbf{r}})}{\partial x} = ik_c\vec{\mathbf{A}}(\vec{\mathbf{r}})e^{ik_c g(\vec{\mathbf{r}})}\frac{\partial g(\vec{\mathbf{r}})}{\partial x} + \frac{\partial\vec{\mathbf{A}}(\vec{\mathbf{r}})}{\partial x}e^{ik_c g(\vec{\mathbf{r}})}. \tag{2.4}$$

Now, for the derivative with respect to y, we have

$$\frac{\partial\vec{\mathbf{E}}(\vec{\mathbf{r}})}{\partial y} = ik_c\vec{\mathbf{A}}(\vec{\mathbf{r}})e^{ik_c g(\vec{\mathbf{r}})}\frac{\partial g(\vec{\mathbf{r}})}{\partial y} + \frac{\partial\vec{\mathbf{A}}(\vec{\mathbf{r}})}{\partial y}e^{ik_c g(\vec{\mathbf{r}})}, \tag{2.5}$$

and finally, for the derivative with respect to z, we have the same pattern as expected,

$$\frac{\partial\vec{\mathbf{E}}(\vec{\mathbf{r}})}{\partial z} = ik_c\vec{\mathbf{A}}(\vec{\mathbf{r}})e^{ik_c g(\vec{\mathbf{r}})}\frac{\partial g(\vec{\mathbf{r}})}{\partial z} + \frac{\partial\vec{\mathbf{A}}(\vec{\mathbf{r}})}{\partial z}e^{ik_c g(\vec{\mathbf{r}})}. \tag{2.6}$$

The next step is to derive E, again with respect to x, y, z, for which we will derive equations (2.4)–(2.6) with respect to x, y, z, respectively. We start with the derivative of equation (2.4) regarding x,

$$\begin{aligned}\frac{\partial^2\vec{\mathbf{E}}(\vec{\mathbf{r}})}{\partial x^2} &= \frac{\partial^2\vec{\mathbf{A}}(\vec{\mathbf{r}})}{\partial x^2}e^{ik_cg(\vec{\mathbf{r}})} - k_c^2\vec{\mathbf{A}}(\vec{\mathbf{r}})e^{ik_cg(\vec{\mathbf{r}})}\left[\frac{\partial g(\vec{\mathbf{r}})}{\partial x}\right]^2 \\ &+ i\left[k_c\vec{\mathbf{A}}(\vec{\mathbf{r}})e^{ik_cg(\vec{\mathbf{r}})}\frac{\partial^2 g(\vec{\mathbf{r}})}{\partial x^2} + 2k_ce^{ik_cg(\vec{\mathbf{r}})}\frac{\partial\vec{\mathbf{A}}(\vec{\mathbf{r}})}{\partial x}\frac{\partial g(\vec{\mathbf{r}})}{\partial x}\right],\end{aligned} \tag{2.7}$$

then, the derivative of equation (2.5) regarding y,

$$\begin{aligned}\frac{\partial^2\vec{\mathbf{E}}(\vec{\mathbf{r}})}{\partial y^2} &= \frac{\partial^2\vec{\mathbf{A}}(\vec{\mathbf{r}})}{\partial y^2}e^{ik_cg(\vec{\mathbf{r}})} - k_c^2\vec{\mathbf{A}}(\vec{\mathbf{r}})e^{ik_cg(\vec{\mathbf{r}})}\left[\frac{\partial g(\vec{\mathbf{r}})}{\partial y}\right]^2 \\ &+ i\left[k_c\vec{\mathbf{A}}(\vec{\mathbf{r}})e^{ik_cg(\vec{\mathbf{r}})}\frac{\partial^2 g(\vec{\mathbf{r}})}{\partial y^2} + 2k_ce^{ik_cg(\vec{\mathbf{r}})}\frac{\partial\vec{\mathbf{A}}(\vec{\mathbf{r}})}{\partial y}\frac{\partial g(\vec{\mathbf{r}})}{\partial y}\right],\end{aligned} \tag{2.8}$$

finally, the derivative of equation (2.6) regarding z has the same pattern,

$$\begin{aligned}\frac{\partial^2\vec{\mathbf{E}}(\vec{\mathbf{r}})}{\partial z^2} &= \frac{\partial^2\vec{\mathbf{A}}(\vec{\mathbf{r}})}{\partial z^2}e^{ik_cg(\vec{\mathbf{r}})} - k_c^2\vec{\mathbf{A}}(\vec{\mathbf{r}})e^{ik_cg(\vec{\mathbf{r}})}\left[\frac{\partial g(\vec{\mathbf{r}})}{\partial z}\right]^2 \\ &+ i\left[k_c\vec{\mathbf{A}}(\vec{\mathbf{r}})e^{ik_cg(\vec{\mathbf{r}})}\frac{\partial^2 g(\vec{\mathbf{r}})}{\partial z^2} + 2k_ce^{ik_cg(\vec{\mathbf{r}})}\frac{\partial\vec{\mathbf{A}}(\vec{\mathbf{r}})}{\partial z}\frac{\partial g(\vec{\mathbf{r}})}{\partial z}\right].\end{aligned} \tag{2.9}$$

Now, we need to insert equations (2.7), (2.8) and (2.9) in equation (2.3). This step is crucial so please notice that we separate the real and imaginary components from equations (2.7), (2.8) and (2.9). Notice that the first real terms in equations (2.7), (2.8) and (2.9) are proportional to the Laplacian of $\vec{\mathbf{A}}(\vec{\mathbf{r}})$.

$$\frac{\partial^2\vec{\mathbf{A}}(\vec{\mathbf{r}})}{\partial x^2}e^{ik_cg(\vec{\mathbf{r}})} + \frac{\partial^2\vec{\mathbf{A}}(\vec{\mathbf{r}})}{\partial y^2}e^{ik_cg(\vec{\mathbf{r}})} + \frac{\partial^2\vec{\mathbf{A}}(\vec{\mathbf{r}})}{\partial z^2}e^{ik_cg(\vec{\mathbf{r}})} = \frac{\nabla^2\vec{\mathbf{A}}(\vec{\mathbf{r}})}{\vec{\mathbf{A}}(\vec{\mathbf{r}})}\vec{\mathbf{E}}(\vec{\mathbf{r}}). \tag{2.10}$$

The second real terms of the aforementioned equations are

$$\begin{aligned}&- k_c^2\vec{\mathbf{A}}(\vec{\mathbf{r}})e^{ik_cg(\vec{\mathbf{r}})}\left[\frac{\partial g(\vec{\mathbf{r}})}{\partial x}\right]^2 - k_c^2\vec{\mathbf{A}}(\vec{\mathbf{r}})e^{ik_cg(\vec{\mathbf{r}})}\left[\frac{\partial g(\vec{\mathbf{r}})}{\partial y}\right]^2 \\ &- k_c^2\vec{\mathbf{A}}(\vec{\mathbf{r}})e^{ik_cg(\vec{\mathbf{r}})}\left[\frac{\partial g(\vec{\mathbf{r}})}{\partial z}\right]^2 = -k_c^2\vec{\mathbf{E}}(\vec{\mathbf{r}})|\nabla g(\vec{\mathbf{r}})|^2.\end{aligned} \tag{2.11}$$

Applying the same methodology to the imaginary terms and pulling it all together in equation (2.3) we obtain the following non-easy solving expression that involves $\vec{\mathbf{A}}(\vec{\mathbf{r}})$ and $g(\vec{\mathbf{r}})$,

$$\left[\frac{\nabla^2\vec{\mathbf{A}}(\vec{\mathbf{r}})}{\vec{\mathbf{A}}(\vec{\mathbf{r}})} - k_c^2|\nabla g(\vec{\mathbf{r}})|^2 + k_c^2 n^2 + i\left(k_c \nabla^2 g(\vec{\mathbf{r}}) + 2k_c \frac{\nabla\vec{\mathbf{A}}(\vec{\mathbf{r}}) \cdot \nabla g(\vec{\mathbf{r}})}{\vec{\mathbf{A}}(\vec{\mathbf{r}})}\right)\right]\vec{\mathbf{E}}(\vec{\mathbf{r}}) = 0. \quad (2.12)$$

From the last expression we eliminate the common factor of $\vec{\mathbf{E}}(\vec{\mathbf{r}})$, which inside the bracket must be zero. Since it is a complex number it should be that the real part is zero as well as the imaginary part. Let's focus on the real part of equation (2.12):

$$\frac{\nabla^2\vec{\mathbf{A}}(\vec{\mathbf{r}})}{\vec{\mathbf{A}}(\vec{\mathbf{r}})} - k_c^2|\nabla g(\vec{\mathbf{r}})|^2 + k_c^2 n^2 = 0. \quad (2.13)$$

Equation (2.13) is still too complicated to obtain a general solution for $\vec{\mathbf{A}}(\vec{\mathbf{r}})$ and $g(\vec{\mathbf{r}})$. If we look at the imaginary part of the scenario, it is even worse. Therefore, we need to sacrifice either $\vec{\mathbf{A}}(\vec{\mathbf{r}})$ or $g(\vec{\mathbf{r}})$ in order to only study the consequence of the other. If we eliminate $g(\vec{\mathbf{r}})$, we will have an answer similar to the one obtained in chapter 1. So let's explore the consequence of eliminating $\vec{\mathbf{A}}(\vec{\mathbf{r}})$. We can't just eliminate $\vec{\mathbf{A}}(\vec{\mathbf{r}})$ without justification since it is the amplitude, but if we observe if $\vec{\mathbf{A}}(\vec{\mathbf{r}})$ does not changes a lot in respect to $\vec{\mathbf{r}}$, then the second derivative of $\vec{\mathbf{A}}(\vec{\mathbf{r}})$ with respect to $\vec{\mathbf{r}}$ will be very small in comparison to $\vec{\mathbf{A}}(\vec{\mathbf{r}})$, which leads to

$$\vec{\mathbf{A}}(\vec{\mathbf{r}}) > > \nabla^2\vec{\mathbf{A}}(\vec{\mathbf{r}}), \quad \frac{\nabla^2\vec{\mathbf{A}}(\vec{\mathbf{r}})}{\vec{\mathbf{A}}(\vec{\mathbf{r}})} \to 0. \quad (2.14)$$

This approximation is called slowly varying envelope approximation (SVEA), or sometimes also called slowly varying amplitude approximation (SVAA). SVEA is the assumption that the amplitude slowly varies, therefore the derivatives are small enough that we can neglect their effect.

The slowly varying envelope approximation is often used because the resulting equations are, in many cases, easier to solve than the original equations, reducing the order of—all or some of—the highest-order partial derivatives. But the validity of the assumptions which are made needs to be justified. In our case, it is justified as equation (2.14) holds. Equation (2.13) under SVEA becomes

$$-k_c^2|\nabla g(\vec{\mathbf{r}})|^2 + k_c^2 n^2 = 0, \quad (2.15)$$

where k_c can easily be dropped out, and we get **the eikonal equation.**

$$\boxed{|\nabla g(\vec{\mathbf{r}})| = n(\vec{\mathbf{r}})} \quad (2.16)$$

where $g(\vec{\mathbf{r}})$ is a function of the position proportional to the wavefronts and related to the optical path length, and n is the refractive index of the medium. The gradient of $g(\vec{\mathbf{r}})$ is directed along the normal to the surface $g(\vec{\mathbf{r}})$ = constant. Therefore, the eikonal function describes the constant-phase surfaces of a scalar wave field.

Equation (2.14) is the fundamental assumption in the geometrical approach of optics, which is that periods and wavelengths of the light waves are much smaller than the time and length scales on which the wave amplitude and the medium vary. Under this approximation, the waves can be regarded locally as planar and monochromatic and, consequently, the electromagnetic propagation can be simplified to a model in which the light is transported along trajectories in space–time called rays.

So far, it is natural to ask if the solutions of the wave equation that we know from the previous chapter hold the SVEA condition of equation (2.14). As an exercise, we take the plane wave

$$E(\vec{\mathbf{r}}, t) = E_0 \cos(kz - wt). \tag{2.17}$$

Notice that the amplitude of the plane wave is constant,

$$\vec{\mathbf{A}}(\vec{\mathbf{r}}) = E_0, \tag{2.18}$$

therefore,

$$\frac{\nabla^2 \vec{\mathbf{A}}(\vec{\mathbf{r}})}{\vec{\mathbf{A}}(\vec{\mathbf{r}})} = 0. \tag{2.19}$$

We conclude that the plane wave holds for SVEA. As an exercise for the reader, check if the spherical wave accomplishes SVEA.

2.3 The ray equation

We have obtained the equation of the eikonal. We know that the normal of $g(\vec{\mathbf{r}})$ is along the path of the light. Everything started because we took equation (2.1)

$$\vec{\mathbf{E}}(\vec{\mathbf{r}}) = \vec{\mathbf{A}}(\vec{\mathbf{r}}) e^{ik_o g(\vec{\mathbf{r}})}, \tag{2.1}$$

and inserted the Helmholtz equation and got the eikonal equation

$$|\nabla g(\vec{\mathbf{r}})| = n(\vec{\mathbf{r}}). \tag{2.16}$$

The question now is what we can infer from the eikonal and path of light? Taking into account that the gradient of the eikonal function gives the direction of propagation of the plane wave predicted by the eikonal equation and that, by definition $|\nabla g(\vec{\mathbf{r}})| = n(\vec{\mathbf{r}})$, this is

$$\frac{d\vec{\mathbf{r}}}{ds} = \frac{\nabla g(\vec{\mathbf{r}})}{|\nabla g(\vec{\mathbf{r}})|}. \tag{2.20}$$

Equivalently, taking into account that if $\vec{\mathbf{r}}(s)$ describes the equation of the ray path parameterized by the arc length, then

$$\nabla g(\vec{\mathbf{r}}) = n\frac{d\vec{\mathbf{r}}}{ds} = n\hat{\mathbf{a}}. \tag{2.21}$$

See figure 2.1, and notice that the trajectory of light in purple is orthogonal to the wavefront in blue. Please notice that we can write equation (2.21) as

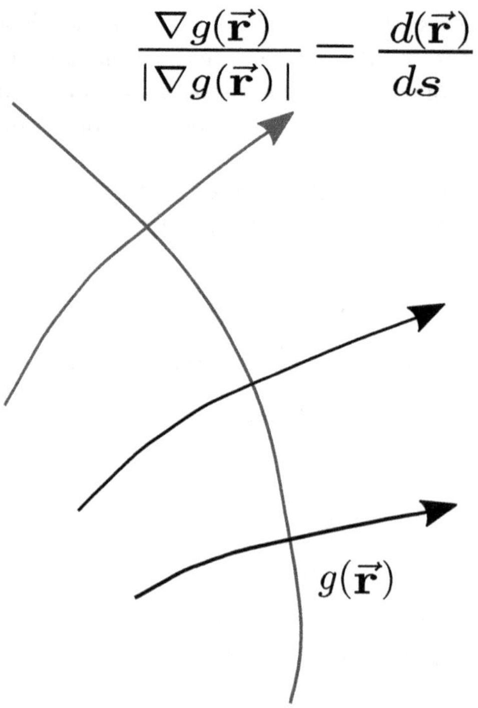

Figure 2.1. The ray is the trajectory of the light under SVEA; in the picture it is colored purple. The ray is perpendicular to the wavefront $g(\vec{\mathbf{r}})$, which is blue.

$$\nabla \times \nabla g(\vec{\mathbf{r}}) = \nabla \times (n\hat{\mathbf{a}}) = 0. \tag{2.22}$$

The curl of a gradient is always zero. Therefore, it can be shown that this equation supports the representation

$$\boxed{\frac{d}{ds}\left(n\frac{d\vec{\mathbf{r}}}{ds}\right) = \nabla n}. \tag{2.23}$$

This is the most standard form of the ray equation. The ray is the imaginary line that represents the direction in which light propagates. The use of this model, widely disclosed in geometric optics, simplifies the calculations due to SVEA. What equations (2.16) and (2.23) tell us is that the ray path is perpendicular to the wavefront. In the following subsections we explore the cases when n is constant or a function, which predicts equation (2.23).

2.3.1 *n* as constant

When in the ray equation we have n as a constant, we get

$$\nabla n \Rightarrow \frac{d}{ds}\left(n\frac{d\vec{\mathbf{r}}}{ds}\right) = 0. \tag{2.24}$$

Applying the derivative operator $\frac{d}{ds}$, we get

$$\frac{d^2\vec{\mathbf{r}}}{ds^2} = 0. \tag{2.25}$$

If we integrate once,

$$\frac{d^2\vec{\mathbf{r}}}{ds^2} = \vec{\mathbf{m}}, \tag{2.26}$$

where $\vec{\mathbf{m}}$ is the constant vector. If we integrate twice,

$$\vec{\mathbf{r}} = \vec{\mathbf{m}}s + \vec{\mathbf{b}}, \tag{2.27}$$

the vector $\vec{\mathbf{b}}$ can be seen as the initial condition and $\vec{\mathbf{m}}$ as the slope of the path ray.

The last expression is just a line. This means that when n is constant, the light propagates in straight lines. This implication is what makes SVEA so useful, since working with straight lines is far more manageable than working with curved lines. We can be sure of this when n is constant.

2.3.2 $n(\vec{\mathbf{r}})$ as a function

$n(\vec{\mathbf{r}})$ is a function; different trajectories may occur, giving rise to interesting phenomena such as mirages, perfectly explainable with geometric optics, and that arising from the variation of the refractive index with temperature so that it can be considered stratified. This is the origin, for example, of the effects that we have all been able to see on the pavement/side of roads on particularly hot days:

$$\frac{d}{ds}\left(n\frac{d\vec{\mathbf{r}}}{ds}\right) = \nabla n. \tag{2.23}$$

In this treatise, we limit ourselves to studying light under constant refractive index mediums, since the main topic of this book is stigmatism. Stigmatism will be formally introduced in chapter 5.

2.4 The Snell law from the eikonal

Another relevant implication of the ray equation is light passing from a medium with refractive index n_1 to another medium with refractive index n_2. To show how the ray path changes, we recall the eikonal equation in its vector form,

$$\nabla g(\vec{\mathbf{r}}) = n\vec{\mathbf{a}}. \tag{2.28}$$

If we integrate along a close loop in both sides, we have

$$\oint \nabla g(\vec{\mathbf{r}}) \cdot d\vec{\mathbf{l}} = \oint n\vec{\mathbf{a}} \cdot d\vec{\mathbf{l}}. \tag{2.29}$$

Notice that the above integrals are equal to zero. This can be proven by Stokes' theorem and equation (2.28).

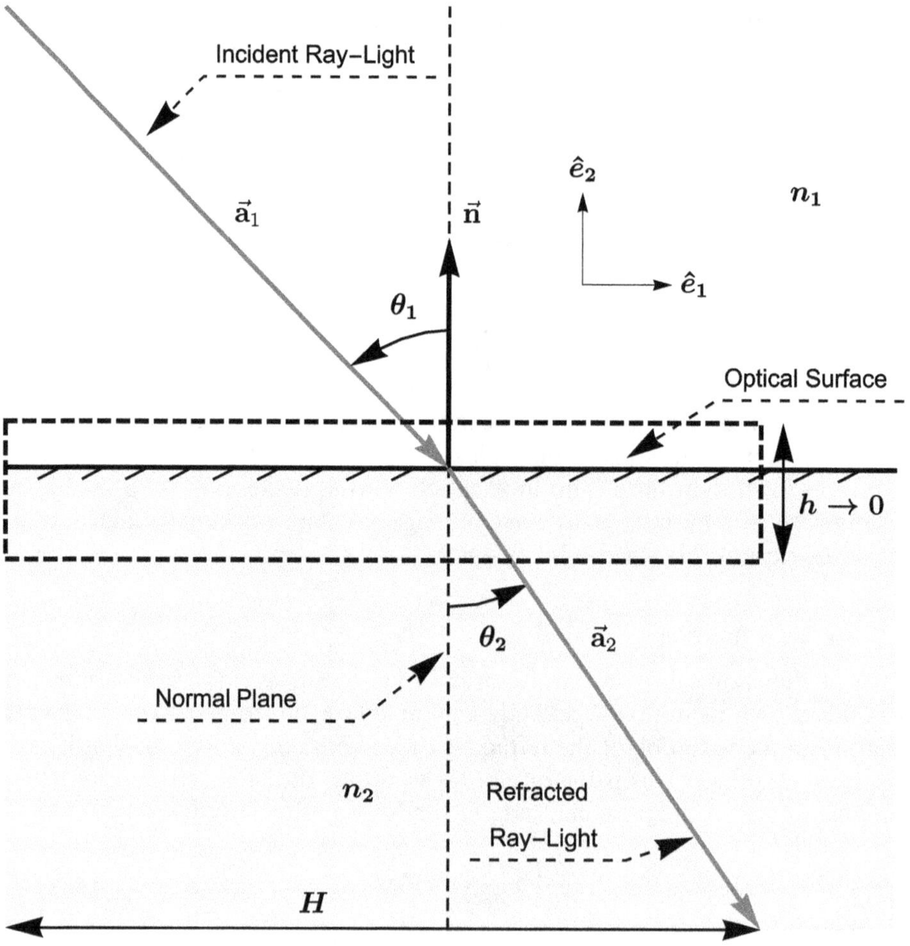

Figure 2.2. Diagram of the refraction of a ray when it passes from one medium with refraction index n_1 to another of refraction index n_2.

$$\oint_{\partial S} \nabla g(\vec{\mathbf{r}}) \cdot d\vec{\mathbf{l}} = \int\int_S \nabla \times \nabla g(\vec{\mathbf{r}}) \cdot dS = 0, \tag{2.30}$$

where ∂S is the border of the surface S. It is zero because the curl of a gradient is zero $\nabla \times \nabla g(\vec{\mathbf{r}}) = 0$. Therefore, as a consequence we have

$$\oint n\vec{\mathbf{a}} \cdot d\vec{\mathbf{l}} = 0. \tag{2.31}$$

Now we are in a position to answer the question presented at the beginning of this section. How does light behave when it passes from a medium with refractive index n_1 to another medium with refractive index n_2? Please notice in figure 2.2 that there is an entrance ray with direction $\vec{\mathbf{a}}_1$ and output $\vec{\mathbf{a}}_2$. Therefore, if we compute the closed integral over the path presented in figure 2.2, taking the result of equation (2.31),

$$n_2\vec{\mathbf{a}}_2 \cdot H\vec{\mathbf{e}}_1 - n_1\vec{\mathbf{a}}_1 \cdot H\vec{\mathbf{e}}_1 = 0. \tag{2.32}$$

Notice that we set $h \to 0$. Manipulating equation (2.32), we have

$$\vec{\mathbf{e}}_1 \cdot (n_2\vec{\mathbf{a}}_2 - n_1\vec{\mathbf{a}}_1) = 0. \tag{2.33}$$

Notice that the normal vector has the same direction as the unit vector of vertical components, $\vec{\mathbf{n}} = \vec{\mathbf{e}}_2$. Therefore,

$$\vec{\mathbf{n}} \times (n_2\vec{\mathbf{a}}_2 - n_1\vec{\mathbf{a}}_1) = 0. \tag{2.34}$$

From the last equation, there are two implications. The first one is

$$n_2\vec{\mathbf{a}}_2 = n_1\vec{\mathbf{a}}_1, \tag{2.35}$$

for this it is necessary that $\vec{\mathbf{a}}_1 = \vec{\mathbf{a}}_2$, which implies that

$$n_2 = n_1. \tag{2.36}$$

This result means that the light does not pass into a change of refraction index. The second implication is expressed in the following equation,

$$\vec{\mathbf{n}} \times n_2\vec{\mathbf{a}}_2 = \vec{\mathbf{n}} \times n_1\vec{\mathbf{a}}_1, \tag{2.37}$$

using the definition of the cross product,

$$\boxed{n_1 \sin\theta_1 = n_2 \sin\theta_2.} \tag{2.38}$$

The last expression is called Snell's law. Snell's law is a formula used to calculate the angle of refraction of light when crossing the separation surface between two means of propagation of light with different refractive index. The name comes from its discoverer, the Dutch mathematician Willebrord Snell van Royen (1580–1626), although it was first discovered by Ibn Sahl in the year 984 AD.

Another way to express the second implication is that the $\vec{\mathbf{n}}$ is parallel to $(n_2\vec{\mathbf{a}}_2 - n_1\vec{\mathbf{a}}_1)$

$$\vec{\mathbf{n}} \times (n_2\vec{\mathbf{a}}_2 - n_1\vec{\mathbf{a}}_1) = 0. \tag{2.39}$$

This implies that $(n_2\vec{\mathbf{a}}_2 - n_1\vec{\mathbf{a}}_1)$ is along with $\vec{\mathbf{n}}$. Therefore, we have

$$(n_2\vec{\mathbf{a}}_2 - n_1\vec{\mathbf{a}}_1) = p\vec{\mathbf{n}}, \tag{2.40}$$

where p is a constant. If we solve for it, we have

$$p = (n_2\vec{\mathbf{a}}_2 \cdot \vec{\mathbf{n}} - n_1\vec{\mathbf{a}}_1 \cdot \vec{\mathbf{n}}). \tag{2.41}$$

Replacing equation (2.41) in equation (2.40) and dividing by n_2,

$$\vec{\mathbf{a}}_2 = \frac{n_1}{n_2}\vec{\mathbf{a}}_1 + \left(\vec{\mathbf{a}}_2 \cdot \vec{\mathbf{n}} - \frac{n_1}{n_2}\vec{\mathbf{a}}_1 \cdot \vec{\mathbf{n}}\right)\vec{\mathbf{n}}, \tag{2.42}$$

and using the definition of the dot product we get

$$\vec{\mathbf{a}}_2 = \frac{n_1}{n_2}\vec{\mathbf{a}}_1 + \left(\cos\theta_2 - \frac{n_1}{n_2}\cos\theta_1\right)\vec{\mathbf{n}}. \tag{2.43}$$

In a more general way, we can express equation (2.43) as

$$\vec{a}_2 = \frac{n_1}{n_2}\vec{a}_1 + \left(\frac{n_1}{n_2}|\vec{n}||\vec{a}_1|\cos\theta_1 - |\vec{n}||\vec{a}_2|\cos\theta_2\right)\vec{n}. \tag{2.44}$$

Looking at figure 2.2, we can see the vertical component of the refracted ray $\vec{\mathbf{a}}_2$ is given by

$$\vec{\mathbf{n}}\cdot\vec{\mathbf{a}}_2 = |\vec{\mathbf{n}}||\vec{\mathbf{a}}_2|\cos\theta_2. \tag{2.45}$$

Notice that since they are unit vectors $|\vec{\mathbf{n}}| = |\vec{\mathbf{a}}_2| \equiv 1$,

$$\vec{\mathbf{n}}\cdot\vec{\mathbf{a}}_2 = \cos\theta_2, \tag{2.46}$$

so let's square it,

$$(\vec{\mathbf{n}}\cdot\vec{\mathbf{a}}_2)^2 = \cos\theta_2^2 = 1 - \sin\theta_2^2 = 1 - \frac{n_1^2}{n_2^2}\sin\theta_1^2. \tag{2.47}$$

From the definition of the cross product we know

$$\vec{\mathbf{n}}\times\vec{\mathbf{a}}_1 = |\vec{\mathbf{n}}||\vec{\mathbf{a}}_1|\sin\theta_1(\vec{\mathbf{e}}_1\times\vec{\mathbf{e}}_2), \tag{2.48}$$

also, $|\vec{\mathbf{n}}| = |\vec{\mathbf{a}}_1| = 1$.

$$\vec{\mathbf{n}}\times\vec{\mathbf{a}}_1 = \sin\theta_1, \tag{2.49}$$

powering to the square,

$$(\vec{\mathbf{n}}\times\vec{\mathbf{a}}_1)^2 = \sin\theta_1^2. \tag{2.50}$$

Replacing the above expression in $(-\vec{\mathbf{n}}\cdot\vec{\mathbf{a}}_2)^2 = 1 - (n_1^2/n_2^2)\sin\theta_1^2$, we get

$$\vec{\mathbf{n}}\cdot\vec{\mathbf{a}}_2 = \sqrt{1 - \frac{n_1^2}{n_2^2}(\vec{\mathbf{n}}\times\vec{\mathbf{a}}_1)^2}. \tag{2.51}$$

It is easy to see that the other terms of equation (2.44) are

$$\frac{n_1}{n_2}[\vec{\mathbf{a}}_1 + |\vec{\mathbf{n}}||\vec{\mathbf{a}}_1|\cos\theta_1\vec{\mathbf{n}}] = \frac{n_1}{n_2}[\vec{\mathbf{a}}_1 - (\vec{\mathbf{n}}\cdot\vec{\mathbf{a}}_1)\vec{\mathbf{n}}]. \tag{2.52}$$

Therefore, Snell's law can be written as

$$\boxed{\vec{\mathbf{a}}_2 = \frac{n_1}{n_2}[\vec{\mathbf{a}}_1 - (\vec{\mathbf{n}}\cdot\vec{\mathbf{a}}_1)\vec{\mathbf{n}}] - \vec{\mathbf{n}}\sqrt{1 - \frac{n_1^2}{n_2^2}(\vec{\mathbf{n}}\times\vec{\mathbf{a}}_1)^2} \quad \text{for } \vec{\mathbf{a}}_2, \vec{\mathbf{a}}_1, \vec{\mathbf{n}} \in \mathbb{R}^2.} \tag{2.53}$$

Equation (2.53) will be very useful in chapters 9 and 10 to express the stigmatic lens equation.

2.5 The Fermat principle from the eikonal

Something very interesting about any theory is that it obeys the principle of least action. When I refer to this theory it is the equation of the eikonal, the ray equation and all their implications. The principle of least action postulates that, for systems of classical physics, the temporal evolution of all physical systems occurred in such a way that an amount called 'action' tended to be the least possible or least energy path. The optics version of the minimum principle is called the Fermat principle.

Fermat's principle says that the path taken between two points by a ray of light is the path that can be passed in the shortest time. Fermat's principle can be stated as:

> The optical length of the path followed by light between two fixed different points is the global minima. The optical length is the physical length multiplied by the refractive index of the medium.

Please pay attention to the word *global minimum*. Remember that the global minimum is the smallest overall value of a set. So, imagine that we have a set, such as its elements are all the possible optical paths from one point to another. These paths have their respective optical length. What the Fermat principle says is that the only physically valid path of our set is the one that has the smallest value of an optical path length.

Mathematically, Fermat's principle can be described as the time T a point of the ray needs to cover a path between the points A and B, given by

$$T = \int_{t_0}^{t_1} dt = \frac{1}{c}\int_{t_0}^{t_1} \frac{c}{v}\frac{ds}{dt}dt = \frac{1}{c}\int_{A}^{B} nds. \tag{2.54}$$

Remember that c is the speed of light in vacuum, ds an infinitesimal displacement along the ray, $v = ds/dt$ the speed of light in a medium and $n = c/v$ the refractive index of that medium. t_0 is the starting time (the ray is in A) and t_1 the arrival time at point B. The optical path length of a ray from A to B is defined by the following integral

$$S = \int_{A}^{B} nds; \tag{2.55}$$

it is related to the travel time by $S = cT$. The optical path length is a purely geometrical quantity since time is not considered in its calculation. The global minimum in the travel time of light between two points A and B is equivalent to the global minimum of the optical path length between A and B.

In the next chapter, we are going to study the variational concepts behind the Fermat principle. In chapter 4 we are going to deduce the eikonal equation, ray equation and their implications from the Fermat principle.

2.6 End notes

In this chapter, we obtained the basis for ray optics, the eikonal equation and the ray equation. From the ray equation, we found that at a constant refractive index, average light travels in a straight line. From the eikonal, we obtained Snell's law that does not say how light changes the trajectory when it is in an interface of constant refractive media.

Finally, we mentioned that all these acts obey the Fermat principle, which we briefly described. In the next chapter, we will see the mathematical foundations of the Fermat principle.

References

Arfken G B and Weber H J 1999 *Mathematical Methods for Physicists* (New York: Academic)

Boas M L 2006 *Mathematical Methods in the Physical Sciences* (New York: Wiley)

Born M and Wolf E 2013 *Principles of Optics: Electromagnetic Theory of Propagation, Interference and Diffraction of Light* (Amsterdam: Elsevier)

Buchdahl H A 1993 *An Introduction to Hamiltonian Optics* (North Chelmsford, MA: Courier Corporation)

Cao S and Greenhalgh S 1994 *Finite-difference solution of the eikonal equation using an efficient, first-arrival, wavefront tracking scheme Geophysics* **59** 632–43

Currier R and Herman M F 1985 Numerical comparison of generalized surface hopping, classical analog, and self-consistent eikonal approximations for nonadiabatic scattering *J. Chem. Phys.* **82** 4509–16

Dacorogna B, Glowinski R and Pan T-W 2003 Numerical methods for the solution of a system of eikonal equations with Dirichlet boundary conditions *C. R. Math.* **336** 511–8

Griffiths D J 2005 *Introduction to Electrodynamics* (Cambridge: Cambridge University Press)

Hecht E 1974 *Schaum's Outline of Optics* (New York: McGraw-Hill)

Hecht E 2012 *Optics* (Noida: Pearson Education India)

Hoffnagle J A and Shealy D L 2011 Refracting the k-function: Stavroudis's solution to the eikonal equation for multielement optical systems *J. Opt. Soc. Am.* **A28** 1312–21

Huang L, Shu C-W and Zhang M 2008 Numerical boundary conditions for the fast sweeping high order WENO methods for solving the Eikonal equation *J. Comput. Math.* **26** 336–46

Hysing S-R and Turek S 2005 The Eikonal equation: numerical efficiency versus algorithmic complexity on quadrilateral grids *Proc. ALGORITMY* **22**

Jackson J D 1999 *Classical Electrodynamics* (Hoboken, NJ: Wiley)

Lakshminarayanan V, Ghatak A and Thyagarajan K 2002 *Lagrangian Optics* (Berlin: Springer)

Lax M, Louisell W H and McKnight W B 1975 From Maxwell to paraxial wave optics *Phys. Rev.* **A11** 1365

Luneburg R K 1964 *Mathematical Theory of Optics* (Berkeley, CA: University of California Press)

Pegis R J 1961 I The modern development of Hamiltonian optics *Progress in Optics* **vol 1** (Amsterdam: Elsevier) pp 1–29

Qian J, Zhang Y-T and Zhao H-K 2007 Fast sweeping methods for eikonal equations on triangular meshes *SIAM J. Numer. Anal.* **45** 83–107

Ronchi V and Barocas V 1970 *The Nature of Light: An Historical Survey* (Cambridge, MA: Harvard University Press) p 12+ 288

Spira A and Kimmel R 2004 An efficient solution to the eikonal equation on parametric manifolds *Interfaces Free Boundaries* **6** 315–27

Stavroudis O 2012 *The Optics of Rays, Wavefronts, and Caustics* **vol 38** (Amsterdam: Elsevier)

Stavroudis O N 1995 The k function in geometrical optics and its relationship to the archetypal wave front and the caustic surface *J. Opt. Soc. Am.* **A12** 1010–6

Stavroudis O N 2006 *The Mathematics of Geometrical and Physical Optics: The K-function and Its Ramifications* (New York: Wiley)

Stavroudis O N and Hurtado-Ramos J B 2000 Maxwell equations and the k function *J. Opt. Soc. Am.* **A17** 1469–74

Vavryčuk V 2012 On numerically solving the complex eikonal equation using real ray-tracing methods: a comparison with the exact analytical solution *Geophysics* **77** T109–16

Zhao H 2005 A fast sweeping method for eikonal equations *Math. Comput.* **74** 603–27

IOP Publishing

Stigmatic Optics (Second Edition)

Rafael G González-Acuña and Héctor A Chaparro-Romo

Chapter 3

Calculus of variations

In this chapter, we will study the mathematical foundations that lie behind the principle of least action that nature follows, the calculation of variations. The whole theory of geometric optics can be formulated using the calculation of variations, since from this the Fermat principle is deduced. As an example of the robustness of the calculation of variations, we will obtain Newton's second law from pure variational concepts.

3.1 Calculus of variations

The calculus of variations studies the methods that allow finding stationary values of a functional. The functional term is applied to certain functions. A functional is a function that takes functions as its argument; that is, a function whose domain is a set of functions. Stationary values include maximum values, minimum values and values of inflexion with a horizontal tangent.

Since a functional represents the mathematical model of a physical problem, the application of variations methods is essential in areas of knowledge such as theoretical physics, optics, Lagrangian mechanics, quantum mechanics, engineering, etc.

The fundamental equation of the calculus of variations is Euler's equation. This equation relates a functional with its stationary value. In this chapter, we will study and obtain Euler's equation. Later, we will demonstrate the robustness of the calculus of variations by getting Newton's second law from the variations principles.

3.2 The Euler equation

The problem is to find the y such that it makes stationary the following integral,

$$I = \int_{x_1}^{x_2} F(x, y, y')dx, \tag{3.1}$$

where F is a given functional, a function whose domain is a set of functions. $y(x)$ makes I a stationary value, $y(x)$ transforms I to a maximum or minimum.

doi:10.1088/978-0-7503-6423-2ch3

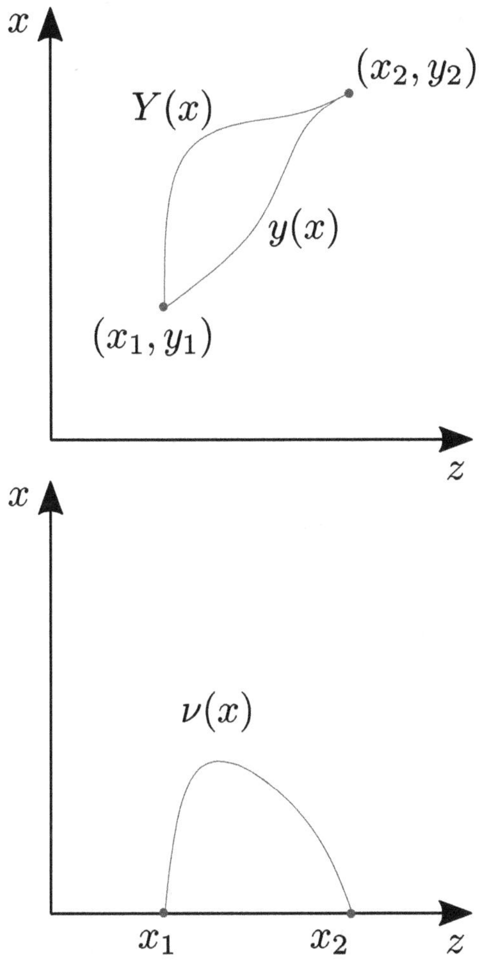

Figure 3.1. *Top:* Two paths given by the curves $y(x)$ and $Y(x)y(x) + \varepsilon\nu(x)$. $y(x)$ is such that I is a minimum and $Y(x)$ is deflected of $y(x)$ by an amount $\nu(x)$. *Bottom:* the deflection $\nu(x)$ is zero at the initial and final points.

The integral I can be seen as the arc length integral. An arc length integral, also called curve rectification, is the measure of the distance or path traveled along a curve or linear dimension. But, we set I as the most general as possible. In our case, we are interested in the shortest path that the light follows. Therefore, we are looking at $y(x)$, which makes I a minimum.

Now, we consider a set of varied curves given by equation (3.1),

$$Y(x) = y(x) + \varepsilon\nu(x). \tag{3.2}$$

Let $\nu(x)$ be a function of x such that it is zero at x_1 and x_2; in other words, $\nu(x_1) = \nu(x_2) = 0$. Also, let $\nu(x)$ have a continuous second derivative in the interval (x_1, x_2). Finally, let ϵ be a completely arbitrary parameter (see figure 3.1). We assume that $Y(x)$ is a set of single-valued curves. From the set of all of the single-valued curves $Y(x)$, we are interested in only picking the curve that makes the following integral a minimum.

$$I(\varepsilon) = \int_{x_1}^{x_2} F(x,\ Y,\ Y')\, dx. \tag{3.3}$$

As expressed in equation (3.3), we can see I as a function of ϵ, $I(\varepsilon)$. The ideal is that if we set $\varepsilon = 0$, we have $Y = y(x)$, the desired extremal. In other words, what we want to have is

$$\left. \frac{dI(\varepsilon)}{d\varepsilon} \right|_{\varepsilon=0} = 0. \tag{3.4}$$

Differentiating equation (3.3) with respect to ϵ,

$$\frac{dI}{d\varepsilon} = \int_{x_1}^{x_2} \left(\frac{\partial F}{\partial Y}\frac{\partial Y}{\partial \varepsilon} + \frac{\partial F}{\partial Y'}\frac{\partial Y}{\partial \varepsilon'} \right) dx. \tag{3.5}$$

Notice that

$$\frac{\partial Y}{\partial \varepsilon} = \nu(x), \qquad \frac{\partial Y'}{\partial \varepsilon} = \nu'(x). \tag{3.6}$$

Replacing equation (3.6) in equation (3.5),

$$\frac{dI}{d\varepsilon} = \int_{x_1}^{x_2} \left[\frac{\partial F}{\partial Y}\nu(x) + \frac{\partial F}{\partial Y'}\nu'(x) \right] dx. \tag{3.7}$$

If we evaluate $\varepsilon = 0$, $Y|_{\varepsilon=0} = y$ and equation (3.7) it becomes

$$\left(\frac{dI}{d\varepsilon} \right)_{\varepsilon=0} = \int_{x_1}^{x_2} \left[\frac{\partial F}{\partial y}\nu(x) + \frac{\partial F}{\partial y'}\nu'(x) \right] dx = 0. \tag{3.8}$$

Integrating the first term of the right side of equation (3.8),

$$\int_{x_1}^{x_2} \frac{\partial F}{\partial y'}\nu'(x) dx = \left. \frac{\partial F}{\partial y'}\nu(x) \right|_{x_1}^{x_2} - \int_{x_1}^{x_2} \frac{d}{dx}\left(\frac{\partial F}{\partial y'} \right) \nu(x) dx. \tag{3.9}$$

Remember that we choose $\nu(x)$ such that $\nu(x_1) = \nu(x_2) = 0$. Therefore, the first term of the right side of equation (3.9) is zero. Replacing equation (3.9) in equation (3.8) we have

$$\left(\frac{dI}{d\varepsilon} \right)_{\varepsilon=0} = \int_{x_1}^{x_2} \left[\frac{\partial F}{\partial y} - \frac{d}{dx}\frac{\partial F}{\partial y'} \right] \nu(x) dx = 0. \tag{3.10}$$

There are two options here, to choose $\nu(x)$ as zero or what is inside the brackets of equation (3.10) to be zero. If $\nu(x)$ is zero, we gained nothing since $\nu(x) = 0$ does not relate F and y such that F is a minimum. Because in $\nu(x) = 0$ there's not even a presence of F and y. Therefore, we choose the second option, and we get Euler's equation

$$\boxed{\frac{\partial F}{\partial y} - \frac{d}{dx}\frac{\partial F}{\partial y'} = 0.} \tag{3.11}$$

Every problem in the calculus of variations is solved by setting up the integral which is to be stationary, addressing what the function F is, replacing it into the Euler equation, and computing the solution of the differential equation obtained from the Euler equation.

The idea is to use the Euler equation to see which path the light uses under certain conditions.

3.3 Newton's second law

In chapter 2, we mentioned that historically, the principle of least action postulated —for systems of classical physics—that the temporal evolution of the physical system as a whole occurred in such a way that a quantity called *action* tended to be the least possible. We also mentioned that the principle of least action in optics is known as the Fermat principle.

The calculus of variations and the principle of least action are not only present in optics. To demonstrate the power of the calculus of variations and the principle of least action, in this section, we will deduce Newton's second law of the calculus of variations and the principle of least action.

So, lets assume that at time t_1 a particle is in position x_1, and at time t_2 the same particle is now in position x_2. Which is the path of least action? We can consider all paths. Some examples of paths can be seen in figure 3.2.

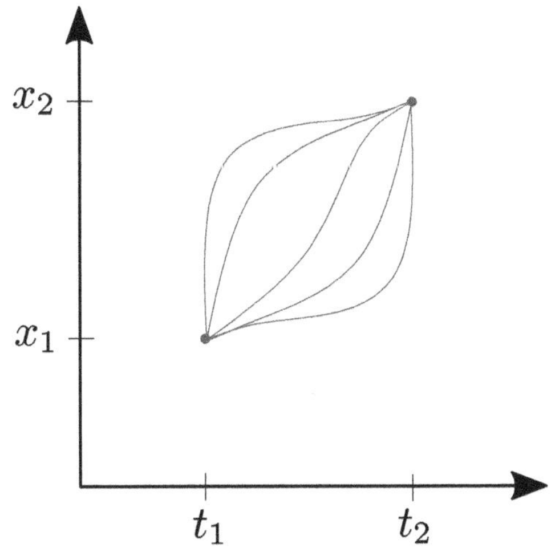

Figure 3.2. Several paths from x_1 to x_2, starting at time t_1 and finishing at time t_2.

If we calculate the difference between kinetic energy and potential energy at each point on the path, then add all of these contributions together to get a final number, we find the path for which this final number is the least. Then, we will know that this is the path that the particle really takes.

We can formalize this by defining something called action S. The action is defined as the time integral from t_1 to t_2 of the difference between the kinetic energy and the potential energy,

$$S = \int_{t_1}^{t_2} (K - P)dt. \tag{3.12}$$

So the idea is to find a path where S is the minimum. But for the moment we are going to explore the differences in single variable functions, because we are going to use the differences in single variable functions in the deduction of the path such that S is the minimum.

If we consider a function $f(x)$ of a single variable x, then one way of knowing whether we are located at the x value, which corresponds to the minimum of $f(x)$, is by moving x a little,

$$x \rightarrow x + \varepsilon, \tag{3.13}$$

so,

$$f(x) \rightarrow f(x + \varepsilon). \tag{3.14}$$

Therefore, using the Taylor series,

$$f(x + \varepsilon) = f(x) + \varepsilon\frac{df}{dx} + \varepsilon^2\frac{d^2f}{dx^2} + \cdots \tag{3.15}$$

then, the change of f is given by

$$\Delta f = f(x + \varepsilon) - f(x) \tag{3.16}$$

$$\Delta f = f(x) + \varepsilon\frac{df}{dx} + \varepsilon^2\frac{d^2f}{dx^2} + \cdots - f(x) \tag{3.17}$$

$$\Delta f = \varepsilon\frac{df}{dx} + \varepsilon^2\frac{d^2f}{dx^2} + \cdots \tag{3.18}$$

So, at minimum value, in other words $df/dx = 0$, the change of f is given by

$$\Delta f = \varepsilon^2\frac{d^2f}{dx^2} + \cdots \tag{3.19}$$

So, we see at the minimum if we make a small change in our x position then our function changes at second order in the small step changes, so Δf is proportional to ε^2.

$$\Delta f = f(x + \varepsilon) - f(x) = 0. \tag{3.20}$$

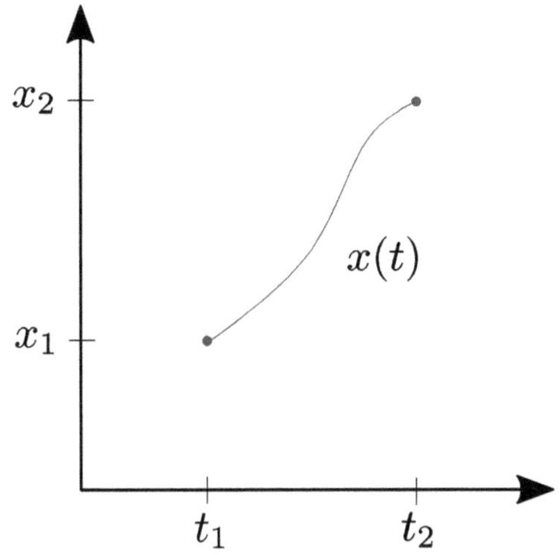

Figure 3.3. A particle moving from x_1 in t_1 to x_2 in t_2.

This means that Δf is zero at first order.

Now, it is important to realize that we are not dealing with the function of a single variable. We are trying to find the minimum of a function, of a function. The task is trying to find a single number to represent the point where it is least, rather than in our case, where what we're trying to do is to find an entire path for which action is least, so we need to use a slightly different approach. So, for example, we imagine the actual path being represented by the function $x(t)$. $x(t)$ represents the moving particle from position x_1 and t_1 to x_2 and t_2 (see figure 3.3).

What we want to do is look at some small deviation away from the actual path. We hope that if we look at the change in the action as the result of that slight deviation away from the actual path, then that change in the action should be zero to first order in the change of the path.

More explicitly, if we deviate the actual path by some amount like the one presented in figure 3.4, the amount changed is $\nu(t)$. So, the actual path is $x(t)$ and the deviated path is $x(t) + \nu(t)$.

Now we recall the action S, equation (3.12),

$$S(x) = \int_{t_1}^{t_2} (K - P)dt. \tag{3.21}$$

If we are working in one dimension we have

$$S = \int_{t_1}^{t_2} \left[\frac{1}{2} m \left(\frac{dx}{dt} \right)^2 - \nu(x) \right] dt, \tag{3.22}$$

where $\frac{1}{2}m\left(\frac{dx}{dt}\right)^2$ is the kinetic and $\nu(x)$ is the potential energy.

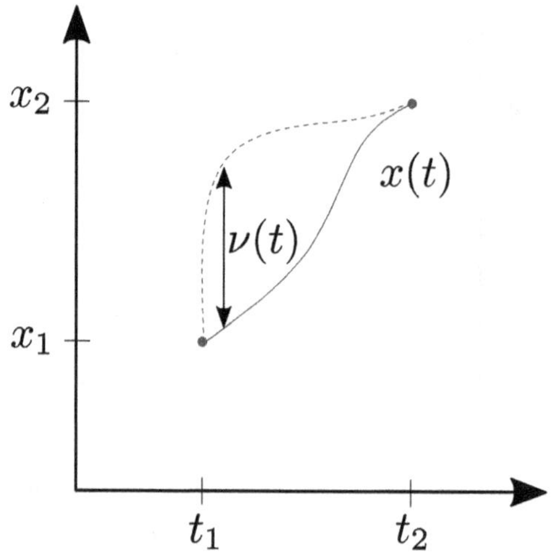

Figure 3.4. The deflection $\nu(t)$ in the path $x(t)$.

What we are going to do is look at how the action changes as a result of the small deviation $\nu(t)$ in the path $x(t)$. Then, we are going to look at the difference between the actual path and the deviated path action.

If x is the path of minimum action then ΔS is zero at first order,

$$\delta S = S(x + \nu) - S(x) = 0, \tag{3.23}$$

thus,

$$S(x + \nu) = \int_{t_1}^{t_2} \left[\frac{1}{2} m \left(\frac{d(x + \nu)}{dt} \right)^2 - V(x + \nu) \right] dt, \tag{3.24}$$

since the derivative is a linear operator, we have

$$S(x + \nu) = \int_{t_1}^{t_2} \left[\frac{1}{2} m \left(\frac{dx}{dt} + \frac{d\nu}{dt} \right)^2 - V(x + \nu) \right] dt. \tag{3.25}$$

Let's focus on the squared term inside the integral of equation (3.25), expanding it,

$$\left(\frac{dx}{dt} + \frac{d\nu}{dt} \right)^2 = \left(\frac{dx}{dt} \right)^2 + \left(\frac{d\nu}{dt} \right)^2 + 2 \frac{dx}{dt} \frac{d\nu}{dt}, \tag{3.26}$$

then, let's focus on the second term inside the integral of equation (3.25), expanding it using the Taylor series,

$$V(x + \nu) = \nu(x) + \nu \frac{dV}{dt} + \frac{\nu^2}{2!} \frac{d^2 V}{dt^2} + \cdots \tag{3.27}$$

We will ignore the higher terms,

$$V(x+\nu)=\nu(x)+\nu\frac{dV}{dt}, \tag{3.28}$$

replacing equations (3.26) and (3.28) in equation (3.25),

$$S(x+\nu)=\int_{t_1}^{t_2}\left[\frac{1}{2}m\left(\frac{dx}{dt}\right)^2-\nu(x)+\frac{1}{2}m\left(2\frac{dx}{dt}\frac{d\nu}{dt}\right)-\nu\frac{dV}{dx}\right]dt. \tag{3.29}$$

Now, if we say that x is the path of minimum action, then ΔS is zero at first order. Thus,

$$\delta S=S(x+\nu)-S(x)=0, \tag{3.30}$$

replacing equations (3.22) and (3.29) in equation (3.30),

$$\delta S=\int_{t_1}^{t_2}\left[m\left(\frac{dx}{dt}\frac{d\nu}{dt}\right)-\nu\frac{dV}{dx}\right]dt=0. \tag{3.31}$$

Now, looking at equation (3.31), notice that we have

$$\frac{d}{dt}\left(\frac{dx}{dt}\nu\right)=\frac{d^2x}{dt^2}\nu+\frac{dx}{dt}\frac{dn}{dt}. \tag{3.32}$$

From it we can get an expression of $\frac{dx}{dt}\frac{dn}{dt}$,

$$\frac{dx}{dt}\frac{dn}{dt}=\frac{d}{dt}\left(\frac{dx}{dt}\nu\right)-\frac{d^2x}{dt^2}\nu, \tag{3.33}$$

integrating from t_1 to t_2 equation (3.33),

$$\int_{t_1}^{t_2}\frac{dx}{dt}\frac{dn}{dt}dt=\int_{t_1}^{t_2}\frac{d}{dt}\left(\frac{dx}{dt}\nu\right)dt-\int_{t_1}^{t_2}\frac{d^2x}{dt^2}\nu dt. \tag{3.34}$$

Notice that the first term of equation (3.34) is zero by the fundamental theorem of calculus and because we choose ν such that

$$\nu(t_1)=\nu(t_2)=0, \tag{3.35}$$

thus,

$$\left.\frac{dx}{dt}\nu\right|_{t_1}^{t_2}=0\Rightarrow\int_{t_1}^{t_2}\frac{d}{dt}\left(\frac{dx}{dt}\nu\right)dt=0. \tag{3.36}$$

Therefore, equation (3.34) can be expressed as

$$\int_{t_1}^{t_2}\frac{dx}{dt}\frac{dn}{dt}dt=-\int_{t_1}^{t_2}\frac{d^2x}{dt^2}\nu dt, \tag{3.37}$$

replacing equation (3.37) in equation (3.31),

$$\delta S = \int_{t_1}^{t_2} \left[-m\frac{d^2x}{dt^2} - \frac{dV}{dx} \right] \nu dt = 0. \tag{3.38}$$

Notice that what is inside the brackets of the above equation should be zero, thus,

$$-\frac{dV}{dx} = m\frac{d^2x}{dt^2}. \tag{3.39}$$

The first terms of the last equation are the force, since the force is the negative derivative of potential energy and the second term is the mass multiplied by the acceleration, thus,

$$F = ma. \tag{3.40}$$

Remember that we are in one dimension, thus the quantities of the last equation are scalars rather than vectors.

3.4 End notes

In this chapter, we studied the basic concepts of the calculus of variations. Starting from them, we find the Euler equation that relates its functional to its stationary value. Finally, as a demonstration of the robustness and depth of the least-action principle, we obtained Newton's second law using the calculus of variations method.

In the next chapter, we will use the calculus of variations techniques for our purposes, studying the nature of light and stigmatism. The latter will be presented concretely until chapter 5.

References

Boas M L 2006 *Mathematical Methods in the Physical Sciences* (New York: Wiley)

Buchdahl H A 1993 *An Introduction to Hamiltonian Optics* (North Chelmsford, MA: Courier Corporation)

del Castillo G F T 2018 *An Introduction to Hamiltonian Mechanics* (Berlin: Springer)

Elsgolc L D 2012 *Calculus of Variations* (North Chelmsford, MA: Courier Corporation)

Gelfand I M and Silverman R A *et al* 2000 *Calculus of Variations* (North Chelmsford, MA: Courier Corporation)

Goldstein H, Poole C and Safko J 2002 *Classical Mechanics* (Reading, MA: Addison-Wesley)

Lakshminarayanan V, Ghatak A and Thyagarajan K 2002 *Lagrangian Optics* (Berlin: Springer)

Luneburg R K 1964 *Mathematical Theory of Optics* (Berkeley, CA: University of California Press)

Marion J B 2013 *Classical Dynamics of Particles and Systems* (New York: Academic)

Morrey C B Jr 2009 *Multiple Integrals in the Calculus of Variations* (Berlin: Springer)

Stavroudis O N 2006 *The Mathematics of Geometrical and Physical Optics: The K-function and Its Ramifications* (New York: Wiley)

IOP Publishing

Stigmatic Optics (Second Edition)

Rafael G González-Acuña and Héctor A Chaparro-Romo

Chapter 4

Optics of variations

A study on the foundations of geometric optics with a purely variational perspective is presented. Starting from the variational calculation and the Fermat principle, we obtain the Lagrangian for light, then with the Euler equation we deduce the ray equation. With the Fermat principle the laws of reflection and refraction are obtained. Finally, the Malus–Dupin theorem is studied.

4.1 Introduction

In the first two chapters of this treatise, we studied Maxwell's equations, and from them, we arrived at the eikonal equation and the ray equation. All this formalism used gave us a broader picture of how light can behave if we take the approximations that justify the existence of the eikonal equation. Right at the end of chapter 2, we mentioned that everything we find could be described under a single principle, the Fermat principle. This principle, in turn, is based on the richness and elegance of the principles of variational calculus.

In chapter 3, we explored variational principles. We discovered that using the Euler equation, we can solve many problems posed by variational calculus. We even deduced Newton's second law from the variational calculus. The latter was only for deductive reasons, but this is a treatise of optics and, more accurately, a treatise of stigmatic optics. Although we will define stigmatism in the next chapter, it is essential that in this study, you study the variational aspects of light (see figure 4.1).

So, let's compute the optical path length of a light ray from one point to another; the path length is given by

$$L = \int_{p_1}^{p_2} n(x, y, z) ds = 0, \tag{4.1}$$

where L is the optical path length from the point p_1 to the point p_2 for a given path. $n(x, y, z)$ is the refraction index along the path and s is a parameter of the arc length of the path. Thus, ds is the infinitesimal arc length.

doi:10.1088/978-0-7503-6423-2ch4

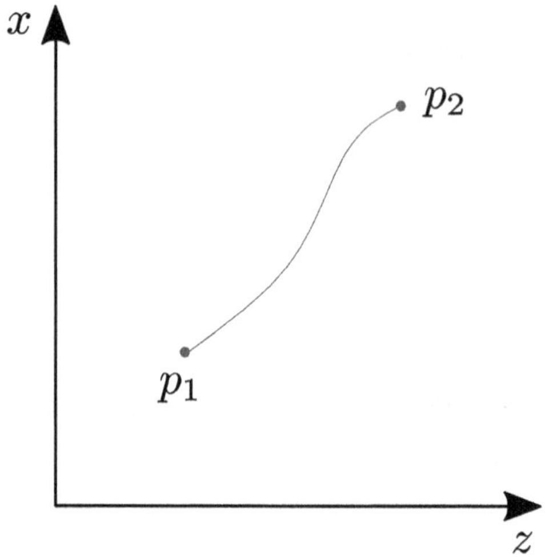

Figure 4.1. A path of light from point p_1 to p_2.

As we have mentioned before, the principle of least action transposed to optics is the principle of Fermat. The Fermat principle tells us that light will travel the least time path, this, expressed in terms of variations, can be seen as the following integral

$$\delta L = \delta \int_{p_1}^{p_2} n(x, y, z) ds = 0, \tag{4.2}$$

where δL can be seen as the variation on L. The light will travel in the path such that the above integral is a minimum.

In this chapter, we are going to study this integral, and from it, we are going to deduce all the things that we know from the behavior of light: The eikonal, the ray equation, Snell's law and the perpendicularity of the rays and waves.

4.2 Lagrangian and Hamiltonian optics

From equation (4.2), we can write the Fermat principle as

$$\delta l = \delta \int_{p_1}^{p_2} L(x, y, z, x', y', z') dz = 0. \tag{4.3}$$

$L(x, y, z, x', y', z')$ is called the Lagrangian, and it is the functional that receives as input parameters the possible paths of light. The above integral is equal to zero, which means that that the path chosen by the light is the minimum. ds is the infinitesimal arc length given by

$$ds = dz\sqrt{1 + x'^2 + y'^2}. \tag{4.4}$$

We use dz since we choose z to be the integration variable. Thus, $x' = \frac{dx}{dz}$ and $y' = \frac{dx}{dz}$. Therefore, we can rewrite equation (4.3) as

$$\delta l = \delta \int_{p_1}^{p_2} L(x, y, x', y', z)dz = 0, \tag{4.5}$$

where the optical Lagrangian is

$$L(x, y, x', y', z) = n(x, y, z)\sqrt{1 + x'^2 + y'^2}. \tag{4.6}$$

Notice that this form of the Lagrangian does not depend on z' because we set z to the independent variable. We choose it by tradition since typically z is set to be the optical axis. Now the Lagrangian has five input parameters that can be divided into three categories. The first one is x and its derivative x', the second one is y, and its derivative y', and the third is the independent variable z. With this in mind, for this system we can get two Euler equations, the first one for x and its derivative x',

$$\frac{d}{dz}\left(\frac{\partial L}{\partial x'}\right) - \frac{\partial L}{\partial x} = 0, \tag{4.7}$$

another one for y and its derivative y'

$$\frac{d}{dz}\left(\frac{\partial L}{\partial y'}\right) - \frac{\partial L}{\partial y} = 0. \tag{4.8}$$

Notice that the role of z is similar to the variable of time in chapter 3, where we found Newton's second law.

Using equation (4.6), we can rewrite the equation as (4.7),

$$\frac{d}{dz}\left(\frac{n\dot{x}}{\sqrt{1 + \dot{x}^2 + \dot{y}^2}}\right) = \sqrt{1 + \dot{x}^2 + \dot{y}^2}\frac{\partial n}{\partial x}, \tag{4.9}$$

and equation (4.8) as

$$\frac{d}{dz}\left(\frac{n\dot{y}}{\sqrt{1 + \dot{x}^2 + \dot{y}^2}}\right) = \sqrt{1 + \dot{x}^2 + \dot{y}^2}\frac{\partial n}{\partial y}. \tag{4.10}$$

Now, manipulating equation (4.9), we get

$$\frac{1}{\sqrt{1 + \dot{x}^2 + \dot{y}^2}}\frac{d}{dz}\left(\frac{n\dot{x}}{\sqrt{1 + \dot{x}^2 + \dot{y}^2}}\right) = \frac{\partial n}{\partial x}. \tag{4.11}$$

Doing the same process, but now in equation (4.10), we get

$$\frac{1}{\sqrt{1 + \dot{x}^2 + \dot{y}^2}}\frac{d}{dz}\left(\frac{n\dot{y}}{\sqrt{1 + \dot{x}^2 + \dot{y}^2}}\right) = \frac{\partial n}{\partial y}. \tag{4.12}$$

Replacing equation (4.6) in equation (4.11),

$$\frac{d}{ds}\left(\frac{n\frac{dx}{dz}}{\frac{ds}{dz}}\right) = \frac{\partial n}{\partial x}, \tag{4.13}$$

the above expression is reduced to

$$\frac{d}{ds}\left(n\frac{dx}{ds}\right) = \frac{\partial n}{\partial x}, \tag{4.14}$$

and we apply the same procedure on equation (4.12),

$$\frac{d}{ds}\left(n\frac{dy}{ds}\right) = \frac{\partial n}{\partial y}. \tag{4.15}$$

For the z direction we can also have a similar equation, since we could set $ds = dz\sqrt{1 + y'^2 + z'^2}$ where now the derivatives, inside the square root, are with respect to x. Thus, we have

$$\frac{d}{ds}\left(n\frac{dz}{ds}\right) = \frac{\partial n}{\partial z}. \tag{4.16}$$

Combining equations (4.14)–(4.16), in vector form we have

$$\boxed{\frac{d}{ds}\left(n\frac{d\vec{\mathbf{r}}}{ds}\right) = \nabla n,} \tag{4.17}$$

where $\vec{\mathbf{r}} = x\hat{\mathbf{e}}_1 + y\hat{\mathbf{e}}_1 + z\hat{\mathbf{e}}_1$. Equation (4.17) is the ray equation; the same ray equation that we found in chapter 2. Different origins, the same result. Absolutely astonishing and beautiful.

Well, there is more; Lagrangian mechanics has an alternative version, which is also based on the calculus of variations, and it is called Hamiltonian mechanics. Hamilton studied the Lagrangian mechanics in terms of what he called generalized momenta. The generalized momenta in direction x is

$$p_x = \frac{\partial L}{\partial x'}, \tag{4.18}$$

and the generalized momenta in direction y is

$$p_y = \frac{\partial L}{\partial y'}. \tag{4.19}$$

Computing the derivative of equation (4.18), p_x can be written as

$$p_x = n\frac{x'}{\sqrt{1 + x'^2 + y'^2}} = n\frac{dx}{ds}, \tag{4.20}$$

and computing the derivative of equation (4.19), for p_y we get

$$p_y = n\frac{y'}{\sqrt{1 + x'^2 + y'^2}} = n\frac{dy}{ds}. \tag{4.21}$$

Equations (4.20) and (4.21) are the generalized momenta. From them it is easy to see that generalized momenta in a given direction is the refraction index multiplied by the cosine director of the path of light in that given direction.

Making an analogy of the Hamiltonian in classical mechanics, the Hamiltonian for the ray is given by

$$H(x, y, p_x, p_y, z) = p_x x' + p_y y' - L. \tag{4.22}$$

Replacing equations (4.20), (4.21) and (4.6), we have

$$H(x, y, p_x, p_y, z) = -\sqrt{n^2 - p_x^2 - p_y^2}, \tag{4.23}$$

therefore, the equations of motion are the following

$$\begin{cases} \dfrac{\partial q_x}{\partial z} = \dfrac{\partial H}{\partial p_x} \\ \dfrac{\partial q_y}{\partial z} = \dfrac{\partial H}{\partial p_y} \\ \dfrac{\partial p_x}{\partial z} = -\dfrac{\partial H}{\partial q_x}, \\ \dfrac{\partial p_y}{\partial z} = -\dfrac{\partial H}{\partial q_y} \end{cases} \tag{4.24}$$

where we take $x \to q_x$ and $y \to q_y$.

Given the coordinates point (q_x, q_y) and the generalized momenta (p_x, p_y), it is possible to solve the equations of motion for the ray position.

Lagrangian optics and Hamiltonian optics differ in certain aspects of notation but without encapsulating the same behavior of light, which is the behavior of light under SVEA, as the ray equation. In the next section, we will see more tangible applications of the Fermat principle. The Fermat principle is the fundamental basis of the concept called stigmatism and of all known stigmatic systems.

4.3 Law of reflection

In chapter 2, we found Snell's law for refraction starting from implications of the eikonal equation. Well, as the name of this expression suggests, such an equation is for when light is refracted, in the case of how much light is reflected. Reflection is the change in the direction of a wavefront at an interface between two different media so that the wavefront returns into the medium from which it originated.

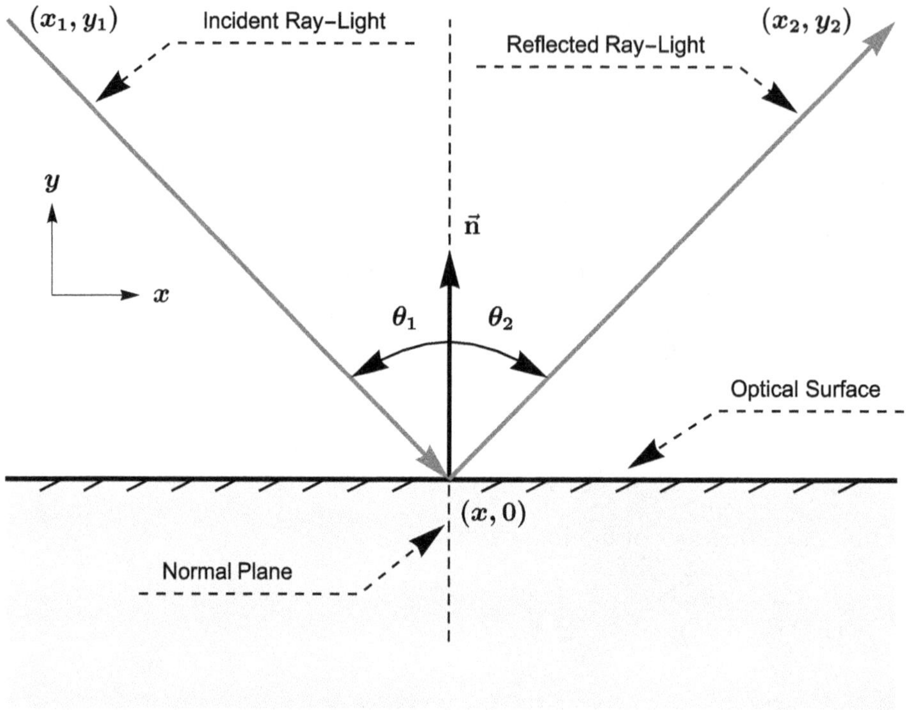

Figure 4.2. Diagram of the reflection of light.

Fermat's principle predicts how the path of light will behave when it is reflected. Observe figure 4.2; it shows an incident ray of light on a surface. The surface reflects the ray. We can see that if the ray starts from point x_1, y_1, it arrives in a straight line to the point where it is reflected in the surface and then reaches point x_2, y_2. Taking this into account, we can write the time that light takes from point x_1, y_1 to point x_2, y_2 as the following expression

$$T(x) = \frac{1}{c}\left[\sqrt{(x - x_1)^2 + y_1^2} + \sqrt{(x_2 - x)^2 + y_2^2}\right], \tag{4.25}$$

where x is the place on the plane on the horizontal direction where the light strikes. Now we know that $T(x)$ is correct in the way that for places with constant refractive index, the light moves in straight lines; all this from the ray equation.

But we do not know yet how the light reflects. The point $(x, 0)$ is crucial here. The value of x will tell us the slope of the straight lines. From the Fermat principle, we know that light moves in the least time. Thus, we can derive with respect to x and equal to zero

$$\frac{\partial T(x)}{\partial x} = 0, \tag{4.26}$$

computing the derivative,

$$\frac{x - x_1}{\sqrt{(x - x_1)^2 + y_1^2}} = \frac{x_2 - x}{\sqrt{(x_2 - x)^2 + y_2^2}}. \tag{4.27}$$

From figure 4.2, we can see that the last equation can be re-formulated as,

$$\sin\theta_1 = \sin\theta_2. \tag{4.28}$$

The last expression leads to

$$\boxed{\theta_1 = \theta_2}. \tag{4.29}$$

In figure 4.2, we can see a simple diagram of a reflection taking place on a flat mirror, the incident angle θ_1 is equal to the reflected angle θ_2; in other words $\theta_1 = \theta_2$.

We can show that it is a real minimum by computing the second derivative

$$\begin{aligned} c\frac{\partial^2 T}{\partial x^2} &= \frac{1 - \sin^2\theta_1}{\sqrt{(x - x_1)^2 + y_1^2}} + \frac{1 - \sin^2\theta_2}{\sqrt{(x_2 - x)^2 + y_2^2}} \\ &= (1 - \sin^2\theta_2)\left[\frac{1}{\sqrt{(x - x_1)^2 + y_1^2}} + \frac{1}{\sqrt{(x_2 - x)^2 + y_2^2}}\right]. \end{aligned} \tag{4.30}$$

Since $\sin\theta_2^2 \leqslant 1$, the above expression is positive for all values of x, which means we have an absolute minimum.

4.4 Law of refraction

Refraction occurs when light travels through a region that has a changing index of refraction. The simplest case of refraction occurs when there is an interface between a uniform medium with an index of refraction n_1 and another medium with an index of refraction n_2. In such situations, **Snell's Law**, also called the **Law of Refraction**, describes the resulting deflection of the light ray. We get Snell's law from the eikonal equation in chapter 2. Now, in this section, we deduce it from Fermat's principle.

In figure 4.3, we can see a simple diagram of a reflection taking place on a flat interface between two mediums with constant refraction indexes. From the figure we can put the trajectory of light as

$$R(x) = n_1\sqrt{d_1^2 + x^2} + n_2\sqrt{d_2^2 + (L - x)^2}, \tag{4.31}$$

where d_1 is the height of the initial position of the light ray and x is the horizontal distance from the initial position of the light ray to the origin of the coordinate system. L is the horizontal distance from the initial position to the final position and d_2 is the height of the final position of the light ray.

To take the path which the light completes with the least time we deviate the equation with respect to x.

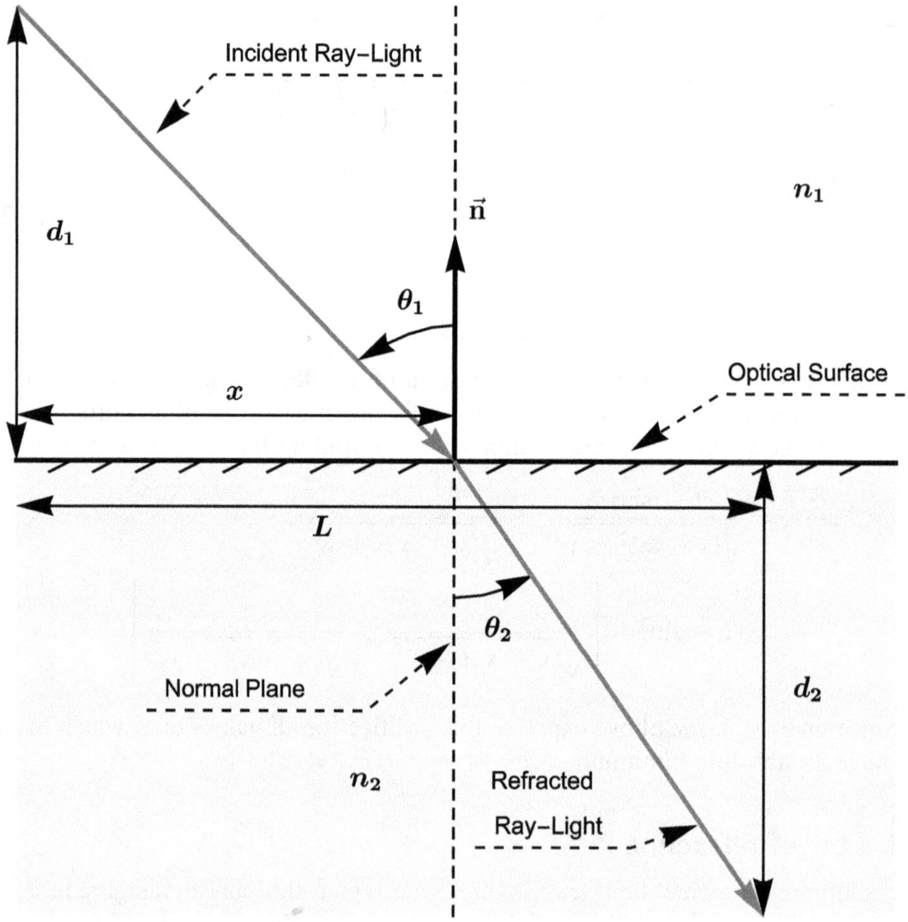

Figure 4.3. Diagram of the refraction of light.

$$\frac{dR(x)}{dx} = n_1 \frac{x}{\sqrt{d_1^2 + x^2}} - n_2 \frac{(L - x)}{\sqrt{d_2^2 + (L - x)^2}}. \tag{4.32}$$

From figure 4.3, we can see that precisely $\sin \theta_1$ is given by

$$\sin \theta_1 = \frac{x}{\sqrt{d_1^2 + x^2}}. \tag{4.33}$$

Also from the figure, we see that $\sin \theta_2$ is

$$\sin \theta_2 = \frac{L - x}{\sqrt{d_2^2 + (L - x)^2}}. \tag{4.34}$$

Therefore, replacing equations (4.33) and (4.34) in equation (4.32),

$$\boxed{n_1 \sin \theta_1 = n_2 \sin \theta_2.} \tag{4.35}$$

When the light rays are in a homogeneous medium, their paths are straight lines. When the light changes from one homogeneous medium to another medium, the refraction (deviation) is expressed by the above equation, which is Snell's law for refraction in angle notations.

4.5 Fermat's principle and Snell's law

Fermat's principle states that light travels along the path that takes the least time. Consider the refraction index as a measure of the speed of light in a material. Then you can use Fermat's principle to derive the same angular relationship between the incident and refracted rays as Snell's law. Fermat's principle and Snell's law share information in common, but they are not the same.

4.6 The Malus–Dupin theorem

The Malus–Dupin theorem is one of the fundamental theorems of geometric optics that relates to wave optics. The Malus–Dupin theorem tells us: if on each ray emitted by a source, we travel the same optical paths, then the points that delimit them form a surface normal to all rays. We call this surface a wavefront. It coincides with the wavefront given by the oscillatory theory under SVEA. When deduced from the Fermat principle, it is valid despite the number of reflections or refractions that the ray may undergo before reaching its destination. The proof is made from the Fermat principle.

We take two distinct infinitesimally separated paths, $[AB]$ and $[AB']$, where A is the focus and B and B' are the arrival points separated by equal optical paths. Then we define the respective optical paths as

$$L_B = \int_A^B n(\vec{\mathbf{r}})ds, \tag{4.36}$$

and

$$L_{B'} = \int_A^{B'} n(\vec{\mathbf{r}}')ds', \tag{4.37}$$

$\vec{\mathbf{r}}$ and $\vec{\mathbf{r}}'$ being the respective position vectors, ds and ds' the respective differentials of space and $n(\vec{\mathbf{r}})$ the refractive index. Notice that we admit that the refractive index is differentiable.

Now, for the derivation, we will use a first-order Taylor series development, so we recall it,

$$f(\vec{\mathbf{r}}) = f(\vec{\mathbf{r}}_0) + \nabla f(\vec{\mathbf{r}}_0) \cdot (\vec{\mathbf{r}} - \vec{\mathbf{r}}_0). \tag{4.38}$$

The equation of the path of a light ray (deduced from the Fermat principle, see chapter 2), is

$$\frac{d}{ds}(n\vec{\mathbf{u}}) = \nabla n, \tag{4.39}$$

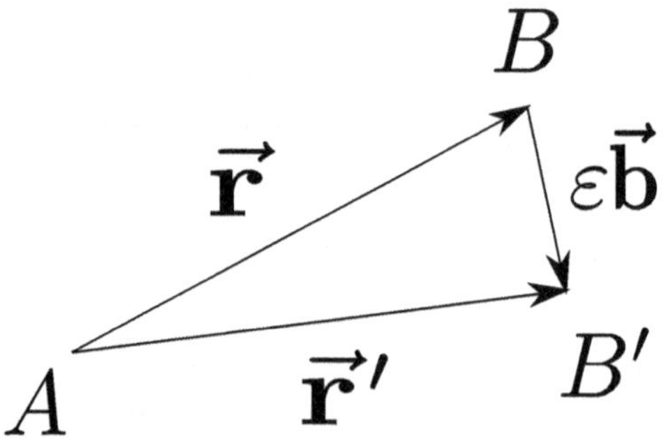

Figure 4.4. A diagram of the relationship **r** with $\vec{\mathbf{r}}'$.

hence,

$$(n\vec{\mathbf{u}}) = \nabla n ds. \tag{4.40}$$

The relationship $\vec{\mathbf{r}}' = \vec{\mathbf{r}} + \varepsilon\vec{\mathbf{b}}$, of which $\vec{\mathbf{dr}}' = \vec{\mathbf{dr}} + \varepsilon\vec{\mathbf{db}}$, can be seen in figure 4.4.

We admit that the refractive index admits a Taylor series development of order 1. Then, we find that

$$n(\vec{\mathbf{r}}') = n(\vec{\mathbf{r}}) + \nabla n(\vec{\mathbf{r}}) \cdot (\vec{\mathbf{r}}' - \vec{\mathbf{r}}) = n(\vec{\mathbf{r}}) + \nabla n(\vec{\mathbf{r}}) \cdot \varepsilon\vec{\mathbf{b}}. \tag{4.41}$$

On the other hand, we will do the same with the module of the position vector:

$$|\vec{\mathbf{r}}'| = |\vec{\mathbf{r}}| + \nabla|\vec{\mathbf{r}}| \cdot (\vec{\mathbf{r}}' - \vec{\mathbf{r}}) \tag{4.42}$$

$$= |\vec{\mathbf{r}}| + \frac{\partial}{\partial r} r\,\vec{\mathbf{e}}_r \cdot (\vec{\mathbf{r}}' - \vec{\mathbf{r}}) \tag{4.43}$$

$$= |\vec{\mathbf{r}}| + \vec{\mathbf{u}} \cdot (\vec{\mathbf{r}}' - \vec{\mathbf{r}}) \tag{4.44}$$

$$= |\vec{\mathbf{r}}| + \vec{\mathbf{u}} \cdot \vec{\mathbf{r}}' - \vec{\mathbf{u}} \cdot \vec{\mathbf{r}} \tag{4.45}$$

$$= |\vec{\mathbf{r}}| + \vec{\mathbf{u}} \cdot \vec{\mathbf{r}}' - |\vec{\mathbf{r}}| \tag{4.46}$$

$$= \vec{\mathbf{u}} \cdot \vec{\mathbf{r}}'. \tag{4.47}$$

Notice that $\vec{\mathbf{r}} = \vec{\mathbf{e}}_r = \vec{\mathbf{u}}$. Thus,

$$ds' = |\vec{\mathbf{dr}}'| = \vec{\mathbf{u}} \cdot \vec{\mathbf{dr}}' = \vec{\mathbf{u}} \cdot \vec{\mathbf{dr}} + \vec{\mathbf{u}} \cdot \varepsilon\vec{\mathbf{db}} = ds + \varepsilon\vec{\mathbf{u}} \cdot \vec{\mathbf{db}}. \tag{4.48}$$

ds' is replaced in the optical path, equation (4.37),

$$L_{B'} = \int_A^{B'} n(\vec{\mathbf{r}}')ds' = \int_A^{B'} [n(\vec{\mathbf{r}}) + \nabla n \cdot \varepsilon\vec{\mathbf{b}}](ds + \varepsilon\vec{\mathbf{u}} \cdot \vec{\mathbf{db}}), \tag{4.49}$$

expanding,

$$L_{B'} = \int_A^{B'} n(\vec{\mathbf{r}})ds + \int_A^{B'} \nabla n \cdot \varepsilon \vec{\mathbf{db}} + \int_A^{B'} n(\vec{\mathbf{r}})\varepsilon\vec{\mathbf{u}} \cdot \vec{\mathbf{db}} + \int_A^{B'} \nabla n \cdot \varepsilon\vec{\mathbf{b}}\varepsilon\vec{\mathbf{u}} \cdot \vec{\mathbf{db}}. \quad (4.50)$$

We remove the order elements from ε^2 from equation (4.50)

$$L_{B'} = \int_A^{B'} n(\vec{\mathbf{r}})ds + \int_A^{B'} \nabla n \cdot \varepsilon \vec{\mathbf{db}} + \int_A^{B'} n(\vec{\mathbf{r}})\varepsilon\vec{\mathbf{u}} \cdot \vec{\mathbf{db}}. \quad (4.51)$$

We calculate ΔL, equation (4.37) less equation (4.51),

$$\Delta L = \int_A^{B'} n(\vec{\mathbf{r}})ds + \int_A^{B'} \nabla n \cdot \varepsilon\vec{\mathbf{b}}ds + \int_A^{B'} n(\vec{\mathbf{r}})\varepsilon\vec{\mathbf{u}} \cdot \vec{\mathbf{db}} - \int_A^{B} n(\vec{\mathbf{r}})ds. \quad (4.52)$$

We factor the terms by powers of ε,

$$\Delta L = \int_A^{B'} n(\vec{\mathbf{r}})ds - \int_A^{B} n(\vec{\mathbf{r}})ds + \varepsilon\left[\int_A^{B'} \nabla n \cdot \vec{\mathbf{b}}ds + \int_A^{B'} n(\vec{\mathbf{r}})\vec{\mathbf{u}} \cdot \vec{\mathbf{db}}.\right] \quad (4.53)$$

By the equation of the trajectories we have $d(n\vec{\mathbf{u}} \cdot \vec{\mathbf{b}}) = d(n\vec{\mathbf{u}}) \cdot \vec{\mathbf{b}} + n\vec{\mathbf{u}} \cdot \vec{\mathbf{db}} = \nabla n \cdot \vec{\mathbf{b}}ds + n\vec{\mathbf{u}} \cdot \vec{\mathbf{db}}$, so,

$$\Delta L = \int_A^{B'} n(\vec{\mathbf{r}})ds - \int_A^{B} n(\vec{\mathbf{r}})ds + \varepsilon\left[\int_A^{B'} d(n\vec{\mathbf{b}} \cdot \vec{\mathbf{u}})\right]. \quad (4.54)$$

If we assume that we have chosen $B \equiv B'$, then

$$\Delta L = \int_A^{B'} n(\vec{\mathbf{r}})ds - \int_A^{B'} n(\vec{\mathbf{r}})ds + \varepsilon\left[\int_A^{B'} d(n\vec{\mathbf{b}} \cdot \vec{\mathbf{u}})\right] = 0, \quad (4.55)$$

turns to

$$\Delta L = +\varepsilon\left[\int_A^{B'} d(n\vec{\mathbf{b}} \cdot \vec{\mathbf{u}})\right] = 0. \quad (4.56)$$

Thus,

$$\int_A^{B'} d(n\vec{\mathbf{b}} \cdot \vec{\mathbf{u}}) = 0, \quad (4.57)$$

which implies

$$n(\vec{\mathbf{r}}_{B'})\vec{\mathbf{b}}(\vec{\mathbf{r}}_{B'}) \cdot \vec{\mathbf{u}}(\vec{\mathbf{r}}_{B'}) - n(\vec{\mathbf{r}}_A)\vec{\mathbf{b}}(\vec{\mathbf{r}}_A) \cdot \vec{\mathbf{u}}(\vec{\mathbf{r}}_A) = 0, \quad (4.58)$$

where $\vec{\mathbf{r}}_{B'}$ is the vector $\vec{\mathbf{r}}$ at point $B \equiv B'$ and $\vec{\mathbf{r}}_A$ is the vector $\vec{\mathbf{r}}$ at A. Since point A is the focus, the separation is always null, consequently,

$$n(\vec{\mathbf{r}}_{B'})\vec{\mathbf{b}}(\vec{\mathbf{r}}_{B'}) \cdot \vec{\mathbf{u}}(\vec{\mathbf{r}}_{B'}) = 0. \quad (4.59)$$

Therefore, the first term of equation should be zero as well,

$$\vec{\mathbf{b}}(\vec{\mathbf{r}}_{B'}) \cdot \vec{\mathbf{u}}(\vec{\mathbf{r}}_{B'}) = 0. \tag{4.60}$$

The only way for the last expression to be zero is when both vectors are perpendicular,

$$\vec{\mathbf{b}}(\vec{\mathbf{r}}_{B'}) \perp \vec{\mathbf{u}}(\vec{\mathbf{r}}_{B'}). \tag{4.61}$$

We have $\vec{\mathbf{r}}' = \vec{\mathbf{r}} + \varepsilon\vec{\mathbf{b}}$, so $\varepsilon\vec{\mathbf{b}}(\mathbf{r})$ joins the points on the surface, from which the formed surface is orthogonal to each ray. We can justify that the points form a surface for continuity.

4.7 End notes

In this chapter, we focused on the variational behavior of light. From it, we successfully found the ray equation from the Euler equation.

Then, we formalized our notation of the Fermat principle with examples, and we found Snell's law of reflection and refraction.

Then, we studied the Lagrangian formalism under the Fermat principles. This procedure helps us to state and prove the Malus–Dupin theorem.

References

Arfken G B and Weber H J 1999 *Mathematical Methods for Physicists* (New York: Academic)

Boas M L 2006 *Mathematical Methods in the Physical Sciences* (New York: Wiley)

Born M and Wolf E 2013 Principles of Optics: Electromagnetic Theory of Propagation *Interference and Diffraction of Light* (Amsterdam: Elsevier)

Buchdahl H A 1993 *An Introduction to Hamiltonian Optics* (North Chelmsford, MA: Courier Corporation)

Cao S and Greenhalgh S 1994 Finite-difference solution of the eikonal equation using an efficient, first-arrival, wavefront tracking scheme *Geophysics* **59** 632–43

Currier R and Herman M F 1985 Numerical comparison of generalized surface hopping, classical analog, and self-consistent eikonal approximations for nonadiabatic scattering *J. Chem. Phys.* **82** 4509–16

Dacorogna B, Glowinski R and Pan T-W 2003 Numerical methods for the solution of a system of eikonal equations with Dirichlet boundary conditions *C. R. Math.* **336** 511–8

Elsgolc L D 2012 *Calculus of Variations* (North Chelmsford, MA: Courier Corporation)

Gelfand I M and Silverman R A *et al* 2000 *Calculus of Variations* (North Chelmsford, MA: Courier Corporation)

Goldstein H, Poole C and Safko J 2002 *Classical Mechanics* (Reading, MA: Addison-Wesley)

Griffiths D J 2005 *Introduction to Electrodynamics* (Cambridge: Cambridge University Press)

Hecht E 1974 *Schaum's Outline of Optics* (New York: McGraw-Hill)

Hecht E 2012 *Optics* (Noida: Pearson Education India)

Hoffnagle J A and Shealy D L 2011 Refracting the k-function: Stavroudis's solution to the eikonal equation for multielement optical systems *J. Opt. Soc. Am.* **A28** 1312–21

Huang L, Shu C-W and Zhang M 2008 Numerical boundary conditions for the fast sweeping high order WENO methods for solving the Eikonal equation *J. Comput. Math.* **26** 336–46

Hysing S-R and Turek S 2005 The Eikonal equation: numerical efficiency versus algorithmic complexity on quadrilateral grids *Proc. ALGORITMY* **vol 22**

Jackson J D 1999 *Classical Electrodynamics* (Hoboken, NJ: Wiley)

Lakshminarayanan V, Ghatak A and Thyagarajan K 2002 *Lagrangian Optics* (Berlin: Springer)

Lax M, Louisell W H and McKnight W B 1975 From Maxwell to paraxial wave optics *Phys. Rev.* **A11** 1365

Luneburg R K 1964 *Mathematical Theory of Optics* (Berkeley, CA: University of California Press)

Marion J B 2013 *Classical Dynamics of Particles and Systems* (New York: Academic)

Morrey C B Jr 2009 *Multiple Integrals in the Calculus of Variations* (Berlin: Springer)

Pegis R J 1961 I The modern development of Hamiltonian optics *Progress in Optics* **vol 1** (Amsterdam: Elsevier) pp 1–29

Qian J, Zhang Y-T and Zhao H-K 2007 Fast sweeping methods for eikonal equations on triangular meshes *SIAM J. Numer. Anal.* **45** 83–107

Ronchi V and Barocas V 1970 *The Nature of Light: An Historical Survey* (Cambridge, MA: Harvard University Press) p 12+ 288

Spira A and Kimmel R 2004 An efficient solution to the eikonal equation on parametric manifolds *Interfaces Free Boundaries* **6** 315–27

Stavroudis O 2012 The Optics of Rays *Wavefronts, and Caustics* **vol 38** (Amsterdam: Elsevier)

Stavroudis O N 1995 The k function in geometrical optics and its relationship to the archetypal wave front and the caustic surface *J. Opt. Soc. Am.* **A12** 1010–6

Stavroudis O N 2006 *The Mathematics of Geometrical and Physical Optics: The K-function and Its Ramifications* (New York: Wiley)

Stavroudis O N and Hurtado-Ramos J B 2000 Maxwell equations and the k function *J. Opt. Soc. Am.* **A17** 1469–74

Vavryčuk V 2012 On numerically solving the complex eikonal equation using real ray-tracing methods: a comparison with the exact analytical solution *Geophysics* **77** T109–16

Zhao H 2005 A fast sweeping method for eikonal equations *Math. Comput.* **74** 603–27

IOP Publishing

Stigmatic Optics (Second Edition)

Rafael G González-Acuña and Héctor A Chaparro-Romo

Chapter 5

Stigmatism and stigmatic reflective surfaces

Having laid the foundations of geometric optics, the equations of the eikonal and the ray, we can focus on the sub-branches of geometric optics, optical design, aberration theory and stigmatism. In this chapter, we address the aforementioned topics and, in addition, we study the simplest stigmatic systems, conical mirrors.

5.1 Introduction

Imagine that you have a solid connected body with a refractive index different than the environment where the body is located. This body is a lens if it has the function of focusing or dispersing rays through the refraction that arises from the difference between the refraction index of the mentioned body and the medium. **Lenses** are fundamental elements of study in geometric optics.

If the body has the function of redirecting rays such that there is no dispersion or focus on them, then the element in question is a **Prism**. If the body does not have a function, then it is just a translucent rock.

We have mentioned that lenses have the function of focusing the light or scattering it. It is also true that mirrors have such features.

The law of reflection and the law of refraction depend on the normal of the surface. Therefore, the shape of the lens/mirror is essential to fulfilling its predetermined function.

When it comes to focusing the light, what is wanted, in principle, is that the rays that come from a point object converge into a point image. Therefore, what is wanted is to have stigmatic lenses and stigmatic mirrors. **Stigmatism** refers to the image-formation property of an optical system which focuses a point object into a point image. Such points are called a stigmatic pair of the optical system.

Stigmatic lenses and stigmatic mirrors need to have a very particular shape. In the case of the mirrors, the reflective surfaces that form the stigmatic mirrors are the conic sections.

doi:10.1088/978-0-7503-6423-2ch5

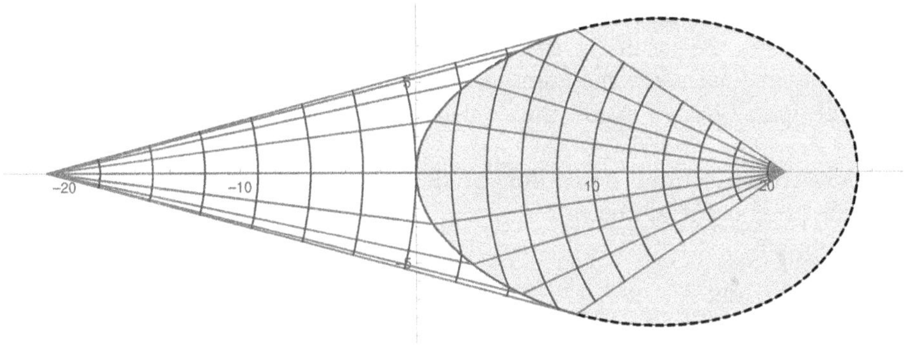

Figure 5.1. Cartesian oval.

- For mirrors with parabolic surfaces, parallel rays that hit the mirror produce reflected rays that converge into a common focus[1].
- For mirrors with spherical surfaces, all rays that emerge from a point object located at a finite distance from the reflecting surface are reflected in the same point; if and only if the object is situated in the center of the circumference.
- For mirrors with elliptic surfaces, all rays that emerge from a point object are reflected in another point, the point image.
- For mirrors with hyperbolic surfaces, all rays that arise from a point object are reflected in a single virtual point image.

In this chapter, we are going to study the aforementioned conical mirrors. The beauty of conical mirrors lies in the intrinsic geometric properties of conic sections.

Other curved surfaces may also focus light, but not in a single point. The stigmatic refractive surface is the Cartesian oval, which is a fourth-order function. In other words, the Cartesian oval is a surface such that all the rays that emerge from a point object are focused on a single image point, once they are refracted (see figure 5.1).

Conic mirrors and the Cartesian oval are results of interest that have been preserved and will be preserved over time by their analytic nature. Cartesian ovals will be studied in chapters 6–8.

In the case of stigmatic lenses, it took more than two thousand years to have a general equation that described their surfaces. The general equation that describes the stigmatic lenses is not trivial. The aforementioned equation will be reviewed in chapter 9.

At this point, it is convenient to define **Stigmatic Optics**. It is the design of an optical system based exclusively on the premise of geometrical optics without

---

[1] Diocles (240 BC–ca. 180 BC), in his work *Burning Mirrors*, was the first person to report this property of the parabolic mirror.

counting any paraxial approximation and without having any optimization process. The relevance of the results given by the analytical optical design is that they are preserved over time, since they are general and not particular cases. Examples of analytical optical design results are conical mirrors, Cartesian ovals and stigmatic lenses.

Before we present the derivation of the shapes of the conic mirror we will introduce the concept of aberrations.

5.2 Aberrations

When a system is not stigmatic for all points of the object, then the system has aberrations. In this section, we will show the terminology implemented throughout this treatise on optical aberrations.

If the optical system has aberration, the point object is projected in a region in the image space instead of at a single point. The nature of the region of space where the image is formed depends on the type of aberration. The optical aberrations of the optical system distort the image formed by the optical system.

Aberrations fall into two classes: chromatic and monochromatic. The variation of a lens's refractive index concerning the wavelength causes chromatic aberrations.

The geometry of the optical system causes monochromatic aberrations. In general, it occurs both when light is reflected or refracted so the reflection law, Snell's law and Fermat's principle are involved in the phenomenon. They have information about the geometry of the imaging system. Therefore, the shape of the optical system is crucial for monochromatic aberrations to vanish. The five basic types of monochromatic aberrations are spherical aberration, coma, astigmatism, field curvature and image distortion.

5.2.1.1 Spherical aberration

Spherical aberration is the phenomenon that exists in an optical system when a point object located on the optical axis does not have a stigmatic correspondence with a point image. In other words, the rays that leave the point object on the optical axis do not converge on a point image on the optical axis. It is called spherical aberration because the spherical lens has this phenomenon. There are lenses that are called aspherical because their shape is different from the sphere. In most cases, their main goal is to reduce spherical aberration. In the following chapters, we will see that a stigmatic lens is aspherical.

In figure 5.2, we can see an example of spherical aberration generated by a spherical surface with constant refraction index n along with the material.

Another example of spherical aberration is presented in figure 5.3. This time the surface is parabolic.

5.2.1.2 Coma

Coma aberration in an optical system refers to the aberration suffered by the image of a point object outside the axis. Coma makes the image appear distorted, with a tail, like a comma or a comet. In other words, an optical system with coma has no

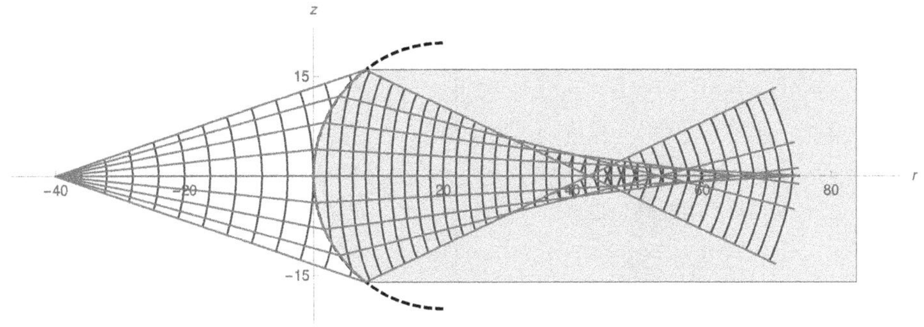

Figure 5.2. Spherical aberration in a spherical surface.

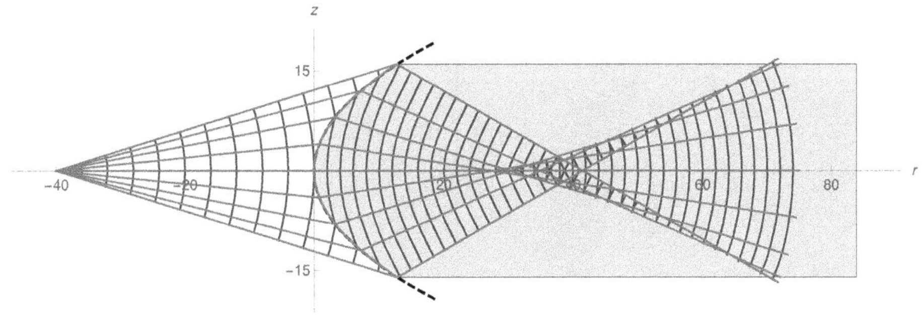

Figure 5.3. Spherical aberration in a parabolic surface.

stigmatic relationship between a point object outside the optical axis and a point image, since this image is not a point but a region (see figures 5.4 and 5.5).

As another way to explain how coma looks, in figure 5.6 we show an interface with rays that come from an off-axis object, then the rays cross the surface of the index refraction n, and they are refracted. The refraction causes an inversion of some of the rays, as can be seen in the same figure. This inversion in the images looks like a comma or a comet.

5.2.1.3 Astigmatism

The astigmatism of a point object takes place when two perpendicular planes have different image points. In figure 5.7 we show a lens with astigmatism.

The distortion in a forming image system is measured with a rectilinear projection. The rectilinear projection is passed through the system, a projection in which straight lines in a scene remain straight in the image if there is no distortion in the system. The most common distortions are shown in figures 5.8 and 5.9.

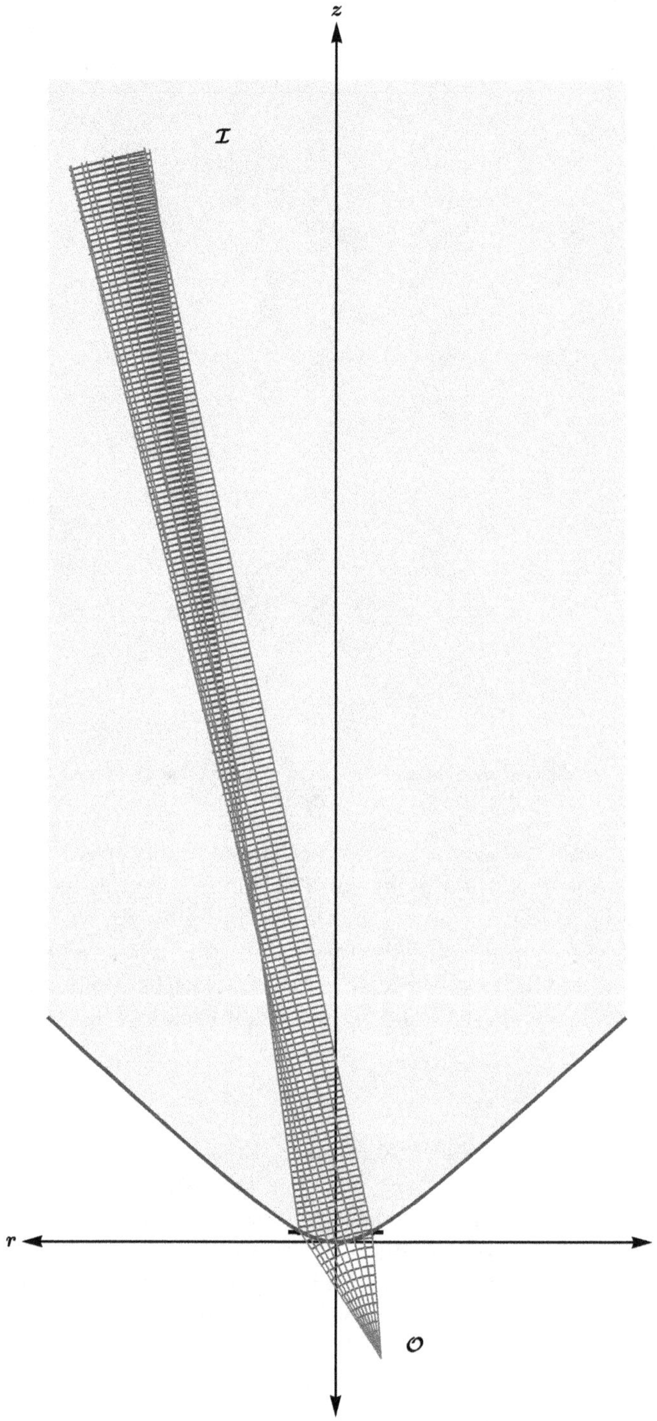

Figure 5.4. Polynomial refractive surface with coma.

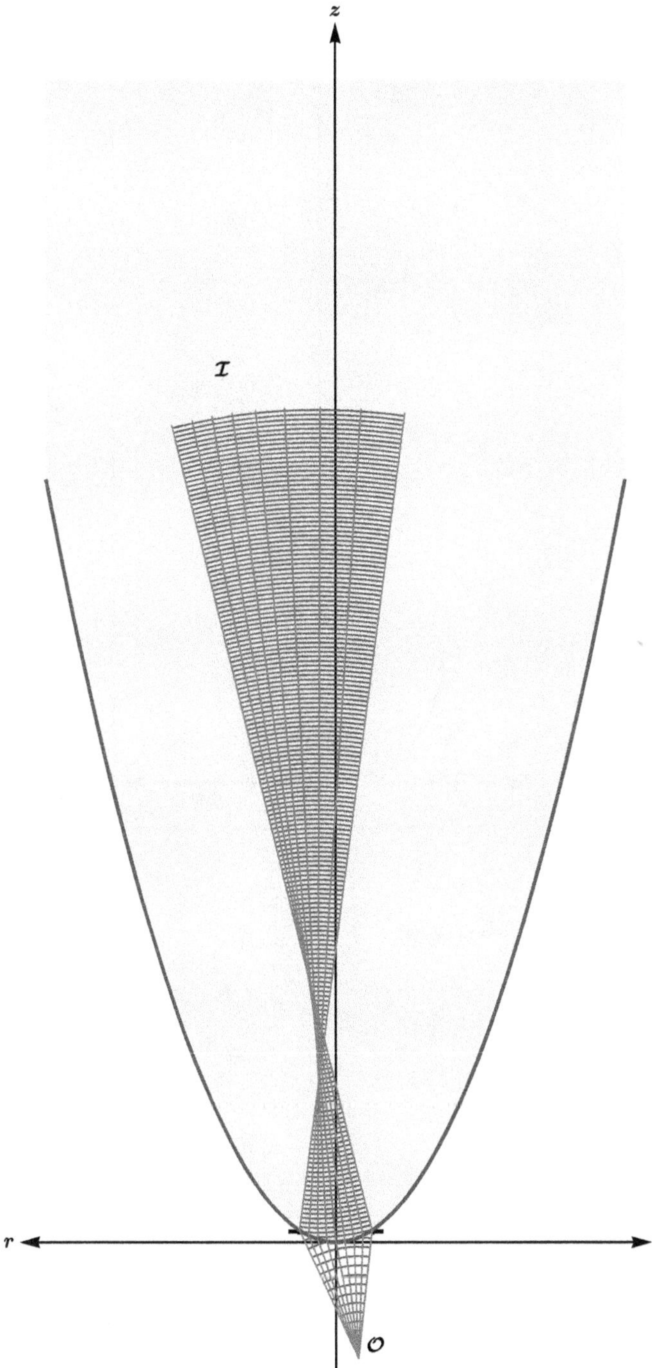

Figure 5.5. Parabolic refractive surface with coma.

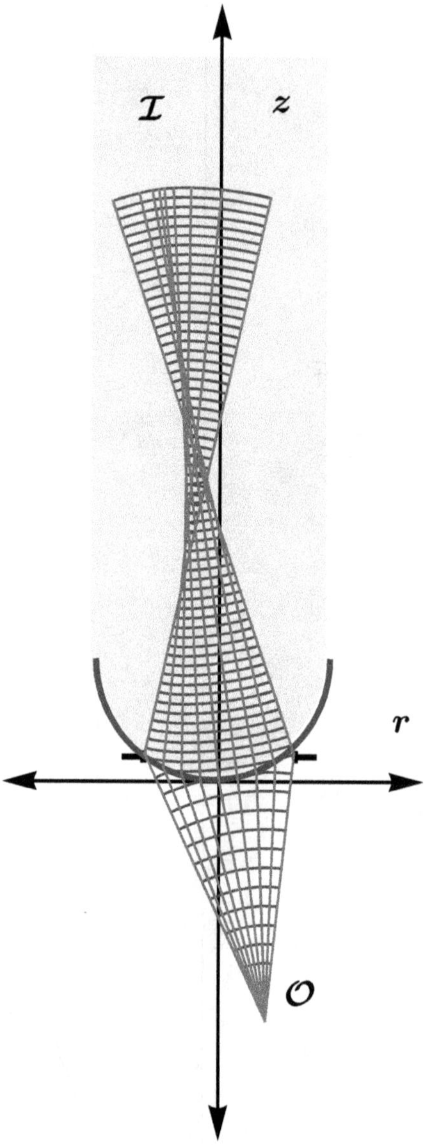

Figure 5.6. Spherical refractive surface with coma.

5.2.1.4 Field curvature

Petzval field curvature, named after Joseph Petzval, describes the optical aberration in which a flat object normal to the optical axis is focused as a curved image. Figure 5.10 shows the field curvature in a Cartesian oval.

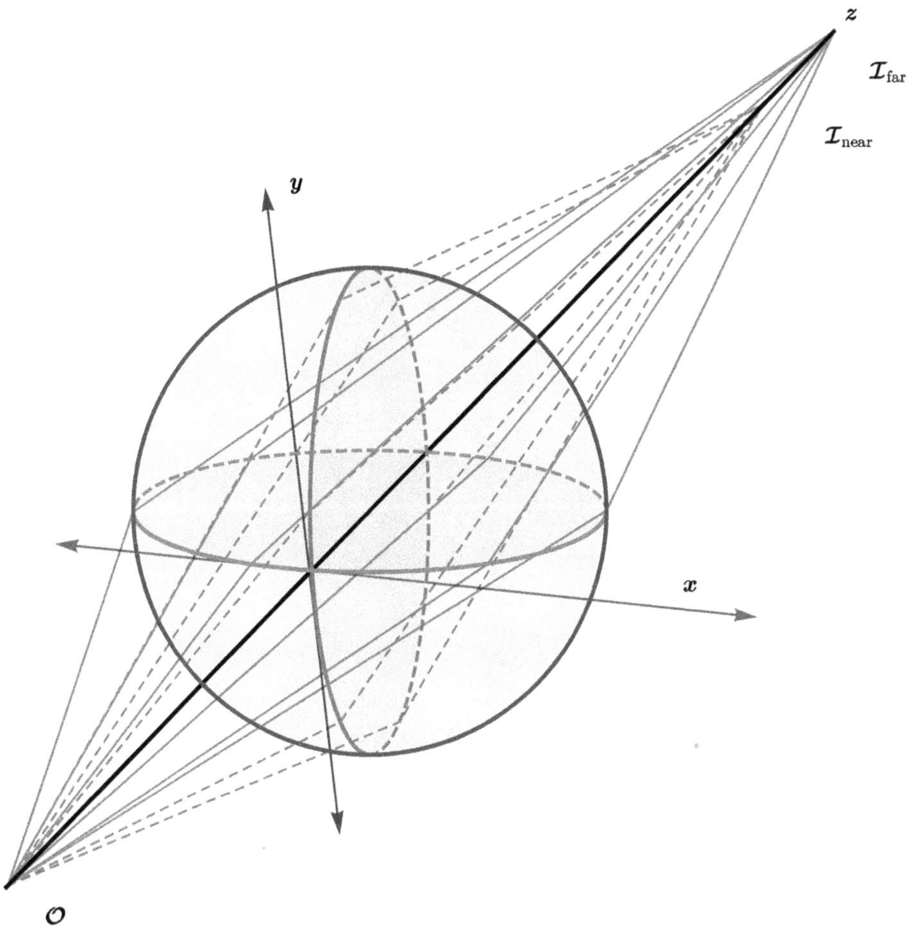

Figure 5.7. An astigmatism presented in a lens.

5.3 Conic mirrors

Once we have introduced the concepts of stigmatism and aberrations, we can start with the design of stigmatic optical systems. The first optical systems that we are going to study are conic mirrors. Ancient Greeks reviewed conic mirrors more than two millennia ago. To obtain the shapes of the stigmatic mirrors, we are going to use the Fermat principle already studied in chapter 4.

5.4 Elliptic mirror

The Fermat principle predicts that the light follows the least time path. If the refraction index of the medium is constant, the least time path from an object placed in z_o is a straight line to the image placed in z_i. This line—we can call it the optical axis if z_o and z_i are aligned—and the ray that follows it, is the axial ray. Hence, if all

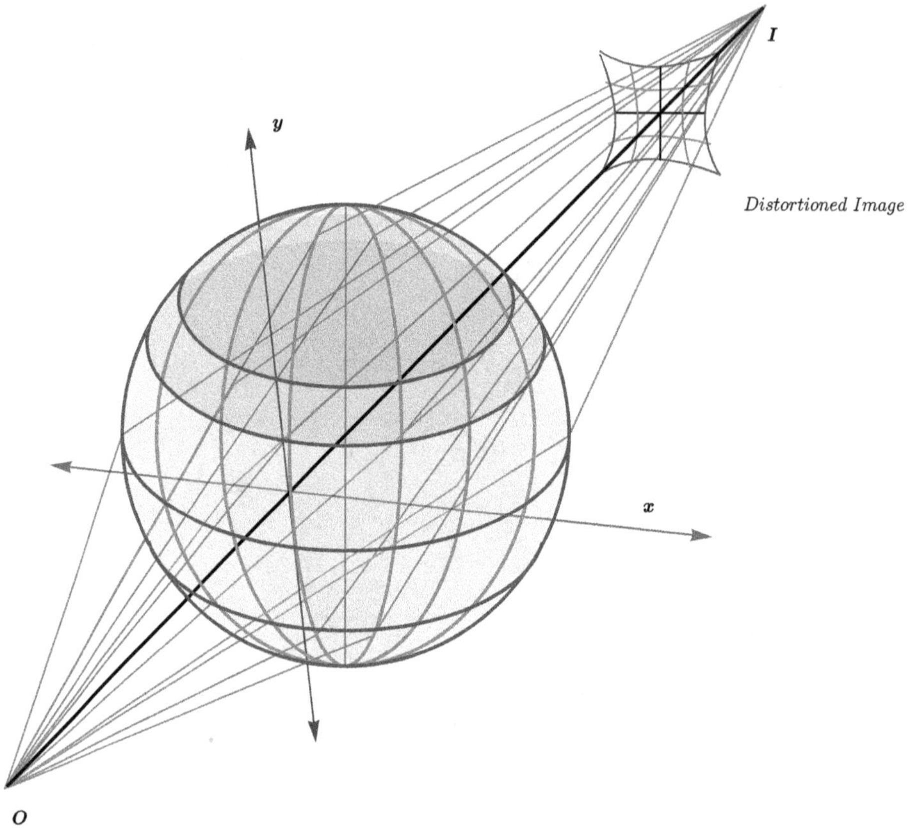

Figure 5.8. Pincushion distortion.

the rays follow the least time path, then the optical path of a non-axial ray is the same as the path of the axial ray, The optical path of the axial ray is

$$-z_o - z_i, \tag{5.1}$$

while the optical path of the non-axial ray is

$$\sqrt{(z_a - z_i)^2 + r_a^2} + \sqrt{r_a^2 + (z_a - z_o)^2}, \tag{5.2}$$

where r_a is the radius of the surface z_a, and it works as the independent variable. The idea is how must z_a be such that the mirror is stigmatic. Taking into account what we have deduced in the first paragraph of this chapter, we have

$$-z_o - z_i = \sqrt{(z_a - z_i)^2 + r_a^2} + \sqrt{r_a^2 + (z_a - z_o)^2}, \tag{5.3}$$

where in the last expression we have $-z_o - z_i$ in the left side because the coordinate system is placed at the vertex of the mirror.

Solving for z_a,

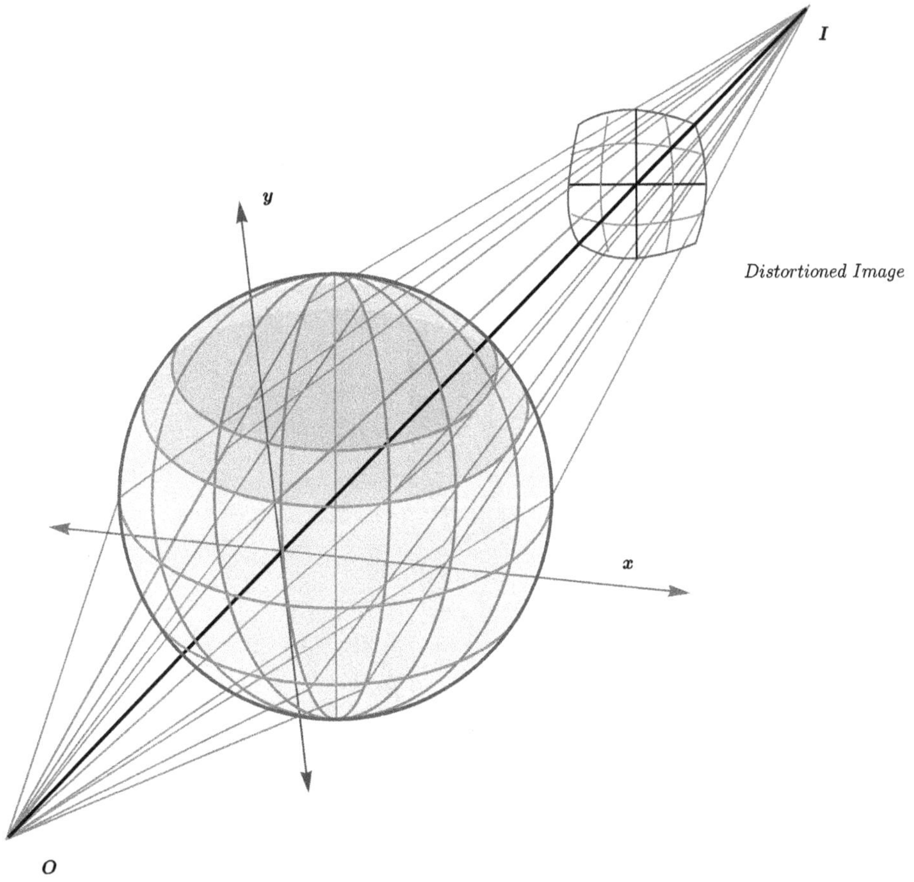

Figure 5.9. Barrel distortion.

$$z_a = \frac{1}{2}\left[z_o + z_i \pm \frac{\sqrt{z_o z_i (z_o + z_i)^2 (z_o z_i - r_a^2)}}{z_o z_i}\right]. \tag{5.4}$$

From the last expression, from the ± we take the positive sign, in order to have an elliptic mirror, hence,

$$z_a = \frac{1}{2}\left[z_o + z_i + \frac{\sqrt{z_o z_i (z_o + z_i)^2 (z_o z_i - r_a^2)}}{z_o z_i}\right]. \tag{5.5}$$

The last expression gives us the shape of a mirror that for a point object and a point image located at z_o and z_i, respectively, the system is stigmatic.

An example of an elliptic mirror can be seen in figure 5.11. The configuration design is captured in the caption of the image.

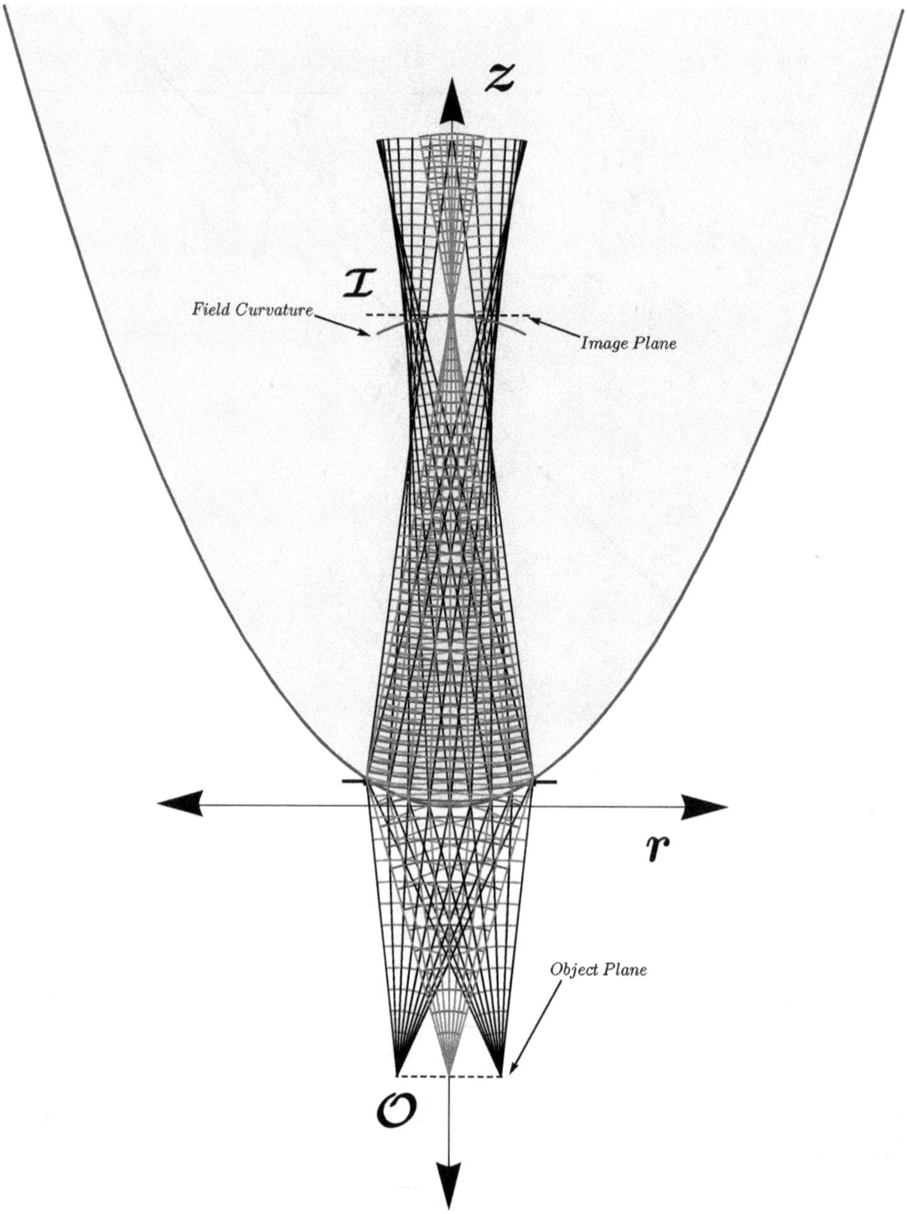

Figure 5.10. Petzval field curvature in a Cartesian oval.

5.5 Circular mirror

The circle is an exceptional case of the ellipse when the two focuses of the ellipse are placed together. Therefore, if we want a stigmatic optical system with a circular mirror, we should take $z_o = z_i$. Replacing $z_o = z_i$ in equation (5.5) we have

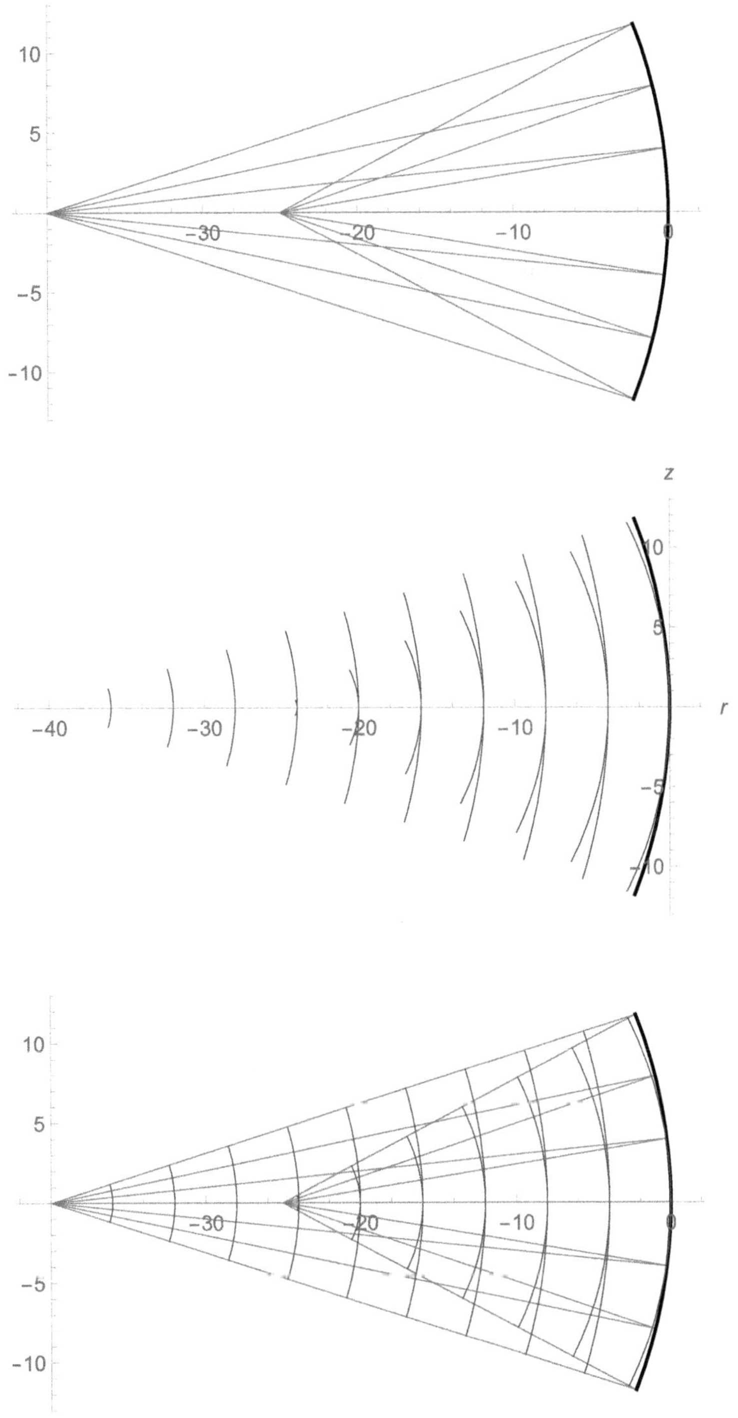

Figure 5.11. Specifications of the design: $z_o = -40$, $z_i = -25$ mm $z=$ equation (5.5).

$$z_a = \sqrt{z_i^2 - r_a^2} + z_i, \tag{5.6}$$

where z_a perfectly matches with the shape of a circle.

An example of a circular mirror is shown in figure 5.12 (see the caption for the design parameters).

5.6 Hyperbolic mirror

When the object is virtual (in the right side of the coordinate system), we should use equation (5.4) with the negative sign. This procedure led us to the hyperbolic mirror

$$z_a = \frac{1}{2}\left[z_o + z_i - \frac{\sqrt{z_o z_i (z_o + z_i)^2 (z_o z_i - r_a^2)}}{z_o z_i}\right]. \tag{5.7}$$

For an example of a hyperbolic mirror, see figure 5.13.

5.7 Parabolic mirror

If the object is very far away from the image $z_o \to \infty$ and we want a stigmatic mirror, we should manipulate equation (5.3) in order to get the following expression

$$1 = \frac{\sqrt{r_a^2 + (z_a - z_o)^2} + z_o}{-\sqrt{(z_a - z_i)^2 + r_a^2} - z_i}. \tag{5.8}$$

Then, we can compute the limit when $z_o \to \infty$,

$$\lim_{z_o \to -\infty}\left[\frac{\sqrt{r_a^2 + (z_a - z_o)^2} + z_o}{-\sqrt{(z_a - z_i)^2 + r_a^2} - z_i}\right] = -\frac{z_a}{\sqrt{(z_i - z_a)^2 + r_a^2} + z_i}. \tag{5.9}$$

Substituting equation (5.9) in equation (5.8), we have

$$1 = -\frac{z_a}{\sqrt{(z_i - z_a)^2 + r_a^2} + z_i}. \tag{5.10}$$

And finally, solving for z_a,

$$z_a = \frac{r_a^2}{4z_i}. \tag{5.11}$$

It is clear that the last expression is a parabola. An example of a parabolic mirror can be found in figure 5.14.

5.8 End notes

In this chapter, we demonstrated several essential concepts, such as stigmatism and how the absence of stigmatism in a system forms aberrations. We present the most common types of aberrations and their diagrams.

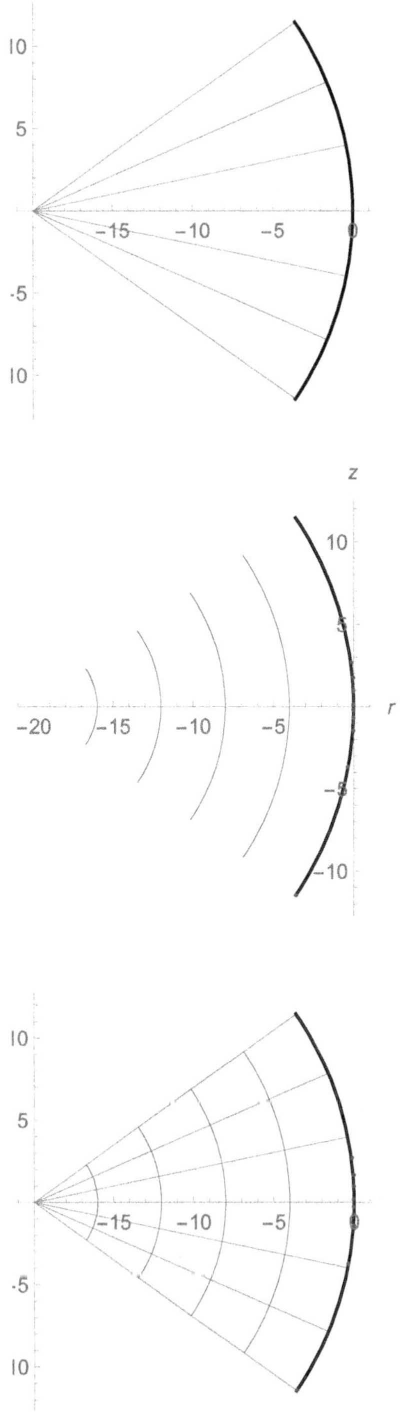

Figure 5.12. Specifications of the design: $z_o = -40$, $z_i = -25$ mm $z_a = \sqrt{z_i^2 - r_a^2} + z_i$.

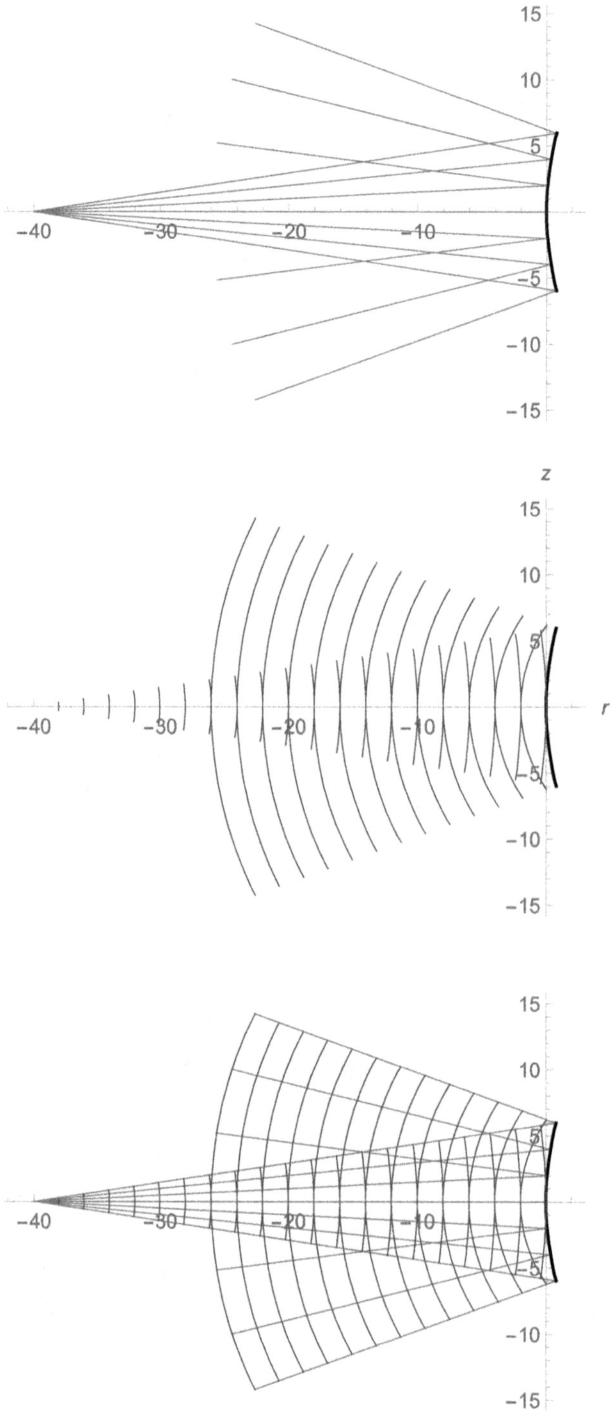

Figure 5.13. Specifications of the design: $z_o = -40$, $z_i = 16$ mm and $z_a =$ equation (5.7).

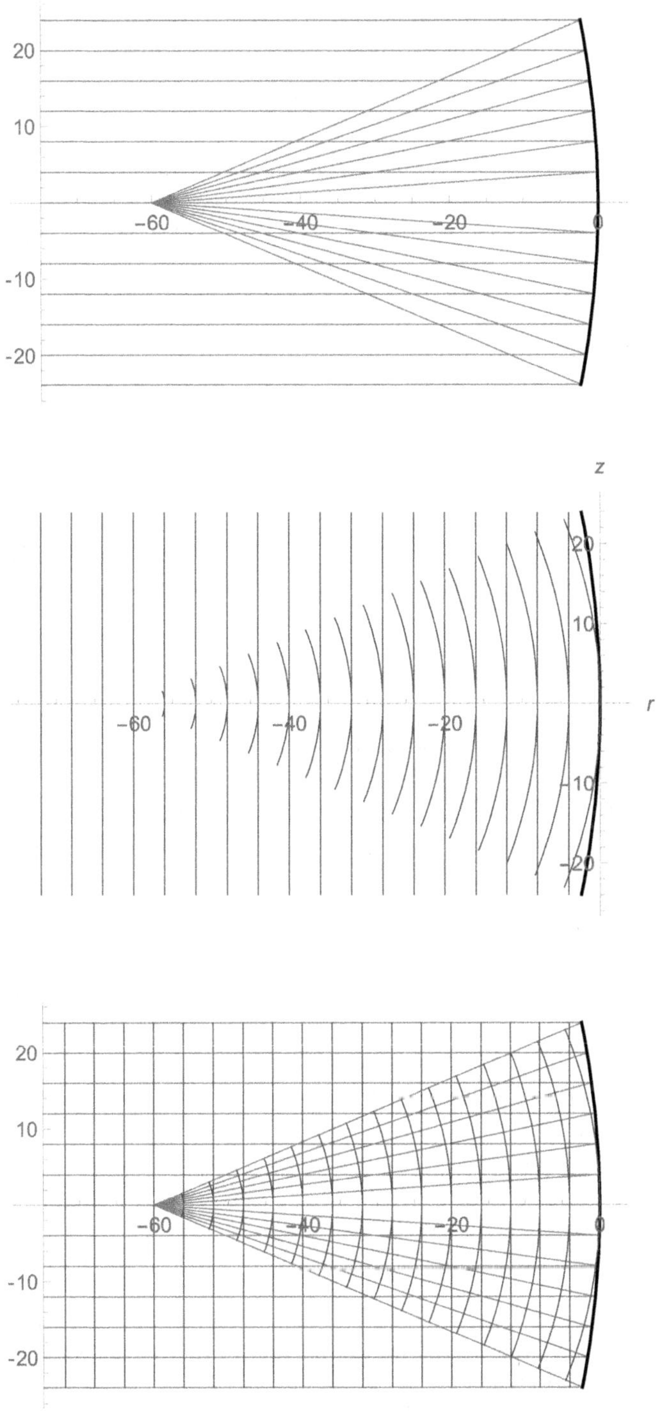

Figure 5.14. Specifications of the design: $z_o = -\infty$, $z_i = -60$ mm and $z_a = \frac{r_a^2}{4z_i}$.

Finally, using the Fermat principle, we find that static mirrors have conical curve shapes. For each case, we show the deduction and an illustrative example.

References

Bass M 1995 *Handbook of Optics, Volume I: Fundamentals Techniques and Design* (New York: McGraw-Hill)

Bellosta H 2002 Burning instruments: from Diocles to Ibn Sahl *Arabic Sci. Phil.* **12** 285–303

Born M and Wolf E 2013 *Principles of Optics: Electromagnetic Theory of Propagation, Interference and Diffraction of Light* (Amsterdam: Elsevier)

Braunecker B, Hentschel R and Tiziani H J 2008 *Advanced Optics Using Aspherical Elements* **vol 173** (Bellingham, WA: SPIE Press)

Chaves J 2016 *Introduction to Nonimaging Optics* 2nd edn (Boca Raton, FL: CRC Press)

Daumas M 1972 *Scientific Instruments of the Seventeenth and Eighteenth Centuries and Their Makers* (London: Batsford) p 361 12+ 142 plates

Descartes R 2012 *The Geometry of Rene Descartes: With a Facsimile of the* First Edition (North Chelmsford, MA: Courier Corporation))

Duerr F, Benítez P, Minano J C, Meuret Y and Thienpont H 2012 Analytic design method for optimal imaging: coupling three ray sets using two free-form lens profiles *Opt. Express* **20** 5576–85

Glassner A S 1989 *An Introduction to Ray Tracing* (Amsterdam: Elsevier)

Gross H 2005 *Handbook of Optical Systems, Volume 1: Fundamentals of Technical Optics* (New York: Wiley) p 848

Hogendijk J P 2002 The burning mirrors of Diocles: reflections on the methodology and purpose of the history of pre-modern science *Early Sci. Med.* **7** 181–97

Kingslake R and Johnson R B 2009 *Lens Design Fundamentals* (New York: Academic)

Lefaivre J 1951 A new approach in the analytical study of the spherical aberrations of any order *J. Opt. Soc. Am.* **41** 647

Lin W, Benítez P, Miñano J C, Infante J and Biot G 2011 *Advances in the SMS design method for imaging optics* **vol 8167** (Bellingham, WA: International Society for Optics and Photonics) p 81670M

Luneburg R K 1964 *Mathematical Theory of Optics* (Berkeley, CA: University of California Press)

Malacara D and Malacara Z 1994 *Handbook of Lens Design* (New York: Marcel Dekker, Inc)

Malacara-Hernández D and Malacara-Hernández Z 2017 *Handbook of Optical Design* (Boca Raton, FL: CRC Press)

Miñano J C, Benítez P, Lin W, Muñoz F, Infante J and Santamaría A 2009 Overview of the SMS design method applied to imaging optics *Novel Optical Systems Design and Optimization XII* **vol 7429** (Bellingham, WA: International Society for Optics and Photonics) p 74290C

Schulz G 1983 Achromatic and sharp real imaging of a point by a single aspheric lens *Appl. Opt.* **22** 3242–8

Scott J F 2016 *The Scientific Work of René Descartes: 1596-1650* (Milton Park: Routledge)

Singer W, Totzeck M and Gross H 2006 Handbook of Optical Systems *Volume 2: Physical Image Formation* (New York: Wiley)

Stavroudis O 2012 *The Optics of Rays, Wavefronts, and Caustics* **vol 38** (Amsterdam: Elsevier)

Stavroudis O N 2006 *The Mathematics of Geometrical and Physical Optics: The K-function and Its Ramifications* (New York: Wiley)

Stavroudis O N and Feder D P 1954 Automatic computation of spot diagrams *J. Opt. Soc. Am.* **44** 163–70

Sun H 2016 *Lens Design: A Practical Guide* (Boca Raton, FL: CRC Press)

Toomer G J 2012 *Diocles, On Burning Mirrors: The Arabic Translation of the Lost Greek Original* **vol 1** (Berlin: Springer)

Vaskas E M 1957 Note on the Wassermann-Wolf method for designing aspheric surfaces *J. Opt. Soc. Am.* **47** 669–70

Wassermann G D and Wolf E 1949 On the theory of aplanatic aspheric systems *Proc. Phys. Soc.* **B62** 2

Winston R, Miñano J C and Benitez P G *et al* 2005 *Nonimaging Optics* (Amsterdam: Elsevier)

IOP Publishing

Stigmatic Optics (Second Edition)

Rafael G González-Acuña and Héctor A Chaparro-Romo

Chapter 6

Stigmatic reflective surfaces: the Cartesian ovals

The next known stigmatic surface is the Cartesian oval. In this chapter, we find different polynomial series for different circumstances of the Cartesian oval. These polynomial series depend on whether the object or image is virtual or real. It is worth mentioning that they are a first approximation of the Cartesian oval.

6.1 Introduction

The classic refractive optical system is based on spherical surfaces; the problem of these surfaces is that they generate a noisy image, because they have spherical aberration, the phenomenon studied in chapter 5. Several techniques reduce spherical aberrations modifying the shape of the surface; most of them are numeric or approximated.

The problem that we are going to solve in this chapter is the following: what is the shape of the surface that does not introduce spherical aberration, once the ray beam has passed through the interface? The shape is called a Cartesian oval. The Cartesian oval has several shapes according to the scenario; for example, if the object and the image are real/virtual, or if one is real and the other is virtual.

This chapter is based on the work of Valencia *et al*, where they presented the functions that describe the Cartesian oval in different scenarios. In the work on which this chapter is based, the authors take some well-known solutions of Cartesian ovals and, in more complex situations, they present polynomial expansions that describe Cartesian ovals. In this approach, it is deduced that all special cases of Cartesian ovals are from Snell's law and not the Fermat principle.

6.2 Stigmatic surfaces

The approach we explore throughout the chapter is the Valencia *et al* approach or the Valencia–Calle approach. Understanding the strategy implemented here will help us to understand the mathematical models coming from the following chapters, the stigmatic lenses.

doi:10.1088/978-0-7503-6423-2ch6

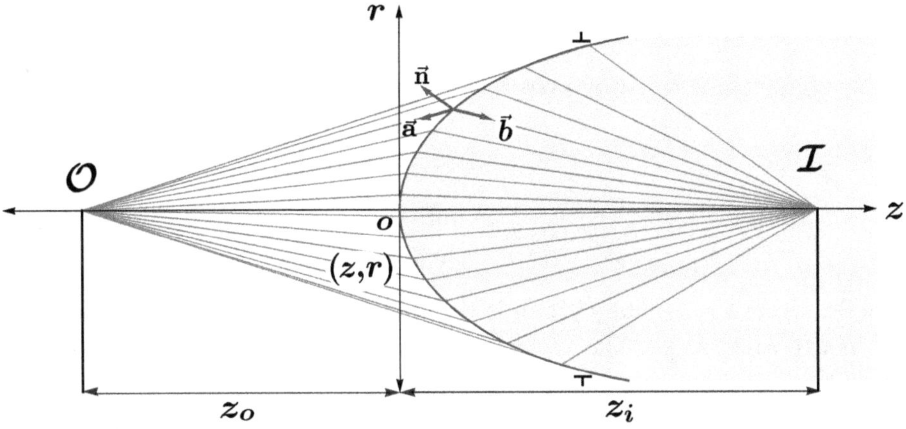

Figure 6.1. Diagram of a Cartesian oval. The origin is placed in the vertex of the Cartesian oval (z, r). The distance from the object to the vertex is z_o. The gap between the origin and the image is z_i. The normal vector of the Cartesian oval is $\vec{n}$, the incident ray is $\vec{a}$ and the refracted ray is $\vec{b}$.

The first assumption is that the interface is between two homogeneous optical materials with a positive refraction index in the object space n_o and the image space n_i. Thus, we calculate the parameter n as

$$n = \frac{n_i}{n_o}. \tag{6.1}$$

Calling Snell's law in the interface, we have

$$\sin \theta_i = n \sin \theta_r, \tag{6.2}$$

where the angles θ_i and θ_r are the incident ray angle and the refracted ray angle, respectively (see figure 6.1). Now, using the following trigonometric identity,

$$\sin \theta^2 + \cos \theta^2 = 1, \tag{6.3}$$

$$\sin \theta = \sqrt{1 - \cos \theta^2}, \tag{6.4}$$

we can express Snell's law, equation (6.2), as

$$\sqrt{1 - \cos \theta_i^2} = n\sqrt{1 - \cos \theta_r^2}. \tag{6.5}$$

Squaring both sides of Snell's law, equation (6.5),

$$1 - \cos \theta_i^2 = n(1 - \cos \theta_r^2). \tag{6.6}$$

Now the cosine of the angle between two vectors $\vec{\mathbf{a}}_1$ and $\vec{\mathbf{a}}_2$ and the dot product of the two vectors are related by

$$\vec{\mathbf{a}}_1 \cdot \vec{\mathbf{a}}_2 = |\vec{\mathbf{a}}_1||\vec{\mathbf{a}}_2| \cos \theta. \tag{6.7}$$

Taking this into account, we express the cosine of Snell's law as

$$\cos\theta_i = \frac{\vec{\mathbf{a}} \cdot \vec{\mathbf{n}}}{|\vec{\mathbf{a}}||\vec{\mathbf{n}}|}, \tag{6.8}$$

and

$$\cos\theta_r = \frac{-\vec{\mathbf{n}} \cdot \vec{\mathbf{b}}}{|\vec{\mathbf{b}}||\vec{\mathbf{n}}|}, \tag{6.9}$$

where the vector $\vec{\mathbf{a}}$ is a unitary vector in the opposite direction to the incident ray, its norm is expressed as $|\vec{\mathbf{a}}|$. The vector $\vec{\mathbf{b}}$ is a unitary vector in the direction of the refracted ray, its norm is expressed as $|\vec{\mathbf{b}}|$. Finally, the vector $\vec{\mathbf{n}}$ is the normal vector of the refractive surface, the surface under study, and its norm is expressed as $|\vec{\mathbf{n}}|$ (see figure 6.1). So, replacing (6.8) and (6.9) in equation (6.6),

$$1 - \left(\frac{\vec{\mathbf{a}} \cdot \vec{\mathbf{n}}}{|\vec{\mathbf{a}}||\vec{\mathbf{n}}|}\right)^2 = n^2\left[1 - \left(\frac{-\vec{\mathbf{n}} \cdot \vec{\mathbf{b}}}{|\vec{\mathbf{b}}||\vec{\mathbf{n}}|}\right)^2\right]. \tag{6.10}$$

Solving for n^2 and simplifying, we get equation (6.10)

$$\frac{|\vec{\mathbf{b}}|^2(|\vec{\mathbf{a}}|^2|\vec{\mathbf{n}}|^2 - (\vec{\mathbf{a}} \cdot \vec{\mathbf{n}})^2)}{|\vec{\mathbf{a}}|^2(|\vec{\mathbf{b}}|^2|\vec{\mathbf{n}}|^2 - (\vec{\mathbf{b}} \cdot \vec{\mathbf{n}})^2)} = n^2, \tag{6.11}$$

where the vectors are given by

$$\vec{\mathbf{a}} = [1, m_i], \tag{6.12}$$

$$\vec{\mathbf{b}} = [1, m_r], \tag{6.13}$$

$$\vec{\mathbf{n}} = \left[\frac{dz}{dr}, -1\right], \tag{6.14}$$

where m_i is the slope of the incident ray, m_r is the slope of the refracted ray and $\frac{dz}{dr}$ is the derivative of the interface with zero spherical aberration with respect to r. Therefore,

$$(\vec{\mathbf{a}} \cdot \vec{\mathbf{n}})^2 = \left(-m_i + \frac{dz}{dr}\right)^2, \tag{6.15}$$

$$(\vec{\mathbf{b}} \cdot \vec{\mathbf{n}})^2 = \left(-m_r + \frac{dz}{dr}\right)^2, \tag{6.16}$$

and

$$|\vec{\mathbf{a}}|^2 = 1 + m_i^2, \tag{6.17}$$

$$|\vec{\mathbf{b}}|^2 = 1 + m_r^2, \tag{6.18}$$

$$|\vec{\mathbf{n}}|^2 = 1 + \left(\frac{dz}{dr}\right)^2. \tag{6.19}$$

Returning to equation (6.10), replacing the expression for $(\vec{\mathbf{a}} \cdot \vec{\mathbf{n}})^2$, $(\vec{\mathbf{b}} \cdot \vec{\mathbf{n}})^2$, $|\vec{\mathbf{a}}|^2$, $|\vec{\mathbf{b}}|^2$ and $|\vec{\mathbf{n}}|^2$, in equation (6.10) we have

$$\frac{(1 + m_r^2)\left[(1 + m_i^2)\left[1 + \left(\frac{dz}{dr}\right)\right] - \left(-m_i + \frac{dz}{dr}\right)^2\right]}{(1 + m_i^2)\left[(1 + m_r^2)\left[1 + \left(\frac{dz}{dr}\right)\right] - \left(-m_r + \frac{dz}{dr}\right)^2\right]} = n^2, \tag{6.20}$$

simplifying and reordering,

$$(1 + m_r^2)\left(1 + m_i\frac{dz}{dr}\right)^2 = n^2(1 + m_i^2)\left(1 + m_r\frac{dz}{dr}\right)^2. \tag{6.21}$$

Now, from equation (6.21), we solve for $\frac{dz}{dr}$ and we get

$$\frac{dz}{dr} = \frac{m_i(1 + m_r^2) - n^2 m_r(1 + m_i^2) \pm n(m_r - m_i)\sqrt{(1 + m_r^2)(1 + m_i^2)}}{n^2 m_r^2(1 + m_i^2) - m_i^2(1 + m_r^2)}. \tag{6.22}$$

From (6.21) we can explore interesting cases, like when the object is at infinity, then the slope of the incident ray is $m_i \to \infty$, therefore,

$$\Rightarrow (m_r^2 + 1)\left(\frac{dz}{dr}\right)^2 = n^2\left(m_r\frac{dz}{dr} + 1\right)^2, \tag{6.23}$$

solving for $\frac{dz}{dr}$,

$$\frac{dz}{dr} = \frac{n}{-nm_r \pm \sqrt{m_r^2 + 1}}. \tag{6.24}$$

The solution of the last differential equation will give us an interface with a surface such that when the rays of an object—placed at minus infinity—cross, it will not introduce any spherical aberration.

Another compelling case is when an image is at infinity, which means that the refracted rays are parallel to the optical axis. Thus, computing the limit $m_r \to \infty$ in equation (6.21),

$$\Rightarrow \left(m_i\frac{dz}{dr} + 1\right)^2 = n^2(m_i^2 + 1)\left(\frac{dz}{dr}\right)^2, \tag{6.25}$$

solving for $\frac{dz}{dr}$,

$$\frac{dz}{dr} = \frac{1}{-m_i \pm \sqrt{m_i^2 + 1}}. \tag{6.26}$$

The solution of the last expression gives us an interface with a surface such that for a finite object, the image is at infinity.

Now, let's focus on the slopes m_i and m_r. First, assume that the light is coming from a point located at (r_o, z_o). From that point, multiple rays emerge that touch the surface under study, z. They cross the surface such that they are refracted and all of them converge in a single point at (r_i, z_i). Therefore, there is no spherical aberration in the image. Taking account of this, we can write the slopes m_i and m_r as

$$\begin{aligned} m_i &= \frac{z - z_o}{r - r_o}, \\ m_r &= \frac{z - z_i}{r - r_i}. \end{aligned} \tag{6.27}$$

Introducing the slopes of equation (6.27) in equation (6.21), we have

$$\left[1 + \left(\frac{z - z_i}{r - r_i}\right)^2\right]\left(1 + \frac{z - z_o}{r - r_o}\frac{dz}{dr}\right)^2 = n^2\left[1 + \left(\frac{z - z_o}{r - r_o}\right)^2\right]\left(1 + \frac{z - z_i}{r - r_i}\frac{dz}{dr}\right)^2. \tag{6.28}$$

Simplifying and reordering the terms we can get to the following expression

$$\frac{\left[r - r_o + (z - z_o)\dfrac{dz}{dr}\right]^2 [(r - r_i)^2 + (z - z_i)^2]}{\left[r - r_i + (z - z_i)\dfrac{dz}{dr}\right]^2 [(r - r_o)^2 + (z - z_o)^2]} = n^2. \tag{6.29}$$

From the last expression, we need to solve for $\frac{dz}{dr}$, so we can get a differential equation of first order. Thus, solving for $\frac{dz}{dr}$ in equation (6.29),

$$\begin{aligned} \frac{dz}{dr} = {} & \frac{(r - r_o)(z - z_o)[(r - r_i)^2 + (z - z_i)^2] - n^2(r - r_i)(z - z_i)[(r - r_o)^2 + (z - z_o)^2]}{n^2(z - z_i)^2[(r - r_o)^2 + (z - z_o)^2] - (z - z_o)^2[(r - r_i)^2 + (z - z_i)^2]}, \\ & \pm \frac{n\sqrt{[(r - r_i)^2 + (z - z_i)^2][(r - r_o)^2 + (z - z_o)^2][(r - r_o)(z - z_i) - (r - r_i)(z - z_o)]^2}}{n^2(z - z_i)^2[(r - r_o)^2 + (z - z_o)^2] - (z - z_o)^2[(r - r_i)^2 + (z - z_i)^2]}. \end{aligned} \tag{6.30}$$

For an arbitrary point, r_o, z_o, r_i and z_i, equation (6.30) is non-linear and the solution has not been obtained yet, but there are several cases where there is a solution and we are going to study them. In the following section we explore particular cases of equation (6.30).

6.2.1 Case I: $r_o = r_i = 0$, $z_o \to -\infty$ and $z_i = f$

In this section, we study the case when $r_0 = r_1 = 0$, which means that the object and the image are along the optical axis. Also, we have $z_o \to -\infty$, which means that the object is far away from the surface.

From equation (6.24), we can replace the slopes of equation (6.27) and obtain the following expression

$$\begin{aligned} \frac{dz}{dr} &= \frac{n}{-nm_r \pm \sqrt{m_r^2 + 1}}, \\ &= \frac{n}{-n\left(\frac{z - z_i}{r - r_i}\right) \pm \sqrt{\left(\frac{z - z_i}{r - r_i}\right)^2 + 1}}. \end{aligned} \tag{6.31}$$

Since the object is far away at infinity, we have $z_o - > - \infty$. So, the rays from the object are parallel and perfectly collimated to the optical axis and the interface under study $z(r)$; therefore, equation (6.31) becomes

$$\begin{aligned} &\lim_{z_0 \to \infty} \left[\frac{dz}{dr} = \frac{n}{-n\left(\frac{z - z_i}{r - r_i}\right) \pm \sqrt{\left(\frac{z - z_i}{r - r_i}\right)^2 + 1}} \right] \\ &\Rightarrow \frac{dz}{dr} = \frac{n(r - r_i)}{-n(z - z_i) \pm \sqrt{(z - z_i)^2 + (r - r_i)^2}}. \end{aligned} \tag{6.32}$$

Now, if the rays converge in the optical axis $r_i = 0$, and $z_i = f$, where f is the focus of the interface, equation (6.32) becomes

$$\begin{aligned} \frac{dz}{dr} &= \frac{n(r - r_i)}{-n(z - z_i) \pm \sqrt{(z - z_i)^2 + (r - r_i)^2}}, \\ &= \frac{n(r)}{-n(z - f) \pm \sqrt{(z - f)^2 + (r)^2}}, \\ &= \frac{r}{f - z \pm \frac{1}{n}\sqrt{(z - f)^2 + r^2}}. \end{aligned} \tag{6.33}$$

In section 6.3 we are going to solve the above equation. For the moment we explore other cases.

6.2.2 Case II: $r_o = r_i = 0$, $z_o = f$ and $z_i \rightarrow -\infty$

Another important case that we need to study is when the refracted rays are collimated and parallel to the optical axis, $z_i \rightarrow \infty$. So, we go back to equation (6.26) and replace the slopes of equation (6.27),

$$\frac{dz}{dr} = \frac{1}{-m_i \pm \sqrt{m_i^2 + 1}}, \tag{6.34}$$

$$= \frac{1}{-\left(\frac{z - z_o}{r - r_o}\right) \pm \sqrt{\left(\frac{z - z_o}{r - r_o}\right)^2 + 1}}. \tag{6.35}$$

Then, we evaluate the limit when $z_i \rightarrow \infty$,

$$\lim_{z_i \rightarrow \infty} \left[\frac{dz}{dr} = \frac{1}{-\left(\frac{z - z_o}{r - r_o}\right) \pm \sqrt{\left(\frac{z - z_o}{r - r_o}\right)^2 + 1}} \right] \tag{6.36}$$
$$\Rightarrow \frac{dz}{dr} = \frac{r - r_o}{-(z - z_o) \pm n\sqrt{(z - z_o)^2 + (r - r_o)^2}}.$$

Again, let's assume that the rays converge at the optical axis, which means $r_i = 0$. Also, we can recall $z_i = f$, where f is the focus of the interface. Thus, equation (6.36) becomes

$$\begin{aligned} \frac{dz}{dr} &= \frac{r - r_o}{-(z - z_o) \pm n\sqrt{(z - z_o)^2 + (r - r_o)^2}}, \\ &= \frac{r}{-(z - f) \pm n\sqrt{(z - f)^2 + (r)^2}}, \\ &= \frac{r}{f - z \pm n\sqrt{(z - f)^2 + r^2}}. \end{aligned} \tag{6.37}$$

In the next section, we are going to solve the above differential equation and other particular cases of equation (6.30).

6.3 Analytical stigmatic refractive surfaces

In this section, we are going to focus on the special cases of equation (6.30). One by one we are going to solve them.

6.3.1 Case A: $r_o = r_i = 0$, $z_o \to -\infty$ and $z_i = f$

In this section, we recall Case I of section 6.2.1. Thus, $r_o = r_i = 0$, $z_o \to -\infty$ and $z_i = f$; when $f > 0$ the focus is real and when $f < 0$ the focus is virtual. So, we recall equation (6.33),

$$\frac{dz}{dr} = \frac{r}{f - z \pm \frac{1}{n}\sqrt{(z - f)^2 + r^2}}. \tag{6.33}$$

The boundary condition is that the Cartesian oval is placed at the origin of the coordinate system,

$$z(0) = 0. \tag{6.38}$$

The four possible solutions of equation (6.33) are the following expressions

$$z = \frac{n(n + 1)f - \text{sign}(f)\sqrt{(n + 1)n^2[(n + 1)f^2 + (-n + 1)r^2]}}{n^2 - 1}, \tag{6.39}$$

$$z = \frac{n(n - 1)f - \text{sign}(f)\sqrt{(n - 1)n^2[(n + 1)f^2 + (-n + 1)r^2]}}{n^2 - 1}, \tag{6.40}$$

$$z = \frac{n(n + 1)f - \text{sign}(f)\sqrt{(n + 1)n^2[(n + 1)f^2 + (-n - 1)r^2]}}{n^2 - 1}, \tag{6.41}$$

$$z = \frac{n(n - 1)f - \text{sign}(f)\sqrt{(n - 1)n^2[(n - 1)f^2 + (-n - 1)r^2]}}{n^2 - 1}. \tag{6.42}$$

In practice, using the ray tracing techniques, we found that the solution is given by

$$z = \frac{n\left[(n - 1)f - \text{sign}(f)\sqrt{(n - 1)[(n - 1)f^2 - (n + 1)r^2]}\right]}{n^2 - 1}. \tag{6.43}$$

Examples of Cartesian ovals of equation (6.43) can be seen in figures 6.2 and 6.3.

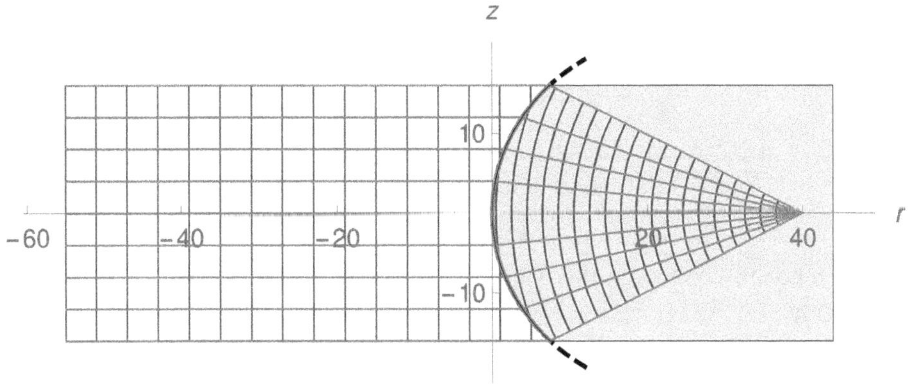

Figure 6.2. Design specifications: $n = 2$, $z_o \to -\infty$, $z_i = 40$ mm and $z=$ equation (6.43).

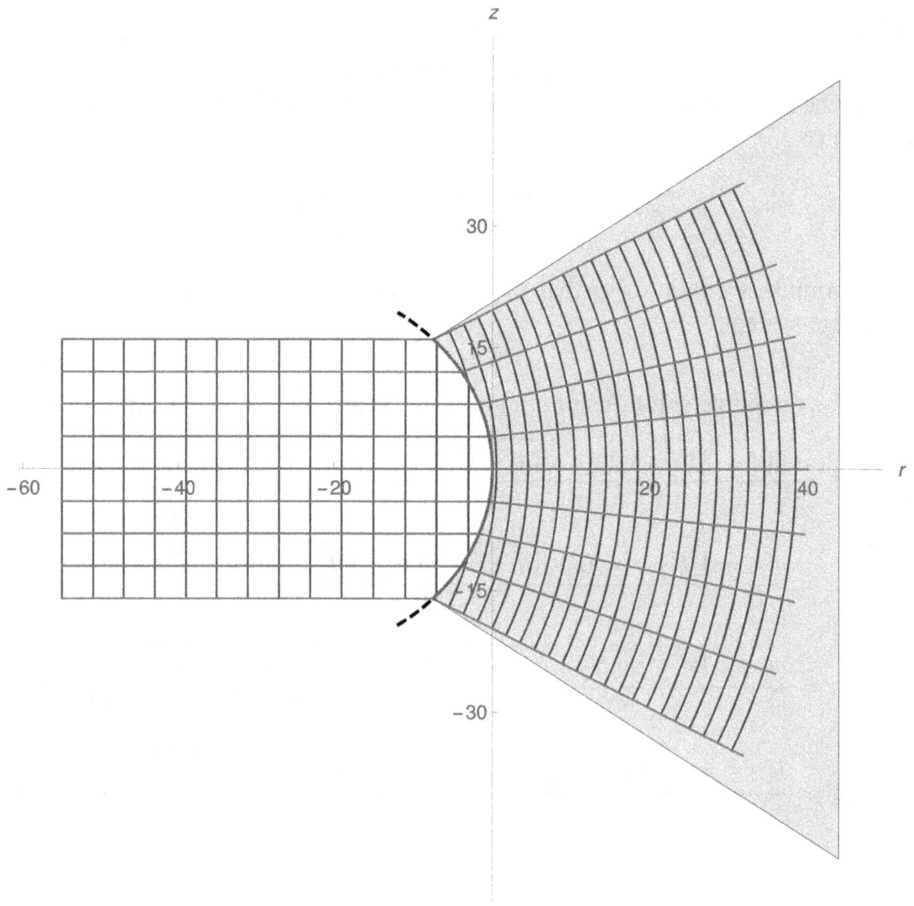

Figure 6.3. Design specifications: $n = 2$, $z_o \to -\infty$, $z_i = 40$ mm and $z =$ equation (6.43).

6.3.2 Case B: $r_o = r_i = 0,\ z_o = f$ and $z_i \to -\infty$

Now, we recall case II of section 6.2.2. Thus, $r_o = r_i = 0$, z_f and $z_i \to -\infty$, so we recall equation (6.37),

$$\frac{dz}{dr} = \frac{r}{f - z \pm n\sqrt{(z - f)^2 + r^2}}. \tag{6.37}$$

Again, we need to solve the equation, taking the boundary condition when the Cartesian oval is placed at the origin,

$$z(0) = 0. \tag{6.44}$$

Mathematically speaking, there are four solutions for equation (6.37), which can be written as

$$z = \frac{(n - 1)f + \text{sign}(f)\sqrt{(n - 1)[(n + 1)f^2 + (n + 1)r^2]}}{n^2 - 1}, \tag{6.45}$$

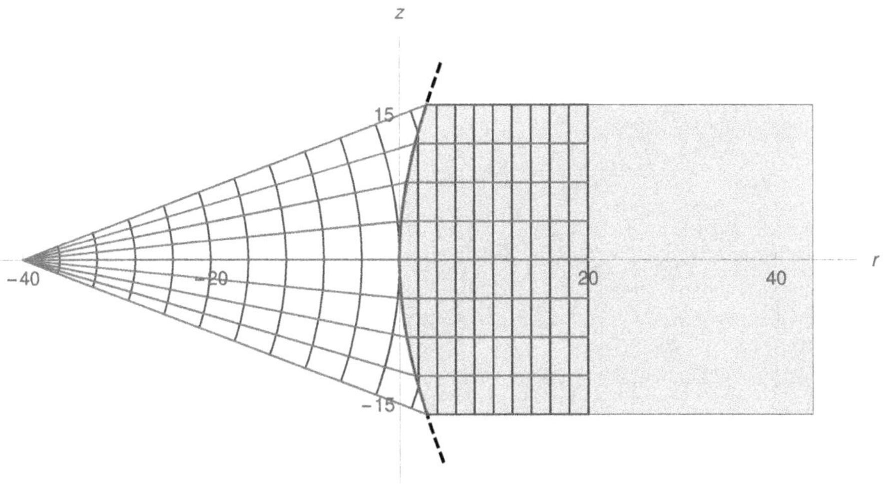

Figure 6.4. Design specifications: $n = 2$, $z_o = -40$ mm, $z_i \to \infty$ and $z =$ equation (6.49).

$$z = \frac{(n+1)f - \mathrm{sign}(f)\sqrt{(n+1)[(n+1)f^2 + (n-1)r^2]}}{n^2 - 1}, \tag{6.46}$$

$$z = \frac{(n-1)f - \mathrm{sign}(f)\sqrt{(n-1)[(n-1)f^2 + (n+1)r^2]}}{n^2 - 1}, \tag{6.47}$$

$$z = \frac{(n+1)f + \mathrm{sign}(f)\sqrt{(n-1)[(n+1)f^2 + (n-1)r^2]}}{n^2 - 1}. \tag{6.48}$$

To find which of the four is valid, we plot the ray tracing. We found that the solution of equation (6.37) we want is

$$z = \frac{(n-1)f - \mathrm{sign}(f)\sqrt{(n-1)[(n-1)f^2 + (n+1)r^2]}}{n^2 - 1}. \tag{6.49}$$

An example of the solution can be seen in figures 6.4 and 6.5.

6.3.3 Case C: $r_o = r_i = 0$, $z_o = \mp f$ and $z_i = \pm f$

In this case, we assume that the object is finite and the image as well. But in this case, both distances are the same value f but a different sign. So, then in this case we have $r_o = r_i = 0$, $z_o = -f$ and $z_i = +f$, taking these values in equation (6.30) we have

$$\begin{aligned}\frac{dz}{dr} &= \frac{(r-r_o)(z-z_o)[(r-r_i)^2+(z-z_i)^2] - n^2(r-r_i)(z-z_i)[(r-r_o)^2+(z-z_o)^2]}{n^2(z-z_i)^2[(r-r_o)^2+(z-z_o)^2] - (z-z_o)^2[(r-r_i)^2+(z-z_i)^2]} \\ &\pm \frac{n\sqrt{[(r-r_i)^2+(z-z_i)^2][(r-r_o)^2+(z-z_o)^2][(r-r_o)(z-z_i)-(r-r_i)(z-z_o)]^2}}{n^2(z-z_i)^2[(r-r_o)^2+(z-z_o)^2] - (z-z_o)^2[(r-r_i)^2+(z-z_i)^2]}.\end{aligned} \tag{6.31}$$

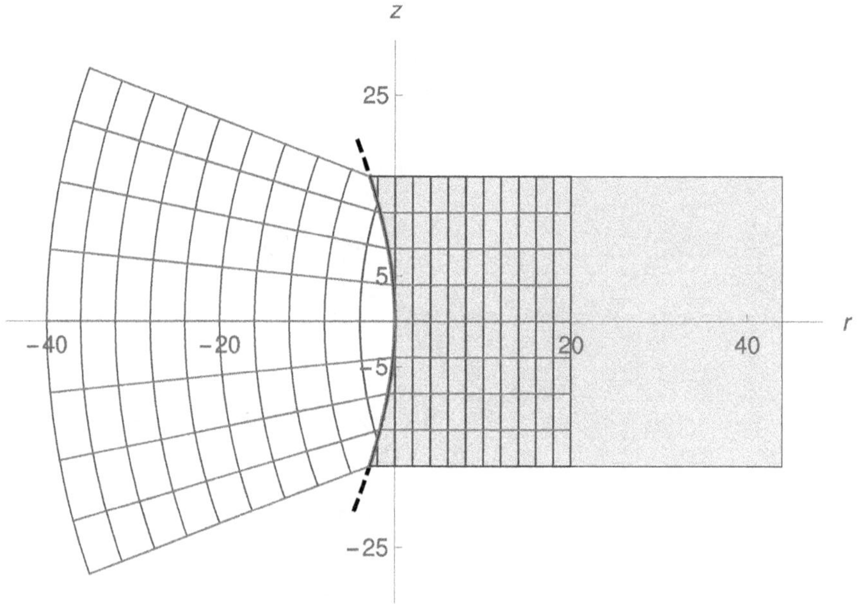

Figure 6.5. Design specifications: $n = 2$, $z_o = 40$ mm, $z_i \to \infty$ and $z =$ equation (6.49).

Thus, we set $r_o = r_i = 0$, $z_o = -f$ and $z_i = f$ in equation (6.30),

$$\frac{dz}{dr} = \frac{(r)(z+f)[(r)^2+(z-f)^2] - n^2(r)(z-f)[(r)^2+(z+f)^2]}{n^2(z-f)^2[(r)^2+(z+f)^2] - (z+f)^2[(r)^2+(z-f)^2]} \pm \frac{n\sqrt{[(r)^2+(z-f)^2][(r)^2+(z+f)^2][(r)(z-f)-(r)(z+f)]^2}}{n^2(z-f)^2[(r)^2+(z+f)^2] - (z+f)^2[(r)^2+(z-f)^2]}, \tag{6.50}$$

manipulating,

$$\frac{dz}{dr} = \frac{r^3(z+f) + r(z+f)(z-f)^2 - n^2 r(z-f)[r^2+(z+f)^2]}{n^2(z-f)^2[r^2+(z+f)^2] - (z+f)^2[r^2+(z-f)^2]} + \frac{\pm 2fnr\sqrt{[r^2+(z-f)^2][r^2+(z+f)^2]}}{n^2(z-f)^2[r^2+(z+f)^2] - (z+f)^2[r^2+(z-f)^2]}. \tag{6.51}$$

The solution of equation (6.61) is obtained by series, which corresponds to

$$\begin{aligned} z &= c_2\frac{r^2}{2f} + c_4\frac{r^4}{8f^3} + c_6\frac{2r^6}{32f^5} + c_8\frac{5r^8}{128f^7} + c_{11}\frac{2r^{10}}{215f^9} + c_{12}\frac{14f^{12}}{2048f^{11}}\cdots \\ &= \sum_{k=1}^{\infty} c_{2k}\frac{I_k r^{2k}}{(2f)^{2k-1}}. \end{aligned} \tag{6.52}$$

The coefficients are given by

$$\begin{cases} c_2 = \dfrac{n \pm 1}{n \mp 1}, \\ c_4 = \dfrac{n \mp 1}{n \pm 1}, \\ c_6 = \dfrac{(n \pm 1)(n^2 + 6n \pm 1)}{(n \mp 1)^3}, \\ c_8 = \dfrac{n \pm 1}{n \mp 1}, \\ c_{10} = \dfrac{(n \pm 1)(7n^4 \pm 124n^3 + 122n^2 \pm 124n + 7)}{(n \mp 1)^5}, \\ c_{12} = \dfrac{(n \pm 1)(3n^4 \mp 44n^3 - 46n^2 \pm 44n + 3)}{(n \pm 1)^5}. \end{cases} \tag{6.53}$$

Using the ray tracing to verify the signs of the coefficients, we find that the coefficient must be the following

$$\begin{cases} c_2 = \dfrac{n + 1}{n - 1}, \\ c_4 = \dfrac{n + 1}{n - 1}, \\ c_6 = \dfrac{(n + 1)(n^2 + 6n + 1)}{(n - 1)^3}, \\ c_8 = \dfrac{n + 1}{n - 1}, \\ c_{10} = \dfrac{(n + 1)(7n^4 + 124n^3 + 122n^2 + 124n + 7)}{(n - 1)^5}, \\ c_{12} = \dfrac{(n + 1)(3n^4 - 44n^3 - 46n^2 - 44n + 3)}{(n - 1)^5}. \end{cases} \tag{6.54}$$

An example of a Cartesian oval of this section can be seen in figure 6.6.

6.3.4 Case D: $r_o = r_i = 0$, $z_o = -\alpha f$ and $z_i = +f$

In this case, we assume that the object has a finite gap between the surface, and the image also has a finite position with respect to the surface. But both distances are not the same. We take the distance from the object to the surface as $-\alpha f$ and the distance from the surface to the image as f. So, the special case D is when $r_o = r_i = 0$, $z_o = -\alpha f$ and $z_i = +f$. Taking these values in equation (6.30), we have

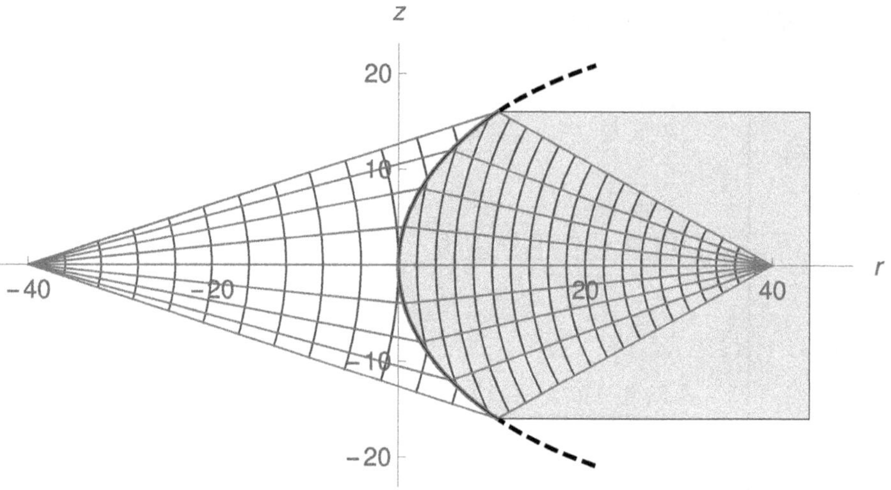

Figure 6.6. Design specifications: $n = 2, f = 40$ mm $z_o = -f$, $z_i = f$ and $z =$ equation (6.52).

$$\frac{dz}{dr} = \frac{(r - r_o)(z - z_o)[(r - r_i)^2 + (z - z_i)^2] - n^2(r - r_i)(z - z_i)[(r - r_o)^2 + (z - z_o)^2]}{n^2(z - z_i)^2[(r - r_o)^2 + (z - z_o)^2] - (z - z_o)^2[(r - r_i)^2 + (z - z_i)^2]} \pm \frac{n\sqrt{[(r - r_i)^2 + (z - z_i)^2][(r - r_o)^2 + (z - z_o)^2][(r - r_o)(z - z_i) - (r - r_i)(z - z_o)]^2}}{n^2(z - z_i)^2[(r - r_o)^2 + (z - z_o)^2] - (z - z_o)^2[(r - r_i)^2 + (z - z_i)^2]}, \tag{6.31}$$

replacing $r_o = r_i = 0$, $z_o = -\alpha f$ and $z_i = +f$ in the last equation,

$$\frac{dz}{dr} = \frac{(r)(z + \alpha f)[(r)^2 + (z - f)^2] - n^2(r)(z - f)[(r)^2 + (z + \alpha f)^2]}{n^2(z - f)^2[(r)^2 + (z + \alpha f)^2] - (z + \alpha f)^2[(r)^2 + (z - f)^2]} \pm \frac{n\sqrt{[(r)^2 + (z - f)^2][(r)^2 + (z + \alpha f)^2][(r)(z - f) - (r)(z + \alpha f)]^2}}{n^2(z - f)^2[(r)^2 + (z + \alpha f)^2] - (z + \alpha f)^2[(r)^2 + (z - f)^2]}, \tag{6.55}$$

reordering terms,

$$\frac{dz}{dr} = \frac{r^3(z + \alpha f) + r(z + \alpha f)(z - f)^2 - n^2 r(z - f)[r^2 + (z + \alpha f)^2]}{n^2(z - f)^2[r^2 + (z + \alpha f)^2] - (z + \alpha f)^2[r^2 + (z - f)^2]} + \frac{\pm fnr(\alpha + 1)\sqrt{(r^2 + (z - f)^2][r^2 + (z + \alpha f)^2]}}{n^2(z - f)^2[r^2 + (z + \alpha f)^2] - (z + \alpha f)^2[r^2 + (z - f)^2]}. \tag{6.56}$$

The solution of equation (6.56) is obtained by series, which corresponds to

$$\begin{aligned} z &= c_2\frac{r^2}{2f} + c_4\frac{r^4}{8f^3} + c_6\frac{2r^6}{32f^5} + c_8\frac{5r^8}{128f^7} + c_{11}\frac{2r^{10}}{215f^9} + c_{12}\frac{14f^{12}}{2048f^{11}}\cdots \\ &= \sum_{k=1}^{\infty} c_{2k}\frac{I_k r^{2k}}{(2f)^{2k-1}}. \end{aligned} \tag{6.57}$$

Mathematically, the coefficients are given by

$$\begin{cases} c_2 = \dfrac{\alpha n \pm 1}{\alpha(n \mp 1)}, \\ c_4 = \dfrac{\alpha^3 n^2 \pm (\alpha^3 + 2\alpha^2 - 2\alpha - 1)n - 1}{\alpha^3(n \mp 1)^2}, \\ c_6 = \dfrac{\alpha^5 n^3 \pm (2\alpha^5 + 3\alpha^4 - 3\alpha^3 + \alpha^2 + 3\alpha + 1)n^2 + (\alpha^5 + 3\alpha^4 + \alpha^3 - 3\alpha^2 + 3\alpha + 2)n \pm 1}{\alpha^5(n \mp 1)^3}, \\ c_8 = \dfrac{1}{\alpha^7(n \pm 1)^4}(\alpha^7 n^4 \pm (3\alpha^7 + 4\alpha^6 - 4\alpha^5 + 2\alpha^4 + 2\alpha^3 - 4\alpha^2 - 4\alpha + 1)n^3), \\ \quad + \dfrac{1}{\alpha^7(n \pm 1)^4}(3\alpha^7 + 8\alpha^6 - 8\alpha^4 + 8\alpha^3 - 8\alpha - 3)n^2 \\ \quad \pm \dfrac{1}{\alpha^7(n \pm 1)^4}[(\alpha^7 + 4\alpha^6 + 4\alpha^5 - 2\alpha^4 - 2\alpha^3 + 4\alpha^2 - 4\alpha - 3)n - 1). \end{cases} \tag{6.58}$$

The optical valid coefficients are

$$\begin{cases} c_2 = \dfrac{\alpha n + 1}{\alpha(n - 1)}, \\ c_4 = \dfrac{\alpha^3 n^2 + (\alpha^3 + 2\alpha^2 - 2\alpha - 1)n - 1}{\alpha^3(n - 1)^2}, \\ c_6 = \dfrac{\alpha^5 n^3 + (2\alpha^5 + 3\alpha^4 - 3\alpha^3 + \alpha^2 + 3\alpha + 1)n^2 + (\alpha^5 + 3\alpha^4 + \alpha^3 - 3\alpha^2 + 3\alpha + 2)n + 1}{\alpha^5(n - 1)^3}, \\ c_8 = \dfrac{1}{\alpha^7(n - 1)^4}(\alpha^7 n^4 + (3\alpha^7 + 4\alpha^6 - 4\alpha^5 + 2\alpha^4 + 2\alpha^3 - 4\alpha^2 - 4\alpha + 1)n^3) \\ + \dfrac{1}{\alpha^7(n - 1)^4}(3\alpha^7 + 8\alpha^6 - 8\alpha^4 + 8\alpha^3 - 8\alpha - 3)n^2 \\ + \dfrac{1}{\alpha^7(n - 1)^4}[(\alpha^7 + 4\alpha^6 + 4\alpha^5 - 2\alpha^4 - 2\alpha^3 + 4\alpha^2 - 4\alpha - 3)n - 1). \end{cases} \tag{6.59}$$

Examples of Cartesian ovals of case D can be seen in figures 6.7 and 6.8.

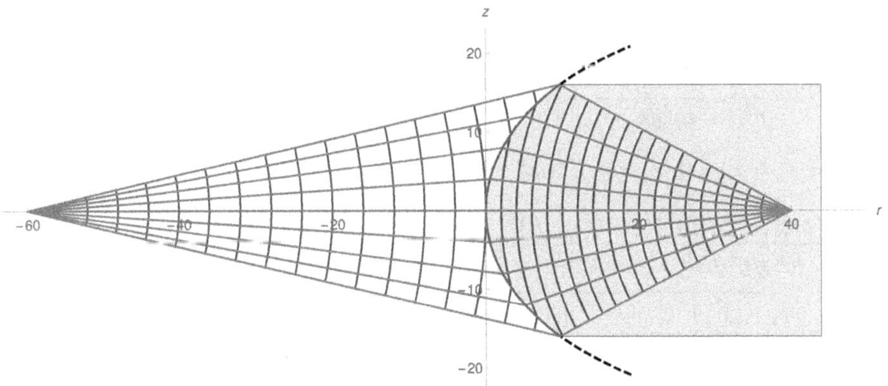

Figure 6.7. Design specifications: $n = 2$, $\alpha = 1.5$, $f = 40$ mm $z_o = -\alpha f$, $z_i = f$ and $z =$ equation (6.57).

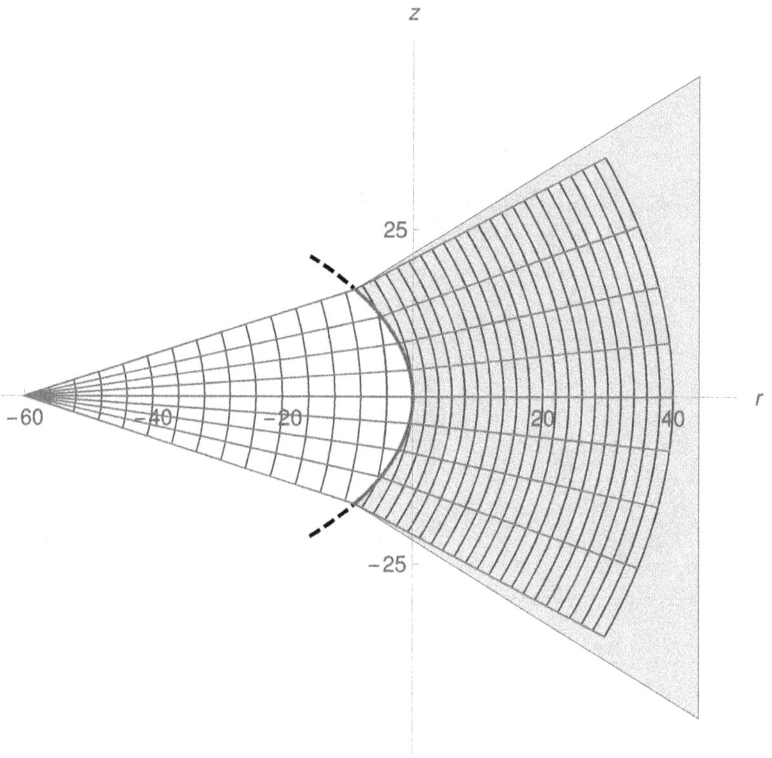

Figure 6.8. Design specifications: $n = 2$, $\alpha = 1.5$, $f = 40$ mm $z_o = -\alpha f$, $z_i = -f$ and $z =$ equation (6.57).

6.3.5 Case E: $r_o = r_i = 0$, $z_o = \alpha f$ and $z_i = -f$

This special case is very similar to the previous special cases. But now we assume that $z_o = \alpha f$, so the technical data are given by $r_o = r_i = 0$, $z_o = \alpha f$ and $z_i = -f$. Recalling equation (6.30),

$$\begin{aligned}\frac{dz}{dr} &= \frac{(r - r_o)(z - z_o)[(r - r_i)^2 + (z - z_i)^2] - n^2(r - r_i)(z - z_i)[(r - r_o)^2 + (z - z_o)^2]}{n^2(z - z_i)^2[(r - r_o)^2 + (z - z_o)^2] - (z - z_o)^2[(r - r_i)^2 + (z - z_i)^2]} \\ &\pm \frac{n\sqrt{[(r - r_i)^2 + (z - z_i)^2][(r - r_o)^2 + (z - z_o)^2][(r - r_o)(z - z_i) - (r - r_i)(z - z_o)]^2}}{n^2(z - z_i)^2[(r - r_o)^2 + (z - z_o)^2] - (z - z_o)^2[(r - r_i)^2 + (z - z_i)^2]},\end{aligned} \tag{6.30}$$

replacing $r_o = r_i = 0$, $z_o = \alpha f$ and $z_i = -f$,

$$\begin{aligned}\frac{dz}{dr} &= \frac{(r)(z - \alpha f)[(r)^2 + (z + f)^2] - n^2(r)(z + f)[(r)^2 + (z - \alpha f)^2]}{n^2(z + f)^2[(r)^2 + (z - \alpha f)^2] - (z - \alpha f)^2[(r)^2 + (z + f)^2]} \\ &\pm \frac{n\sqrt{[(r)^2 + (z + f)^2][(r)^2 + (z - \alpha f)^2][(r)(z + f) - (r)(z - \alpha f)]^2}}{n^2(z + f)^2[(r)^2 + (z - \alpha f)^2] - (z - \alpha f)^2[(r)^2 + (z + f)^2]},\end{aligned} \tag{6.60}$$

simplifying,

$$\begin{aligned}\frac{dz}{dr}=&\frac{r^3(z-\alpha f)+r(z-\alpha f)(z+f)^2-n^2r(z+f)(r^2+(z-\alpha f)^2]}{n^2(z+f)^2[r^2+(z-\alpha f)^2]-(z-\alpha f)^2[r^2+(z+f)^2]}\\&+\frac{\pm fnr(\alpha+1)\sqrt{[r^2+(z+f)^2][r^2+(z-\alpha f)^2]}}{n^2(z+f)^2[r^2+(z-\alpha f)^2]-(z-\alpha f)^2[r^2+(z+f)^2]}.\end{aligned}\tag{6.61}$$

As usual, the solution of equation (6.61) is obtained by series,

$$\begin{aligned}z&=c_2\frac{r^2}{2f}+c_4\frac{r^4}{8f^3}+c_6\frac{2r^6}{32f^5}+c_8\frac{5r^8}{128f^7}+c_{11}\frac{2r^{10}}{215f^9}+c_{12}\frac{14f^{12}}{2048f^{11}}\cdots\\&=\sum_{k=1}^{\infty}c_{2k}\frac{I_kr^{2k}}{(2f)^{2k-1}}.\end{aligned}\tag{6.62}$$

The coefficients are given by

$$\begin{cases}c_2=-\dfrac{\alpha n\pm 1}{\alpha(n\mp 1)},\\c_4=-\dfrac{\alpha^3n^2\pm(\alpha^3+2\alpha^2-2\alpha-1)n-1}{\alpha^3(n\mp 1)^2},\\c_6=-\dfrac{\alpha^5n^3\pm(2\alpha^5+3\alpha^4-3\alpha^3+\alpha^2+3\alpha+1)n^2+(\alpha^5+3\alpha^4+\alpha^3-3\alpha^2+3\alpha+2)n\pm 1}{\alpha^5(n\mp 1)^3},\\c_8=-\dfrac{1}{\alpha^7(n\pm 1)^4}(\alpha^7n^4\pm(3\alpha^7+4\alpha^6-4\alpha^5+2\alpha^4+2\alpha^3-4\alpha^2-4\alpha+1)n^3),\\-\dfrac{1}{\alpha^7(n\pm 1)^4}(3\alpha^7+8\alpha^6-8\alpha^4+8\alpha^3-8\alpha-3)n^2,\\-\dfrac{1}{\alpha^7(n\pm 1)^4}[(\alpha^7+4\alpha^6+4\alpha^5-2\alpha^4-2\alpha^3+4\alpha^2-4\alpha-3)n-1).\end{cases}\tag{6.63}$$

Using the ray tracing we verify the correct coefficients

$$\begin{cases}c_2=-\dfrac{\alpha n+1}{\alpha(n-1)},\\c_4=-\dfrac{\alpha^3n^2+(\alpha^3+2\alpha^2-2\alpha-1)n-1}{\alpha^3(n-1)^2},\\c_6=-\dfrac{\alpha^5n^3+(2\alpha^5+3\alpha^4-3\alpha^3+\alpha^2+3\alpha+1)n^2+(\alpha^5+3\alpha^4+\alpha^3-3\alpha^2+3\alpha+2)n+1}{\alpha^5(n-1)^3},\\c_8=-\dfrac{1}{\alpha^7(n-1)^4}(\alpha^7n^4+(3\alpha^7+4\alpha^6-4\alpha^5+2\alpha^4+2\alpha^3-4\alpha^2-4\alpha+1)n^3)\\-\dfrac{1}{\alpha^7(n-1)^4}(3\alpha^7+8\alpha^6-8\alpha^4+8\alpha^3-8\alpha-3)n^2\\-\dfrac{1}{\alpha^7(n-1)^4}[(\alpha^7+4\alpha^6+4\alpha^5-2\alpha^4-2\alpha^3+4\alpha^2-4\alpha-3)n-1).\end{cases}\tag{6.64}$$

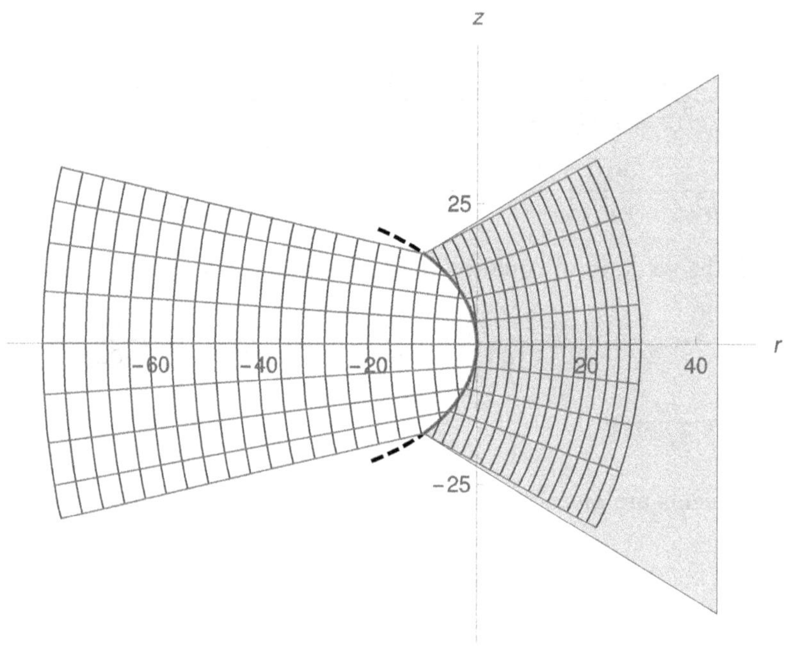

Figure 6.9. Design specifications: $n = 2$, $\alpha = 1.5$, $f = 40$ mm $z_o = -\alpha f$, $z_i = -f$ and $z =$ equation (6.57).

For an example of case E see figure 6.9.

6.4 Conclusions

In this chapter, we studied interfaces that do not generate spherical aberration, which means that all the refracted rays converge in a single point. To construct these surfaces, first we explored the phenomenon described by Snell's law at the surface under study. The first step was to express Snell's law in the form of the vectors involved in the phenomenon instead of the angles. Once Snell's ruling was in its vector form, we solved for the derivative of the surface under study dz/dr. Then, we found a non-linear differential equation, which in some cases could be solved. We studied each instance and presented the solution as well as the plot of the solution.

Studying these interfaces is tremendously helpful in understanding the solution of singlet lenses with zero spherical aberration. In the next chapters, we will focus on similar methods but with two surfaces, not one. However, it was essential to set the basis of this strategy to understand the more complex problems in the following chapters.

References

Avendaño-Alejo M, Román-Hernández E, Castañeda L and Moreno-Oliva V I 2017 Analytic conic constants to reduce the spherical aberration of a single lens used in collimated light *Appl. Opt.* **56** 6244–54

Bass M 1995 *Handbook of Optics, Volume I: Fundamentals Techniques and Design* **vol I** (New York: McGraw-Hill)

Born M and Wolf E 2013 *Principles of Optics: Electromagnetic Theory of Propagation, Interference and Diffraction of Light* (Amsterdam: Elsevier)

Braunecker B, Hentschel R and Tiziani H J 2008 *Advanced Optics Using Aspherical Elements* **vol 173** (Bellingham, WA: SPIE Press)

Castillo-Santiago G, Avendaño-Alejo M, Díaz-Uribe R and Castañeda L 2014 Analytic aspheric coefficients to reduce the spherical aberration of lens elements used in collimated light *Appl. Opt.* **53** 4939–46

Chaves J 2016 *Introduction to Nonimaging Optics* 2nd edn (Boca Raton, FL: CRC Press)

Estrada J C V, Calle Á H B and Hernández D M 2013 Explicit representations of all refractive optical interfaces without spherical aberration *J. Opt. Soc. Am.* **A30** 1814–24

Glassner A S 1989 *An Introduction to Ray Tracing* (Amsterdam: Elsevier)

González-Acuña R G and Chaparro-Romo H A 2018 General formula for bi-aspheric singlet lens design free of spherical aberration *App. Opt.* **57** 9341–5

González-Acuña R G and Guitiérrez-Vega J C 2018 Generalization of the axicon shape: the gaxicon *J. Opt. Soc. Am.* **A35** 1915–8

González-Acuña R G and Gutiérrez-Vega J C 2019 Analytic formulation of a refractive-reflective telescope free of spherical aberration *Opt. Eng.* **58** 085105

González-Acuña R G, Avendaño-Alejo M and Gutiérrez-Vega J C 2019a Singlet lens for generating aberration-free patterns on deformed surfaces *J. Opt. Soc. Am.* **A36** 925–9

González-Acuña R G, Chaparro-Romo H A and Gutiérrez-Vega J C 2019b General formula to design freeform singlet free of spherical aberration and astigmatism *Appl. Opt.* **58** 1010–5

González-Acuña R G, Chaparro-Romo H A and Gutiérrez-Vega J C 2020a Analytic aplanatic singlet lens: setting and design for three-point objects and images in the meridional plane *Opt. Eng.* **59** 055104

González-Acuña R G, Chaparro-Romo H A and Gutiérrez-Vega J C 2020b Chaparro-Romo and J C Gutiérrez-Vega 2020b Analytic solution of the eikonal for a stigmatic singlet lens *Phys. Scr.*

González-Acuña R G, Chaparro-Romo H A and Gutiérrez-Vega J C 2020c *Analytical Lens Design* (Bristol: Institute of Physics Publishing)

González-Acuña R G, Chaparro-Romo H A and Gutíerrez-Vega J C 2019c *Single Lens Telescope* (arXiv:1903.11129)

González-Acuña R G and Gutiérrez-Vega J C 2019a Analytic formulation of a refractive-reflective telescope free of spherical aberration *Opt. Eng.* **58** 1–5

González-Acuña R G and Gutiérrez-Vega J C 2019b General formula to eliminate spherical aberration produced by an arbitrary number of lenses *Opt. Eng.* **58** 1–6

González-Acuña R G and Gutiérrez-Vega J C 2019 General formula for aspheric collimator lens design free of spherical aberration *Current Developments in Lens Design and Optical Engineering XX* **vol 11104** ed R B Johnson, V N Mahajan and S Thibault (Bellingham, WA: International Society for Optics and Photonics, SPIE) pp 181–4

González Acuña R G and Gutiérrez-Vega J C 2019 General formula to design freeform collimator lens free of spherical aberration and astigmatism *Novel Optical Systems, Methods, and Applications XXII* **vol 11105** (Bellingham, WA: International Society for Optics and Photonics) p 111050A

González Acuña R G and Gutiérrez-Vega J C 2019 General formula of the refractive telescope design free spherical aberration *Novel Optical Systems, Methods, and Applications XXII* **vol 11105** ed C F Hahlweg and J R Mulley (Bellingham, WA: International Society for Optics and Photonics, SPIE) pp 162–6

Kingslake R and Johnson R B 2009 *Lens Design Fundamentals* (New York: Academic)

Lefaivre J 1951 A new approach in the analytical study of the spherical aberrations of any order *J. Opt. Soc. Am.* **41** 647

Luneburg R K 1964 *Mathematical Theory of Optics* (Berkeley, CA: University of California Press)

Malacara D 1965 Two lenses to collimate red laser light *Appl. Opt.* **4** 1652–4

Malacara-Hernández D and Malacara-Hernández Z 2016 *Handbook of Optical Design* (Boca Raton, FL: CRC Press)

González-Acuña R G and Gutiérrez-Vega J C 2020 Analytic design of a spherochromatic singlet *J. Opt. Soc. Am.* **A37** 149–53

Schulz G 1983 Achromatic and sharp real imaging of a point by a single aspheric lens *Appl. Opt.* **22** 3242–8

Silva-Lora A and Torres R 2020 Explicit Cartesian oval as a superconic surface for stigmatic imaging optical systems with real or virtual source or image *Proc. R. Soc.* A476

Stavroudis O 2012 *The Optics of Rays, Wavefronts, and Caustics* **vol 38** (Amsterdam: Elsevier)

Stavroudis O N 2006 *The Mathematics of Geometrical and Physical Optics: The K-function and Its Ramifications* (New York: Wiley)

Stavroudis O N and Feder D P 1954 Automatic computation of spot diagrams *J. Opt. Soc. Am.* **44** 163–70

Sun H 2016 *Lens Design: A Practical Guide* (Boca Raton, FL: CRC Press)

Valencia-Estrada J C and Malacara-Doblado D 2014 Parastigmatic corneal surfaces *Appl. Opt.* **53** 3438–47

Valencia-Estrada J C, Flores-Hernández R B and Malacara-Hernández D 2015 Singlet lenses free of all orders of spherical aberration *Proc. R. Soc.* **A471** 20140608

Vaskas E M 1957 Note on the Wassermann-Wolf method for designing aspheric surfaces *J. Opt. Soc. Am.* **47** 669–70

Wassermann G D and Wolf E 1949 On the theory of aplanatic aspheric systems *Proc. Phys. Soc.* **B62** 2

Winston R, Miñano J C and Benitez P G *et al* 2005 *Nonimaging Optics* (Amsterdam: Elsevier)

Wolf E 1948 On the designing of aspheric surfaces *Proc. Phys. Soc.* **61** 494

Wolf E and Preddy W S 1947 On the determination of aspheric profiles *Proc. Phys. Soc.* **59** 704

Yang T, Jin G-F and Zhu J 2017 Automated design of freeform imaging systems *Light Sci. Appl.* **6** e17081

IOP Publishing

Stigmatic Optics (Second Edition)

Rafael G González-Acuña and Héctor A Chaparro-Romo

Chapter 7

The general equation of the Cartesian oval

In this chapter, we will obtain the general equation of the Cartesian oval. Unlike the previous chapter, the presented equation is not a polynomial approximation, but a closed expression. The equation presented is general, because it supports that the object is real or virtual and that the image is real or virtual.

7.1 From Ibn Sahl to René Descartes

Cartesian ovals have been studied for over a thousand years. Although it is named after French mathematician René Descartes, it was actually first considered by Ibn Sahl in 984, in what is now Basra, Iraq.

Ibn Sahl comprehended very well the optics of ancient Greece, but he went much further to the unexplored field of refraction. His treatise *On the Burning Instruments* is so unusual for its time that it makes him the first mathematician understood to have studied stigmatic lens design. In his time, 10th century, the main area of study was based on catoptrics. Ibn Sahl studied burning mirrors, both parabolic and ellipsoidal; he considered hyperbolic plano-convex lenses and hyperbolic biconvex lenses (shown in *On the Burning Instruments*). But his most notable accomplishment is the law of refraction (Snell's law) long before Snell himself existed.

Sahl was the first to discover the stigmatic refractive surface, today known as the Cartesian oval. Sahl suggested a stigmatic lens comprising two Cartesian ovals where the rays refracted are collimated along the optical axis. Sahl could not find a general equation of the Cartesian ovals.

7.2 A generalized problem

In the previous chapter, we presented the Valencia–Calle method of Cartesian ovals. In general, the technique is simple when the object or image are at minus, plus infinite, respectively. Otherwise, if the object and the image are finite, we obtain from Snell's law differential equations that are quite complicated to solve.

doi:10.1088/978-0-7503-6423-2ch7

Of all the representations that we have of Cartesian ovals when the object and the image are finite, we always have long equations, whose deduction procedures are not trivial.

Recently, the scientists Alberto Silva-Lora and Rafael Torres published a parametric equation of the Cartesian ovals in a closed form. Note that in the previous chapter, many of the results proposed by the Valencia–Calle method end with a non-closed polynomial approach.

The Alberto Silva-Lora and Rafael Torres method of obtaining the Cartesian ovals will be studied in this chapter under the name of the Silva–Torres approach or Silva–Torres method.

What makes the Silva–Torres method different from the other procedures is that it has a closed solution that contains all the cases of Cartesian ovals. Also, the Silva–Torres method contains instances of the stigmatic mirrors that we studied in chapter 5.

7.3 Mathematical model

In this section, we will obtain the mathematical derivation of the Cartesian ovals by the Silva–Torres method. The Silva–Torres approach differs from the Valencia–Calle model in that it focuses on the optical path, the Fermat principle rather than Snell's law. Therefore, it avoids differential equations.

The first thing to note is that the Fermat principle predicts that the Cartesian oval is the refractive surface such that the optical path of the axial ray is the same path for all other rays. This is because the axial ray has the minimum optical path length between the object point located at z_o and the image point located at z_i, as shown in figure 7.1.

Therefore, we can match the optical path of any non-axial ray with that of the axial ray, which gives us the following expression

$$n_i\sqrt{(z - z_i)^2 + x^2 + y^2} + n_o\sqrt{(z - z_o)^2 + x^2 + y^2} = n_i z_i - n_o z_o. \tag{7.1}$$

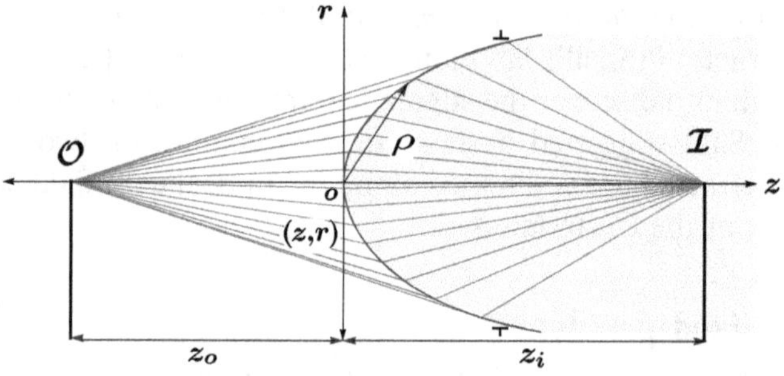

Figure 7.1. Diagram of a Cartesian oval. The origin is placed in the vertex of the Cartesian oval (z, r). The distance from the object to the vertex is z_o. The gap between the origin and the image is z_i. The distance from the origin to a given point of the Cartesian oval is ρ.

Note that n_o is the refractive index of the medium where the object is located and n_i is the corresponding refractive index where the image is.

In our three-dimensional model x, y, z, are points in the three-dimensional space of the Cartesian oval where the non-axial ray passes, as shown in figure 7.1. We take ρ as a distance from the origin of the coordinate system to x, y, z. The origin is placed at the vertex of the Cartesian oval.

The next algebraic steps are quite long, so it is good to assign the following parameters

$$A \equiv \sqrt{(z - z_o)^2 + x^2 + y^2}. \tag{7.2}$$

The other square root is abbreviated as

$$B \equiv \sqrt{(z - z_i)^2 + x^2 + y^2}, \tag{7.3}$$

and finally, the optical path of the axial ray is a constant given by

$$k \equiv n_i z_i - n_o z_o. \tag{7.4}$$

Taking into account the new notation, equation (7.1) has the following form

$$n_i A + n_o B = k. \tag{7.5}$$

Before looking for the Cartesian oval equation, our closest goal is to get rid of the square roots, so we square equation (7.5) and manipulate it,

$$A^2 n_o^2 + B^2 n_i^2 - k^2 = 2ABn_i n_o. \tag{7.6}$$

In the previous equation, the term that has the square roots is on the right side, to eliminate these roots we squared again on two sides of the equation mentioned above,

$$(A^2 n_o^2 + B^2 n_i^2 - k^2)^2 = (2ABn_i n_o)^2. \tag{7.7}$$

So, we expand equation (7.7). The interesting thing is that equation (7.7) no longer has square roots, the result is the following

$$\begin{aligned}
&n_i^4 x^4 + n_o^4 x^4 - 2n_i^2 n_o^2 x^4 + 2y^2 n_i^4 x^2 + 2z^2 n_i^4 x^2 + 2y^2 n_o^4 x^2 + 2z^2 n_o^4 x^2 \\
&- 4y^2 n_i^2 n_o^2 x^2 - 4z^2 n_i^2 n_o^2 x^2 - 4n_i^2 n_o^2 z_i^2 x^2 - 4n_i^2 n_o^2 z_o^2 x^2 - 4zn_i^4 z_i x^2 \\
&+ 4zn_i^2 n_o^2 z_i x^2 - 4zn_o^4 z_o x^2 + 4zn_i^2 n_o^2 z_o x^2 + 4n_i n_o^3 z_i z_o x^2 + 4n_i^3 n_o z_i z_o x^2 \\
&+ y^4 n_i^4 + z^4 n_i^4 + 2y^2 z^2 n_i^4 + y^4 n_o^4 + z^4 n_o^4 + 2y^2 z^2 n_o^4 - 2y^4 n_i^2 n_o^2 - 2z^4 n_i^2 n_o^2 \\
&- 4y^2 z^2 n_i^2 n_o^2 + 4z^2 n_i^4 z_i^2 - 4y^2 n_i^2 n_o^2 z_i^2 - 4z^2 n_i^2 n_o^2 z_i^2 + 4z^2 n_o^4 z_o^2 - 4y^2 n_i^2 n_o^2 z_o^2 \\
&- 4z^2 n_i^2 n_o^2 z_o^2 - 8zn_i n_o^3 z_i z_o^2 + 8zn_i^2 n_o^2 z_i z_o^2 - 4z^3 n_i^4 z_i - 4y^2 zn_i^4 z_i + 4z^3 n_i^2 n_o^2 z_i \\
&+ 4y^2 zn_i^2 n_o^2 z_i - 4z^3 n_o^4 z_o - 4y^2 zn_o^4 z_o + 4z^3 n_i^2 n_o^2 z_o + 4y^2 zn_i^2 n_o^2 z_o + 8zn_i^2 n_o^2 z_i^2 z_o \\
&- 8zn_i^3 n_o z_i^2 z_o + 4y^2 n_i n_o^3 z_i z_o + 4z^2 n_i n_o^3 z_i z_o - 8z^2 n_i^2 n_o^2 z_i z_o + 4y^2 n_i^3 n_o z_i z_o \\
&+ 4z^2 n_i^3 n_o z_i z_o = 0.
\end{aligned} \tag{7.8}$$

In mathematics and theoretical physics, it can become an art in how we simplify or reorder equations for our convenience. The following equation is a rearrangement of equation (7.8). The reason why we order in this way is in the following steps,

$$\begin{aligned}&[(n_i^2 - n_o^2)(x^2 + y^2 + z^2) - 2z(n_i^2 z_i - n_o^2 z_o)]^2\\&- 4n_i n_o(n_i z_i - n_o z_o)[(x^2 + y^2 + z^2)(z_i n_o - n_i z_o) + 2zz_i z_o(n_i - n_o)] = 0.\end{aligned} \tag{7.9}$$

Since ρ is the vector that starts from the origin and ends at points x, y, z, we have $\rho^2 = x^2 + y^2 + z^2$. Substituting $\rho^2 = x^2 + y^2 + z^2$ in equation (7.9) we obtain equation (7.10),

$$\begin{aligned}&[\rho^2(n_i^2 - n_o^2) - 2z(n_i^2 z_i - n_o^2 z_o)]^2\\&- 4n_i n_o(n_i z_i - n_o z_o)[\rho^2(z_i n_o - n_i z_o) + 2zz_i z_o(n_i - n_o)] = 0.\end{aligned} \tag{7.10}$$

Expanding the first term of the previous equation we have

$$\begin{aligned}&\rho^4(n_i^2 - n_o^2)^2 + 4z^2(n_i^2 z_i - n_o^2 z_o)^2 - 4\rho^2 z(n_i^2 - n_o^2)(n_i^2 z_i - n_o^2 z_o)\\&- 4n_i n_o(n_i z_i - n_o z_o)[\rho^2(z_i n_o - n_i z_o) + 2zz_i z_o(n_i - n_o)] = 0.\end{aligned} \tag{7.11}$$

From the previous expression, we want to replace the following algebraic identities

$$(n_i - n_o)^2(n_i + n_o)^2 = -2n_i^2 n_o^2 + n_i^4 + n_o^4, \tag{7.12}$$

and

$$(n_i - n_o)(n_i + n_o) = n_i^2 - n_o^2. \tag{7.13}$$

Therefore, substituting equations (7.12) and (7.13) in equation (7.10) we have equation (7.14),

$$\begin{aligned}&4z^2(n_i^2 z_i - n_o^2 z_o)^2 - 2z(2\rho^2(n_i - n_o)(n_i + n_o)(n_i^2 z_i - n_o^2 z_o)\\&+ 4n_i z_i n_o z_o(n_i - n_o)(n_i z_i - n_o z_o)) + \rho^2[4\rho^2 n_i n_o(n_i - n_o)^2(n_i + n_o)^2(n_i z_i - n_o z_o)] = 0.\end{aligned} \tag{7.14}$$

We can simplify equation (7.14) by dividing it by the parameter D expressed by equation (7.15),

$$D \equiv 4n_i z_i n_o z_o(n_i - n_o)(n_i z_i - n_o z_o). \tag{7.15}$$

Dividing equation (7.14) by equation (7.15), we have

$$c_0 K z^2 - (1 + b_1\rho^2)z + (c_0 + c_1\rho^2)\rho^2 = 0, \tag{7.16}$$

where

$$K \equiv \frac{(n_i^2 z_i - n_o^2 z_o)^2}{n_i n_o(n_i z_i - n_o z_o)(n_i z_o - n_o z_i)}, \tag{7.17}$$

$$c_0 \equiv \frac{n_i z_o - n_o z_i}{z_i z_o (n_i - n_o)}, \tag{7.18}$$

$$c_1 \equiv \frac{(n_i - n_o)(n_i + n_o)^2}{4 n_i n_o z_i z_o (n_i z_i - n_o z_o)}, \tag{7.19}$$

and

$$b_1 \equiv \frac{(n_i + n_o)(n_i^2 z_i - n_o^2 z_i)}{2 n_i n_o z_i z_o (n_i z_i - n_o z_o)}. \tag{7.20}$$

Equation (7.16) is a quadratic equation whose solution is the following expression

$$z = \frac{1}{c_0 K}\left(1 + b_1 \rho^2 \pm \sqrt{1 + (2b_1 - c_0^2 K)\rho^2 + (b_1^2 - c_0 c_1 K)\rho^4}\right). \tag{7.21}$$

Equation (7.21) can be simplified, taking into account that $c_0 c_1 K$ is equal to b_1^2 as demonstrated by the procedure expressed in equation (7.22).

$$\begin{aligned} c_0 c_1 K &= \left[\frac{n_i z_o - n_o z_i}{z_i z_o (n_i - n_o)}\right]\left[\frac{(n_i - n_o)(n_i + n_o)^2}{4 n_i n_o z_i z_o (n_i z_i - n_o z_o)}\right]\left[\frac{(n_i^2 z_i - n_o^2 z_o)^2}{n_i n_o (n_i z_i - n_o z_o)(n_i z_o - n_o z_i)}\right] \\ &= \frac{(n_i + n_o)^2 (n_i^2 z_i - n_o^2 z_i)^2}{4 n_i^2 n_o^2 z_i^2 z_o^2 (n_i z_i - n_o z_o)^2} = b_1^2. \end{aligned} \tag{7.22}$$

Therefore, substituting $b_1^2 = c_0 c_1 K$ in equation (7.21), we have

$$z = \frac{1}{c_0 K}\left(1 + b_1 \rho^2 \pm \sqrt{1 + (2b_1 - c_0^2 K)\rho^2}\right). \tag{7.23}$$

In equation (7.23), multiplying the square root we have a ±; this comes from the fact that the solution of a second-order equation has two roots. Both solutions are mathematically valid. But the only interesting solution for this is the one where the vertex of the Cartesian oval passes through the origin, as shown in figure 7.1.

$$z = \frac{1}{c_0 K}\left(1 + b_1 \rho^2 - \sqrt{1 + (2b_1 - c_0^2 K)\rho^2}\right). \tag{7.24}$$

In the original publication of this method, Silva-Lora and Torres presented the Cartesian oval with the root in the denominator. So, we multiply both sides of equation (7.24) by $\left(1 + b_1 \rho^2 + \sqrt{1 + (2b_1 - c_0^2 K)\rho^2}\right)$ $/\left(1 + b_1 \rho^2 + \sqrt{1 + (2b_1 - c_0^2 K)\rho^2}\right)$,

$$z = \frac{1}{c_0 K}\left(1 + b_1 \rho^2 - \sqrt{1 + (2b_1 - c_0^2 K)\rho^2}\right)\frac{\left(1 + b_1 \rho^2 + \sqrt{1 + (2b_1 - c_0^2 K)\rho^2}\right)}{\left(1 + b_1 \rho^2 + \sqrt{1 + (2b_1 - c_0^2 K)\rho^2}\right)}. \tag{7.25}$$

Simplifying equation (7.25), we obtain the Silva–Torres equation,

$$\boxed{z = \frac{\rho^2(c_0 + c_1\rho^2)}{\sqrt{\rho^2(2b_1 - c_0^2K) + 1} + b_1\rho^2 + 1}.} \tag{7.26}$$

Equation (7.26) is the most important equation in this chapter. It contains information on all the cases of Cartesian ovals. We mean that you can have real or virtual object/images and finite or infinite object/images. Even with equation (7.26) you can get conical mirrors.

In the following section, we show the potential of this equation. But first, it is important to note that equation (7.26) is the sagittal part of the Cartesian oval. To obtain the radial part of the Cartesian oval, from figure 7.1 we know that the radius is

$$\boxed{r = \text{sgn}(\rho)\sqrt{\rho^2 - z^2}.} \tag{7.27}$$

Equation (7.27) is the radial part of the Cartesian oval described by the Silva–Torres model.

7.4 Illustrative examples

In the following examples, we show several Cartesian ovals generated by the Silva–Torres method by directly using equations (7.26) and (7.27). In this section, we discuss all relevant cases, when the object/image is real or virtual. The specifications of each design are shown in the captions for each figure. The figures are 7.2–7.6.

7.5 Collimated input rays

Now we study the case when the object is very far from a Cartesian oval, $z_o \to \infty$. In this case, the Cartesian oval receives collimated rays along the optical axis. To obtain a Cartesian oval in this way, we need to evaluate the limit when $z_o \to \infty$ in all the parameters within the equations (7.26) and (7.27).

We start with the parameter K, and compute the limit when $z_o \to \infty$ in equation (7.17),

$$\lim_{z_o \to -\infty}(K) = \lim_{z_o \to -\infty}\left[\frac{(n_i^2 z_i - n_o^2 z_o)^2}{n_i n_o(n_i z_i - n_o z_o)(n_i z_o - n_o z_i)}\right]. \tag{7.28}$$

Evaluating the limit expressed in the previous equation we have

$$\lim_{z_o \to -\infty}(K) = -\frac{n_o^2}{n^2}. \tag{7.29}$$

As a next step, we apply the limit when $z_o \to \infty$ in c_0, equation (7.18),

$$\lim_{z_o \to -\infty}(c_0) = \lim_{z_o \to -\infty}\left[\frac{n_i z_o - n_o z_i}{z_i z_o(n_i - n_o)}\right]. \tag{7.30}$$

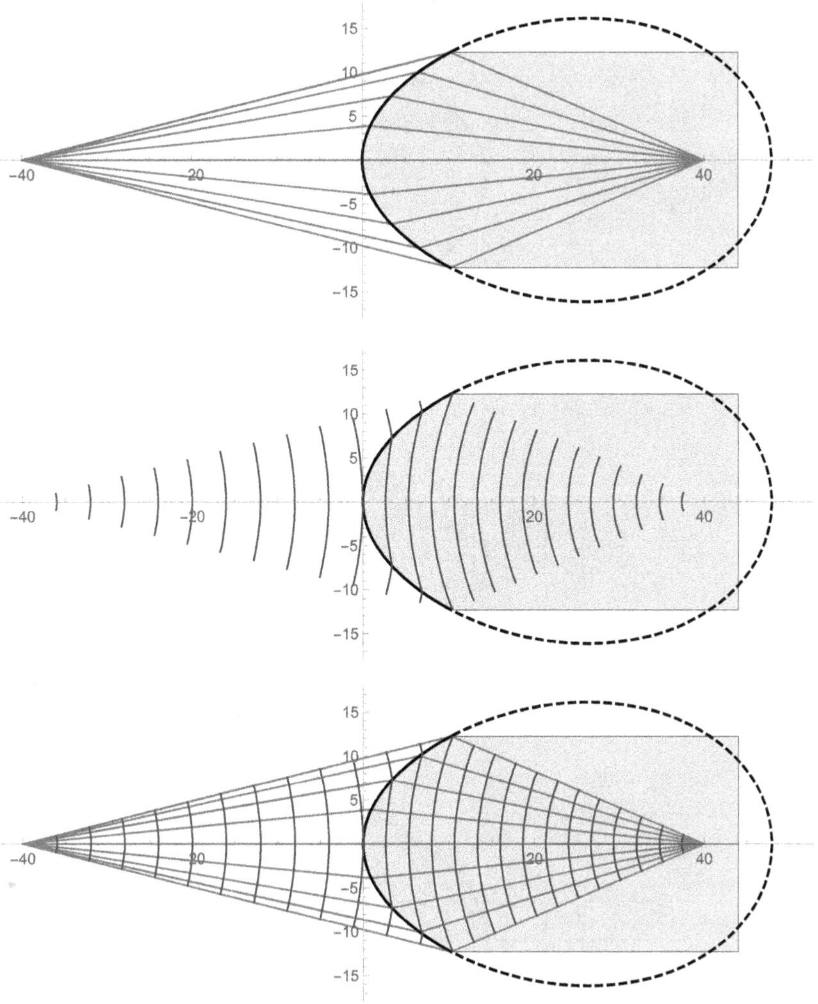

Figure 7.2. Specifications of the design: $n_o = 1$, $n_i = 1.5$, $z_o = -40$ mm, $z_i = 40mm$ $z =$ equation (7.26) and $r =$ equation (7.27).

Computing the limit in equation (7.30), we get

$$\lim_{z_o \to -\infty} (c_0) = \frac{n}{n(z_i - \tau) + n_o(\tau - z_i)}. \tag{7.31}$$

Then, we compute the limit $z_o \to \infty$ in c_1, equation (7.19),

$$\lim_{z_o \to -\infty} (c_1) = \lim_{z_o \to -\infty} \left[\frac{(n_i - n_o)(n_i + n_o)^2}{4n_i n_o z_i z_o(n_i z_i - n_o z_o)} \right] = 0. \tag{7.32}$$

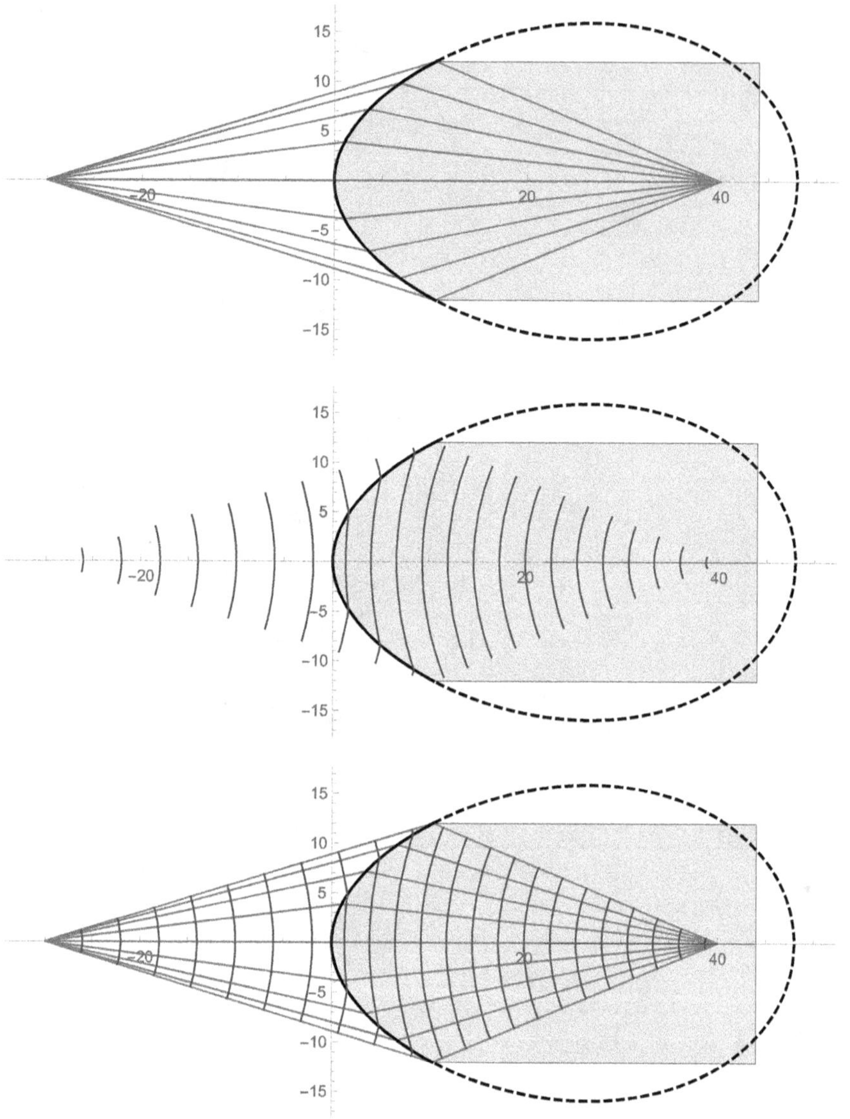

Figure 7.3. Specifications of the design: $n_o = 1$, $n_i = 1.5$, $z_o = -30$ mm, $z_i = 40mm$ $z =$ equation (7.26) and $r =$ equation (7.27).

The same happens when we compute the limit $z_o \to \infty$ in b_1, equation (7.20),

$$\lim_{z_o \to -\infty} (b_1) = \lim_{z_o \to -\infty} \left[\frac{(n_i + n_o)(n_i^2 z_i - n_o^2 z_i)}{2 n_i n_o z_i z_o (n_i z_i - n_o z_o)} \right] = 0. \tag{7.33}$$

Therefore, the sagitta of the Cartesian oval proposed by the Silva–Torres method is given by

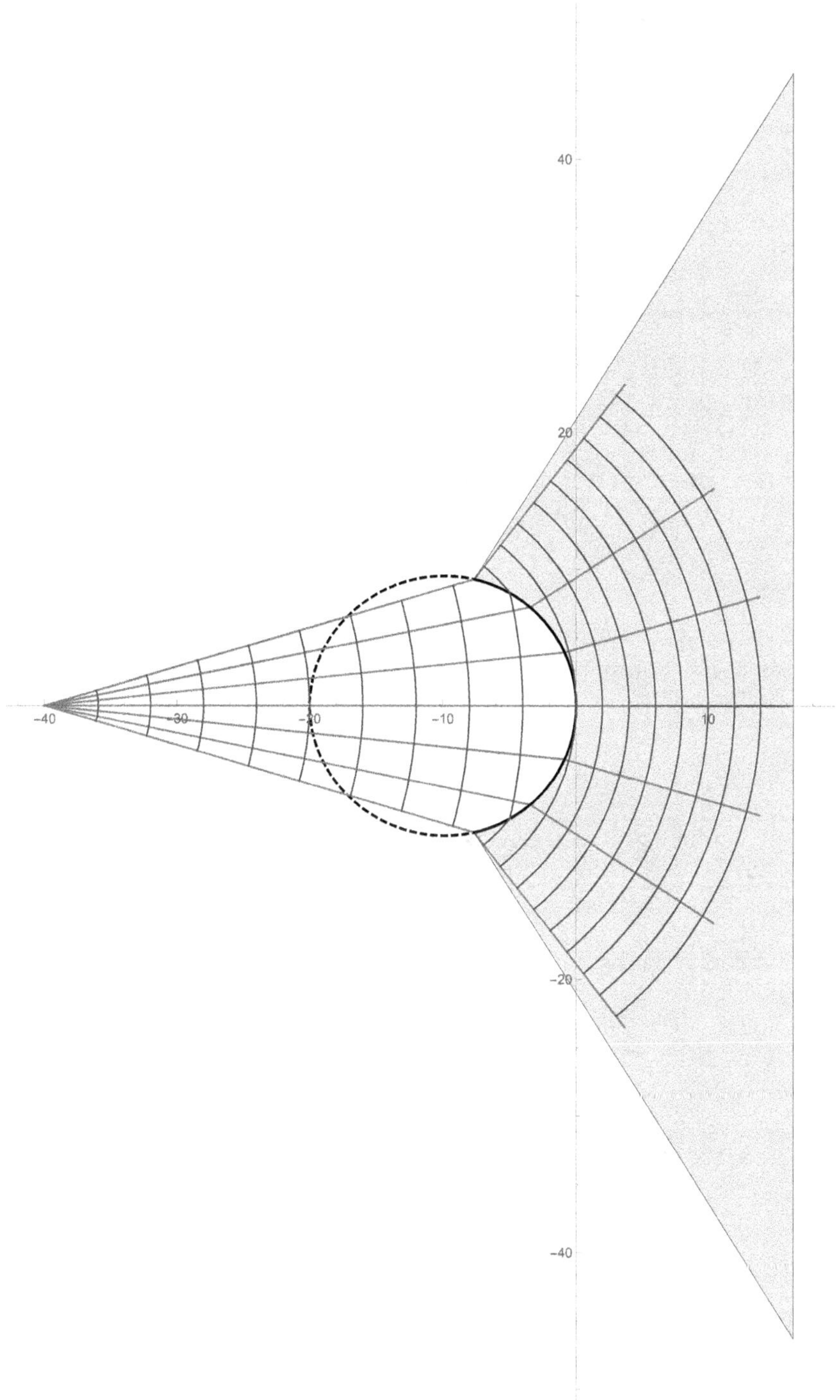

Figure 7.4. Specifications of the design: $n_o = 1$, $n_i = 1.5$, $z_o = -40$ mm, $z_i = -15mm$ $z =$ equation (7.26) and $r =$ equation (7.27).

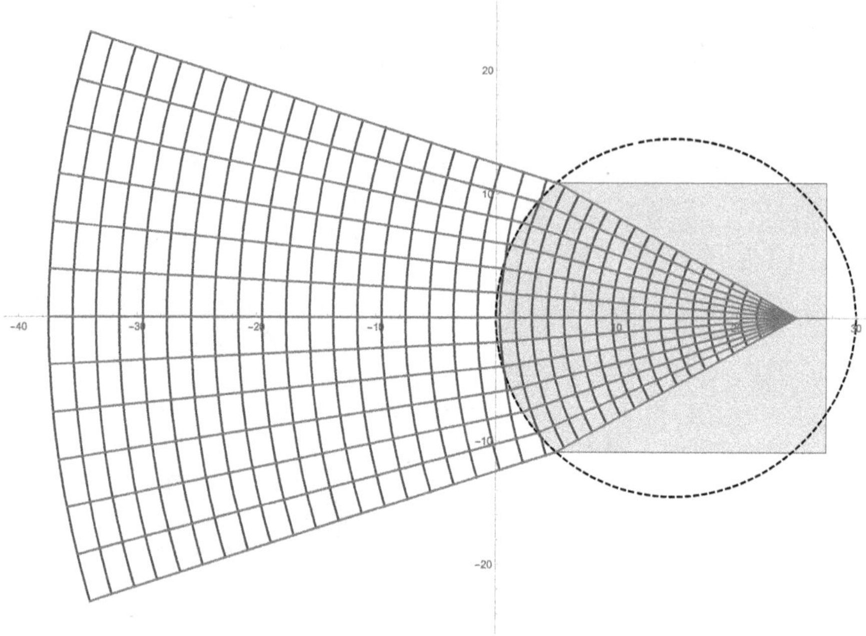

Figure 7.5. Specifications of the design: $n_o = 1, n_i = 1.5, z_o = 40$ mm, $z_i = 25mm$ $z =$ equation (7.26) and $r =$ equation (7.27).

$$\lim_{z_o \to -\infty}(z) = \frac{\left(\lim_{z_o \to -\infty} c_0\right)\rho^2}{\sqrt{1 - \left(\lim_{z_o \to -\infty} c_0\right)^2\left(\lim_{z_o \to -\infty} K\right)\rho^2} + 1}, \tag{7.34}$$

and the radius is given by

$$\lim_{z_o \to -\infty}(r) = \operatorname{sgn}(\rho)\sqrt{\rho^2 - \lim_{z_o \to -\infty}(z^2)}. \tag{7.35}$$

7.6 Illustrative examples

In this section, we will show an example series of Cartesian ovals with the object in minus infinity. The image can be real or virtual. The specifications are shown in each figure (see figures 7.7 and 7.8).

7.7 Collimated output rays

Here we will focus on studying the Cartesian oval that has collimated rays at the system output. For this we have to apply the limit $z_i \to \infty$ in equations (7.17)–(7.20). This model is very similar to the model in the previous section.

Let's start by evaluating the limit $z_i \to \infty$ of K, equation (7.17),

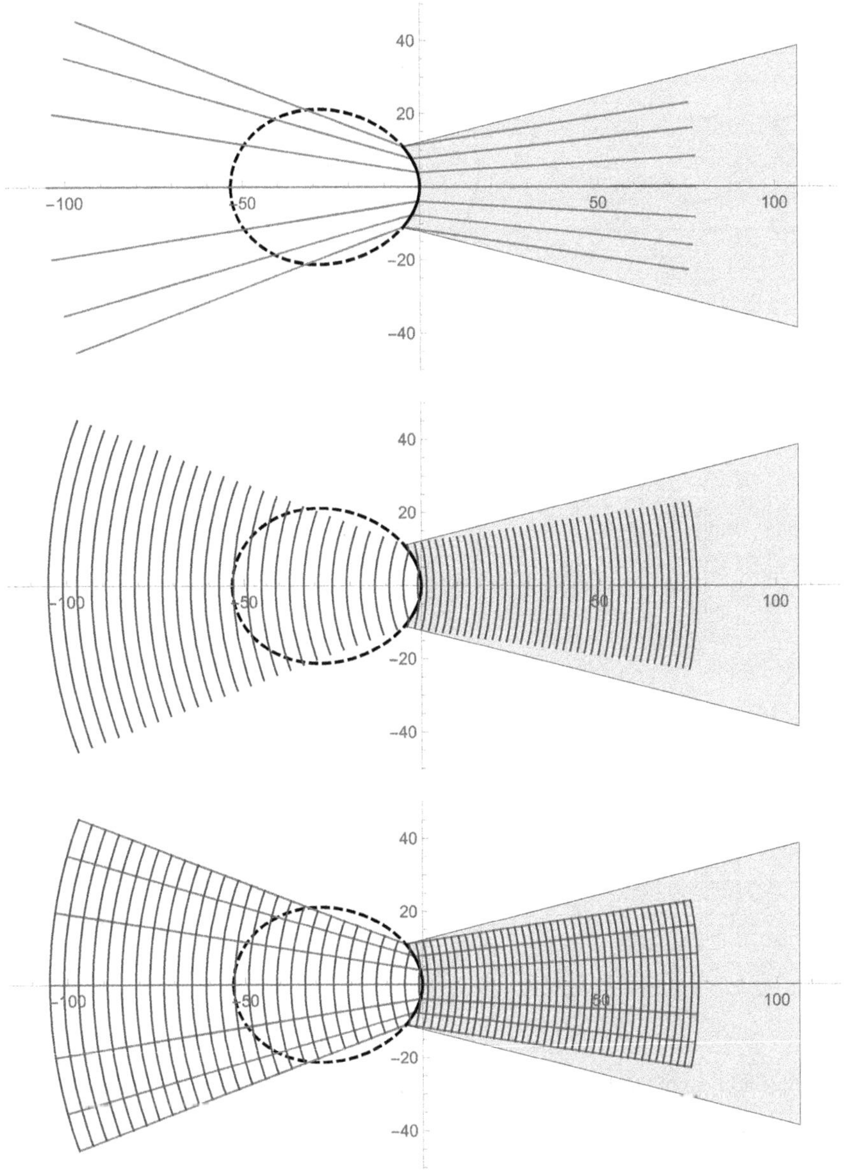

Figure 7.6. Specifications of the design: $n_o = 1$, $n_i = 1.5$, $z_o = 25$ mm, $z_i = -75mm$ $z =$ equation (7.26) and $r =$ equation (7.27).

$$\lim_{z_i \to \infty} (K) = \lim_{z_i \to -\infty} \left[\frac{(n_i^2 z_i - n_o^2 z_o)^2}{n_i n_o (n_i z_i - n_o z_o)(n_i z_o - n_o z_i)} \right] = -\frac{n_o^2}{n^2}. \tag{7.36}$$

Then we have the limit $z_i \to \infty$ of c_0, equation (7.18),

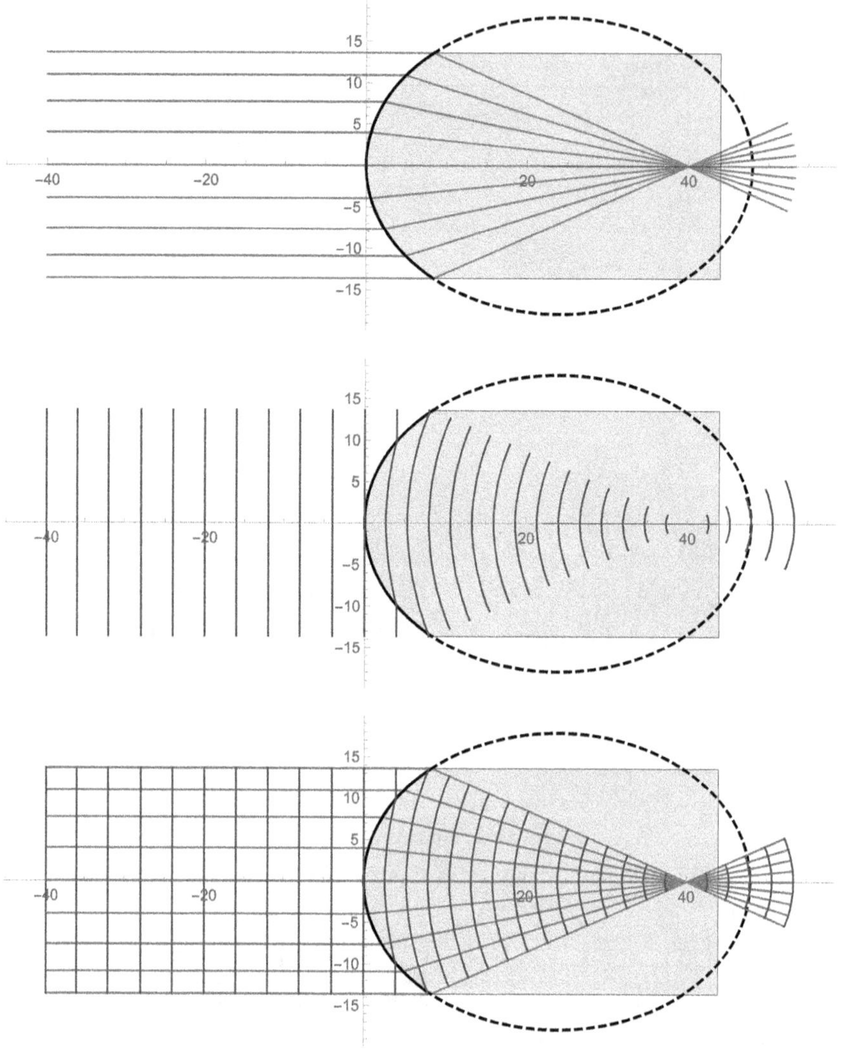

Figure 7.7. Specifications of the design: $n_o = 1$, $n_i = 1.5$, $z_o = -\infty$, $z_i = 40$ mm $z =$ equation (7.34) and $r =$ equation (7.35).

$$\lim_{z_i \to \infty} (c_0) = \lim_{z_i \to -\infty} \left[\frac{n_i z_o - n_o z_i}{z_i z_o (n_i - n_o)} \right], \tag{7.37}$$

computing the limit,

$$\lim_{z_i \to \infty} (c_0) = -\frac{n_o}{n z_o - n_o z_o}. \tag{7.38}$$

Then, we compute the limit $z_i \to \infty$ of c_1, equation (7.19),

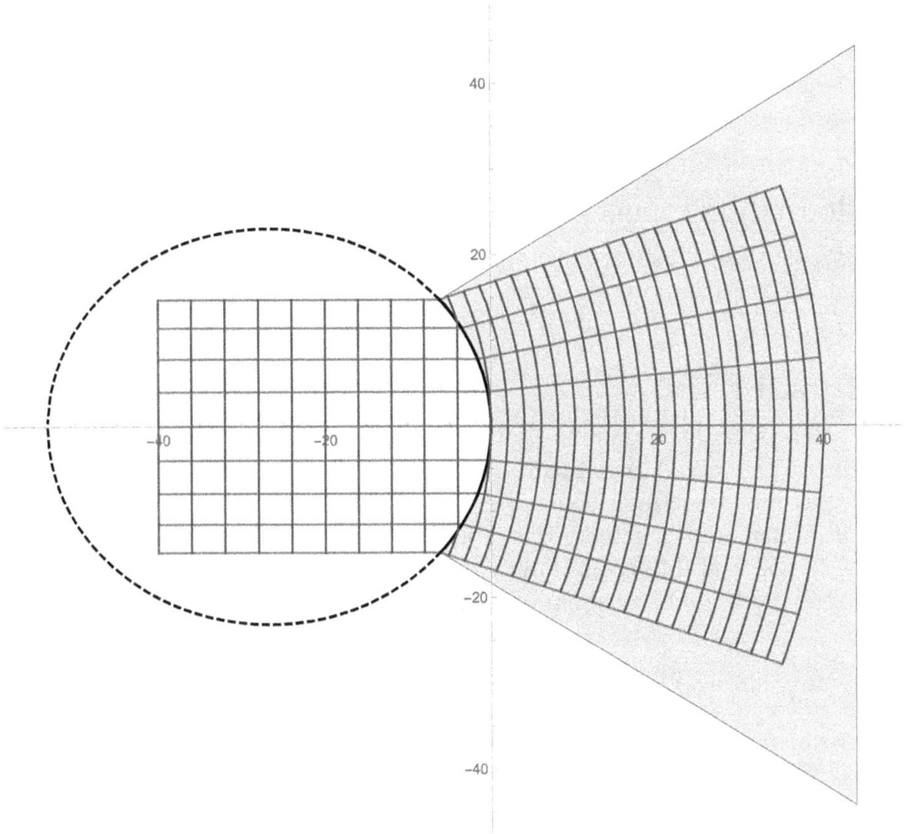

Figure 7.8. Specifications of the design: $n_o = 1, n_i = 1.5, z_o = -\infty, z_i = -40$ mm $z =$ equation (7.34) and $r =$ equation (7.35).

$$\lim_{z_i\to\infty}(c_1) = \lim_{z_i\to\infty}\left[\frac{(n_i - n_o)(n_i + n_o)^2}{4n_i n_o z_i z_o(n_i z_i - n_o z_o)}\right] = 0, \tag{7.39}$$

and the limit $z_i \to \infty$ of b_1, equation (7.20), is given by

$$\lim_{z_i\to\infty}(b_1) = \lim_{z_i\to\infty}\left[\frac{(n_i + n_o)(n_i^2 z_i - n_o^2 z_i)}{2n_i n_o z_i z_o(n_i z_i - n_o z_o)}\right] = 0. \tag{7.40}$$

Once we compute the aforementioned limits, we can express the Cartesian oval proposed by the Silva–Torres method for output collimated rays

$$\lim_{z_i\to\infty} z_a = \frac{\left(\lim\limits_{z_i\to\infty} c_0\right)\rho^2}{\sqrt{1 - \left(\lim\limits_{z_i\to\infty} c_0\right)^2\left(\lim\limits_{z_i\to\infty} K\right)\rho^2} + 1}, \tag{7.41}$$

and radius

$$\lim_{z_i\to\infty} r = \operatorname{sgn}(\rho)\sqrt{\rho^2 - \lim_{z_i\to\infty}(z^2)}\,. \tag{7.42}$$

7.8 Illustrative examples

Now let's see some illustrative examples of Cartesian ovals with images in infinity. As usual, the specifications are given in the caption of each figure.

The figures showing the output collimated rays are 7.9–7.11.

7.9 Refractive surface

We mentioned at the beginning of this chapter that one of the most interesting attributes of the Silva–Torres method is that it is robust enough that it contains all the information concerning conical mirrors. In this section, we demonstrate this mentioned attribute. To achieve this objective, we take two limits. The first limit is when $n_i \to -n_o$ and the second limit is when $n_o \to -1$.

We apply both limits on equation (7.17),

$$\lim_{n_o\to-1}\left[\lim_{n_i\to-n_o}(K)\right] = \lim_{n_o\to-1}\left[\lim_{n_i\to-n_o}\left(\frac{\left(n_i^2 z_i - n_o^2 z_o\right)^2}{n_i n_o(n_i z_i - n_o z_o)(n_i z_o - n_o z_i)}\right)\right]. \tag{7.43}$$

After computing both limits we get

$$\lim_{n_o\to-1}\left[\lim_{n_i\to-n_o}(K)\right] = -\frac{(z_i - z_o)^2}{(z_i + z_o)^2}. \tag{7.44}$$

Then we compute the mentioned limits in c_0, equation (7.18),

$$\lim_{n_o\to-1}\left[\lim_{n_i\to-n_o}(c_0)\right] = \lim_{n_o\to-1}\left[\lim_{n_i\to-n_o}\left(\frac{n_i z_o - n_o z_i}{z_i z_o(n_i - n_o)}\right)\right]. \tag{7.45}$$

The result is the following

$$\lim_{n_o\to-1}\left[\lim_{n_i\to-n_o}(c_0)\right] = \frac{z_i + z_o}{2 z_i z_o}. \tag{7.46}$$

Then, let's focus on c_1, equation (7.19),

$$\lim_{n_o\to-1}\left[\lim_{n_i\to-n_o}(c_1)\right] = \lim_{n_o\to-1}\left[\lim_{n_i\to-n_o}\left(\frac{(n_i - n_o)(n_i + n_o)^2}{4 n_i n_o z_i z_o(n_i z_i - n_o z_o)}\right)\right]. \tag{7.47}$$

The computation gives the following result

$$\lim_{n_o\to-1}\left[\lim_{n_i\to-n_o}(c_1)\right] = 0. \tag{7.48}$$

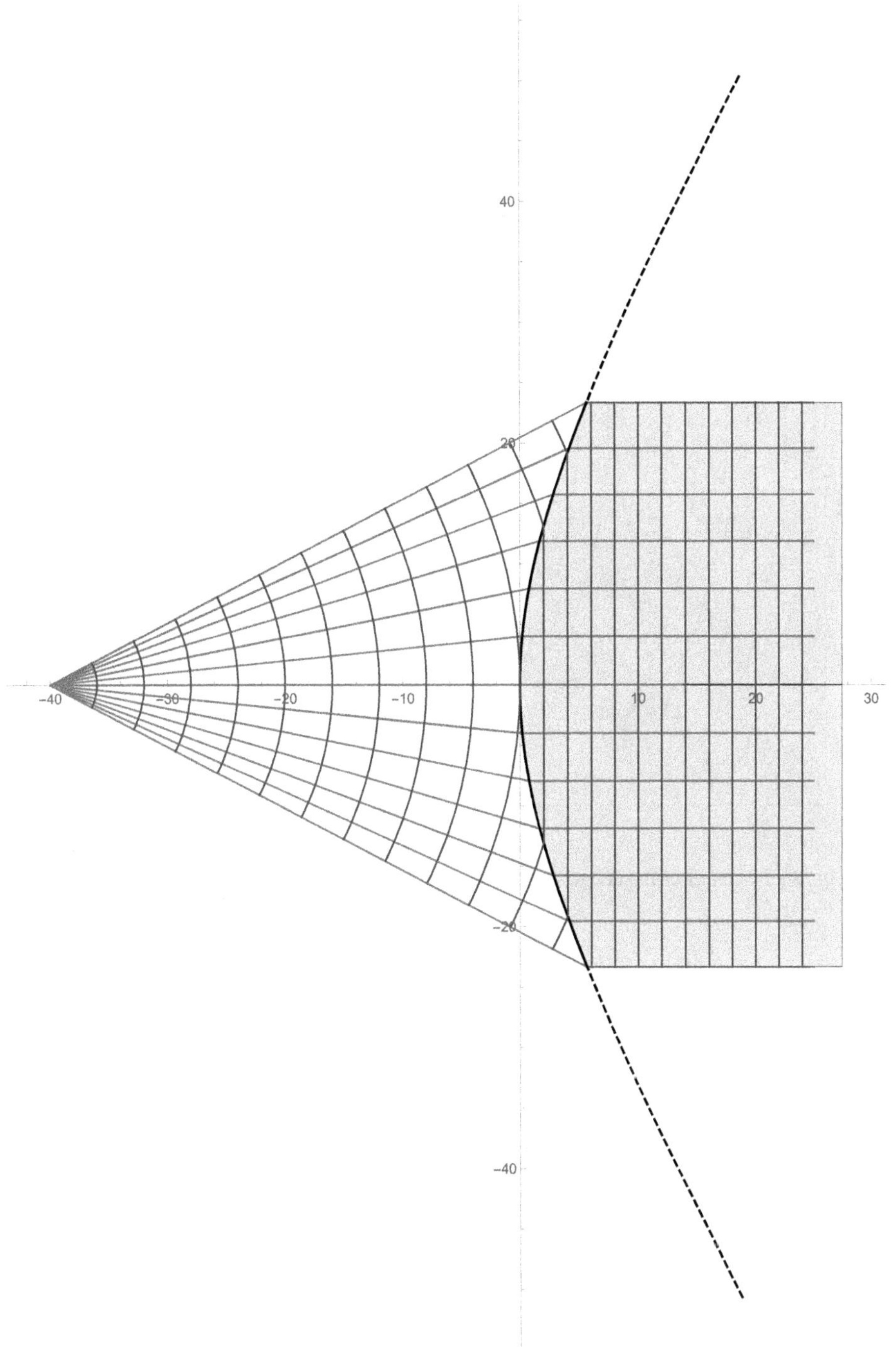

Figure 7.9. Specifications of the design: $n_o = 1$, $n_i = 1.5$, $z_o = -40$ mm, $z_i = \infty$ $z =$ equation (7.41) and $r =$ equation (7.42).

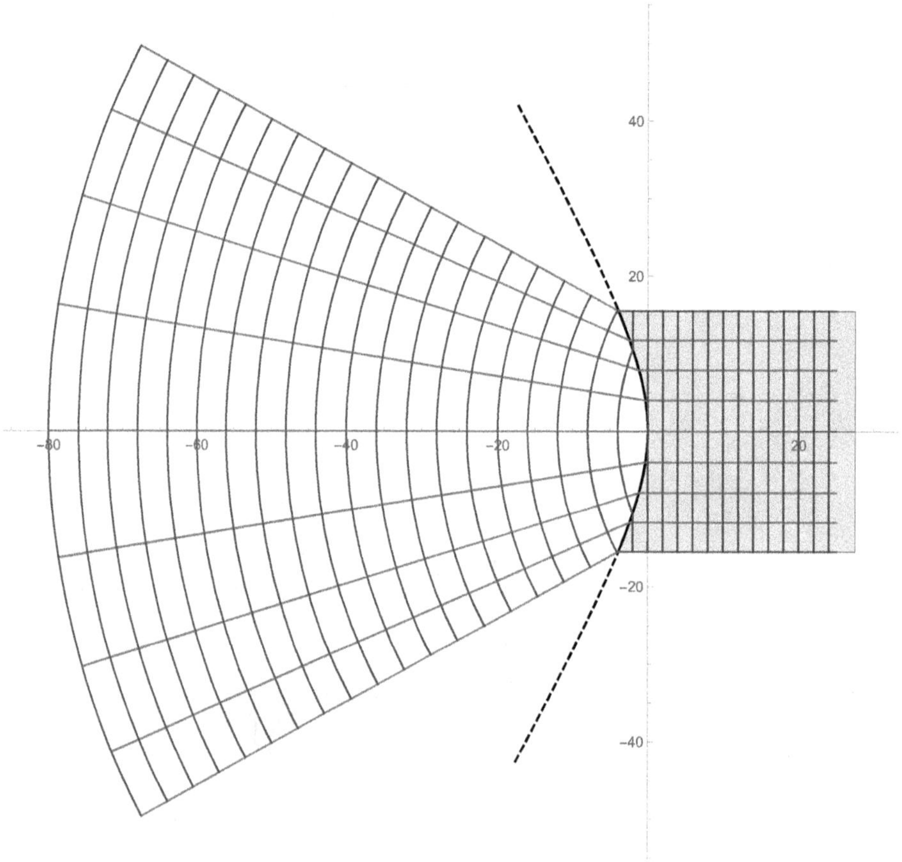

Figure 7.10. Specifications of the design: $n_o = 1$, $n_i = 1.5$, $z_o = 25$ mm, $z_i = \infty$ $z =$ equation (7.41) and $r =$ equation (7.42).

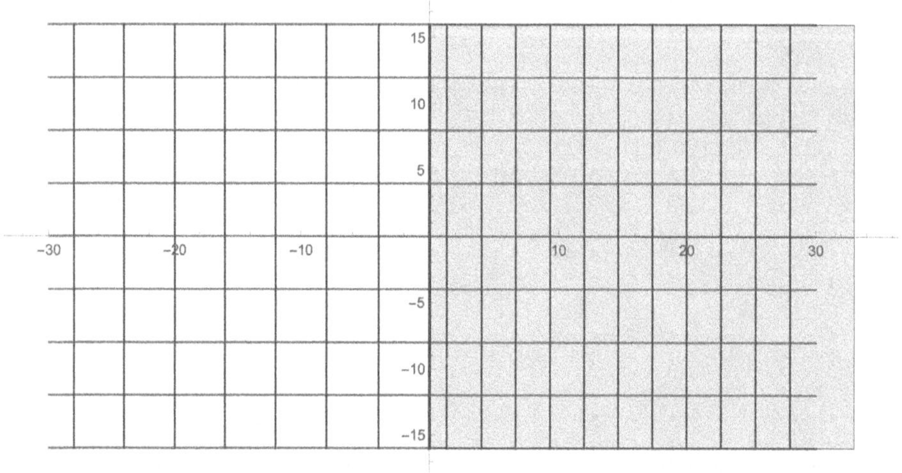

Figure 7.11. Specifications of the design: $n_o = 1$, $n_i = 1.5$, $z_o = -\infty$, $z_{in} = \infty$, $z_i = \infty$ $z =$ equation (7.41) and $r =$ equation (7.42).

Finally, we need to apply both limits on b_1, equation (7.20),

$$\lim_{n_o\to-1}\left[\lim_{n_i\to-n_o}(b_1)\right] = \lim_{n_o\to-1}\left[\lim_{n_i\to-n_o}\left(\frac{(n_i+n_o)(n_i^2 z_i - n_o^2 z_i)}{2n_i n_o z_i z_o(n_i z_i - n_o z_o)}\right)\right]. \tag{7.49}$$

After computing the limits, we can see that b_1 turns to zero, as can be seen in equation (7.49).

$$\lim_{n_o\to-1}\left[\lim_{n_i\to-n_o}(b_1)\right] = 0. \tag{7.50}$$

Therefore, the Cartesian oval proposed by Silva–Torres becomes a conic mirror whose sagitta is given by equation (7.51)

$$\lim_{n_o\to-1}\left[\lim_{n_i\to-n_o}(z)\right] = \frac{\left(\dfrac{z_i+z_o}{2z_i z_o}\right)\rho^2}{\sqrt{1-\left(\dfrac{z_i+z_o}{2z_i z_o}\right)^2\left[-\dfrac{(z_i-z_o)^2}{(z_i+z_o)^2}\right]\rho^2+1}}, \tag{7.51}$$

and its radius is given by

$$\lim_{n_o\to-1}\left[\lim_{n_i\to-n_o}(r)\right] = \mathrm{sgn}(\rho)\sqrt{\rho^2 - \lim_{n_o\to-1}\left[\lim_{n_i\to-n_o}(z)\right]^2}. \tag{7.52}$$

Equations (7.51) and (7.52) can be used to plot spherical, hyperbolic and elliptic mirrors. In the next section, we study the special case of the parabolic mirror.

7.9.1 Parabolic mirror

In order to get the shape of the parabolic mirror from the Silva–Torres method of Cartesian ovals, it is necessary to compute the limit when $z_o \to -\infty$ on equations (7.44), (7.46) (7.51) and (7.52).

We start with equation (7.44),

$$\lim_{z_o\to-\infty}\left\{\lim_{n_o\to-1}\left[\lim_{n_i\to-n_o}(K)\right]\right\} = \lim_{z_o\to-\infty}\left(-\frac{(z_i-z_o)^2}{(z_i+z_o)^2}\right). \tag{7.53}$$

The computation of the limit when $z_o \to -\infty$ gives

$$\lim_{z_o\to-\infty}\left\{\lim_{n_o\to-1}\left[\lim_{n_i\to-n_o}(K)\right]\right\} = -1. \tag{7.54}$$

Then, let's pay attention to equation (7.46),

$$\lim_{z_o\to-\infty}\left\{\lim_{n_o\to-1}\left[\lim_{n_i\to-n_o}(c_0)\right]\right\} = \lim_{z_o\to-\infty}\left(\frac{z_i+z_o}{2z_iz_o}\right), \tag{7.55}$$

applying the limits

$$\lim_{z_o\to-\infty}\left\{\lim_{n_o\to-1}\left[\lim_{n_i\to-n_o}(c_0)\right]\right\} = \frac{1}{2z_i}. \tag{7.56}$$

Therefore, the parabolic mirror from the Silva–Torres perspective is given by

$$\lim_{z\to-\infty}\left\{\lim_{n_o\to-1}\left[\lim_{n_i\to-n_o}(z)\right]\right\} = \frac{\left(\frac{1}{2z_i}\right)\rho^2}{\sqrt{1+\left(\frac{1}{2z_i}\right)^2\rho^2}+1}, \tag{7.57}$$

where

$$\lim_{r\to-\infty}\left\{\lim_{n_o\to-1}\left[\lim_{n_i\to-n_o}(r)\right]\right\} = \mathrm{sgn}(\rho)\sqrt{\rho^2-\lim_{r\to-\infty}\left\{\lim_{n_o\to-1}\left[\lim_{n_i\to-n_o}(z)\right]\right\}^2}. \tag{7.58}$$

For this to be successfully implemented we need to use equations (7.57) and (7.58).

7.10 Illustrative examples

In this section, we present examples of all the conic mirrors using the Silva–Torres approach. The specification of the design is captured in the caption of each example. The examples are in figures 7.12–7.15.

7.11 End notes

In this chapter, we deduced a powerful expression to obtain Cartesian ovals. The first to use this approach were Alberto Silva-Lora and Rafael Torres. Hence, we call this approach the Silva–Torres method. The Silva–Torres method is a closed-form solution of a Cartesian oval. The Silva–Torres method has an all in one single expression, equation (7.26). Equation (7.26) contains all the cases of Cartesian ovals and conic mirrors, and in this chapter we presented examples of these.

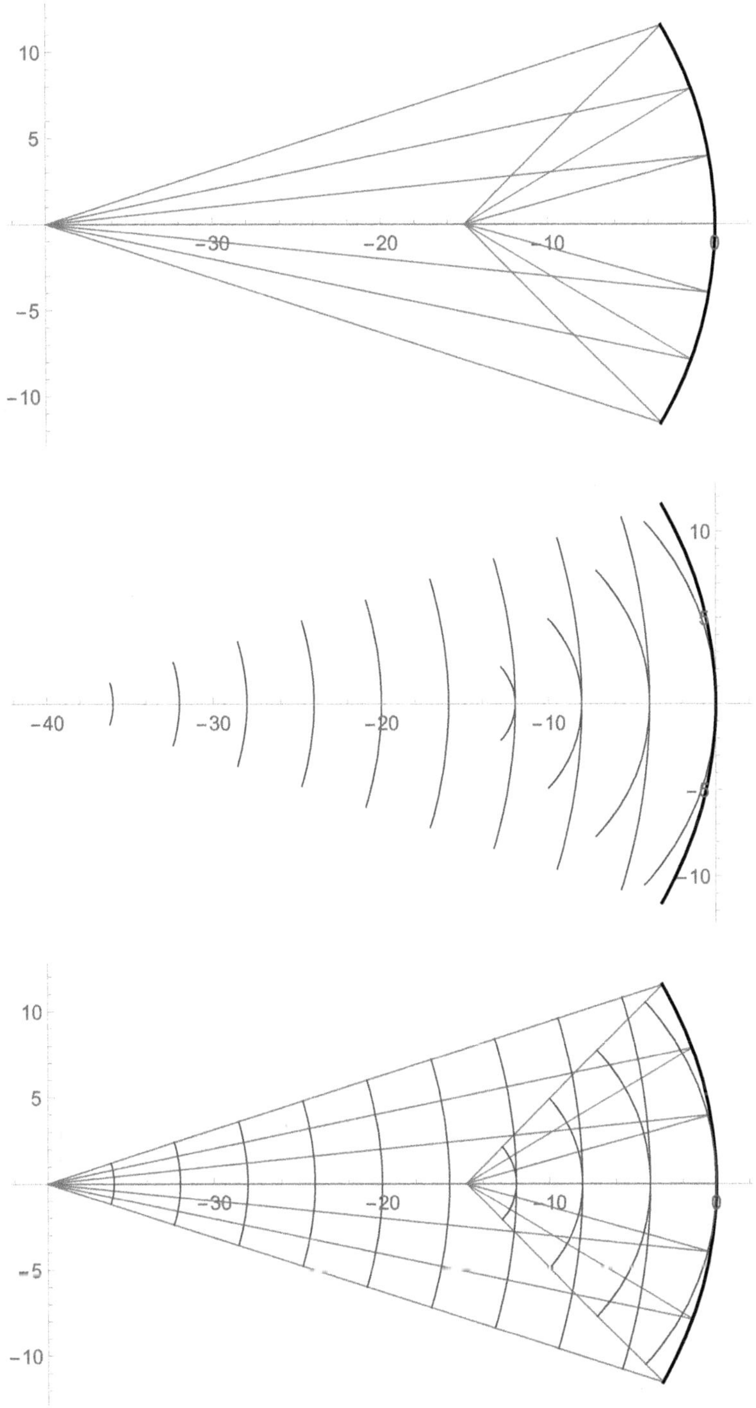

Figure 7.12. Specifications of the design: $n_o = -n_i = 1$, $z_o = -15$, $z_i = -40$ mm $z =$ equation (7.51) and $r =$ equation (7.52).

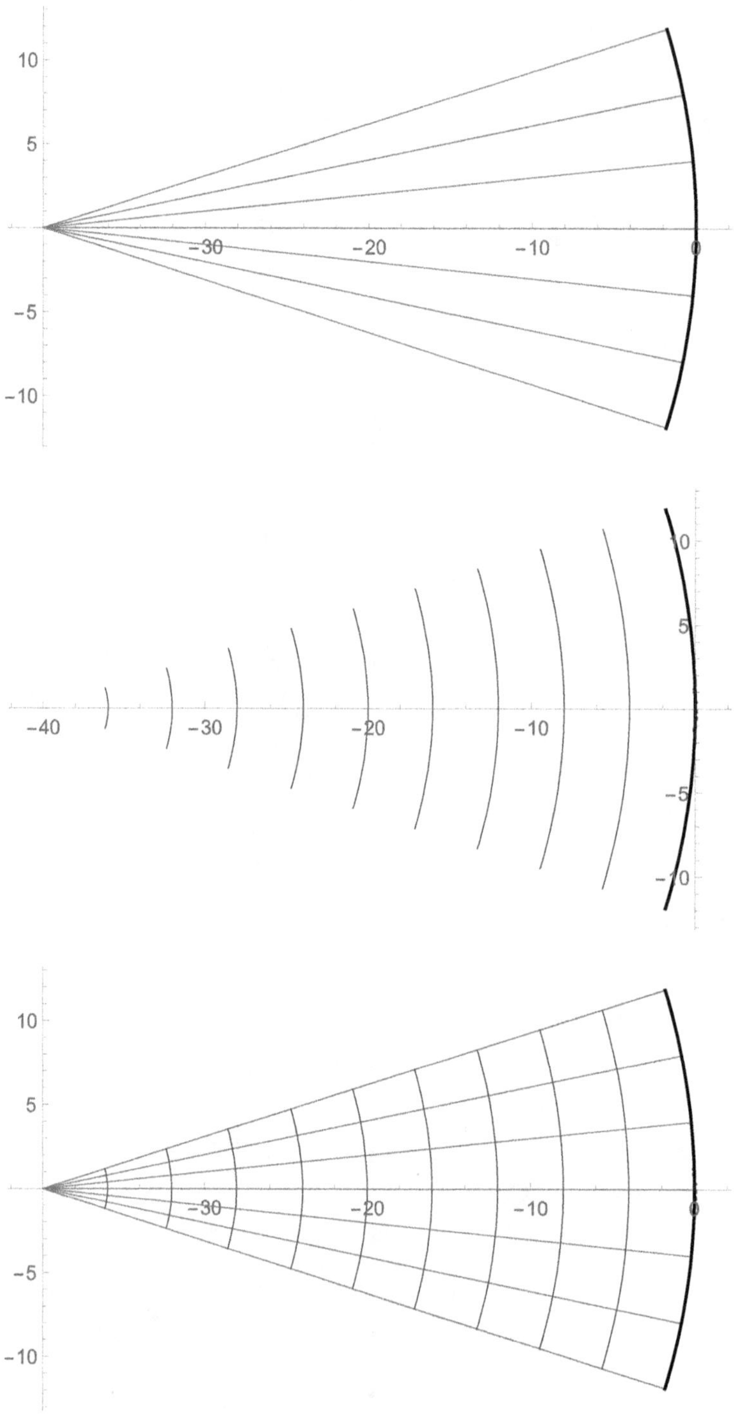

Figure 7.13. Specifications of the design: $n_o = -n_i = 1$, $z_o = -40$, $z_i = -40$ mm $z =$ equation (7.51) and $r =$ equation (7.52).

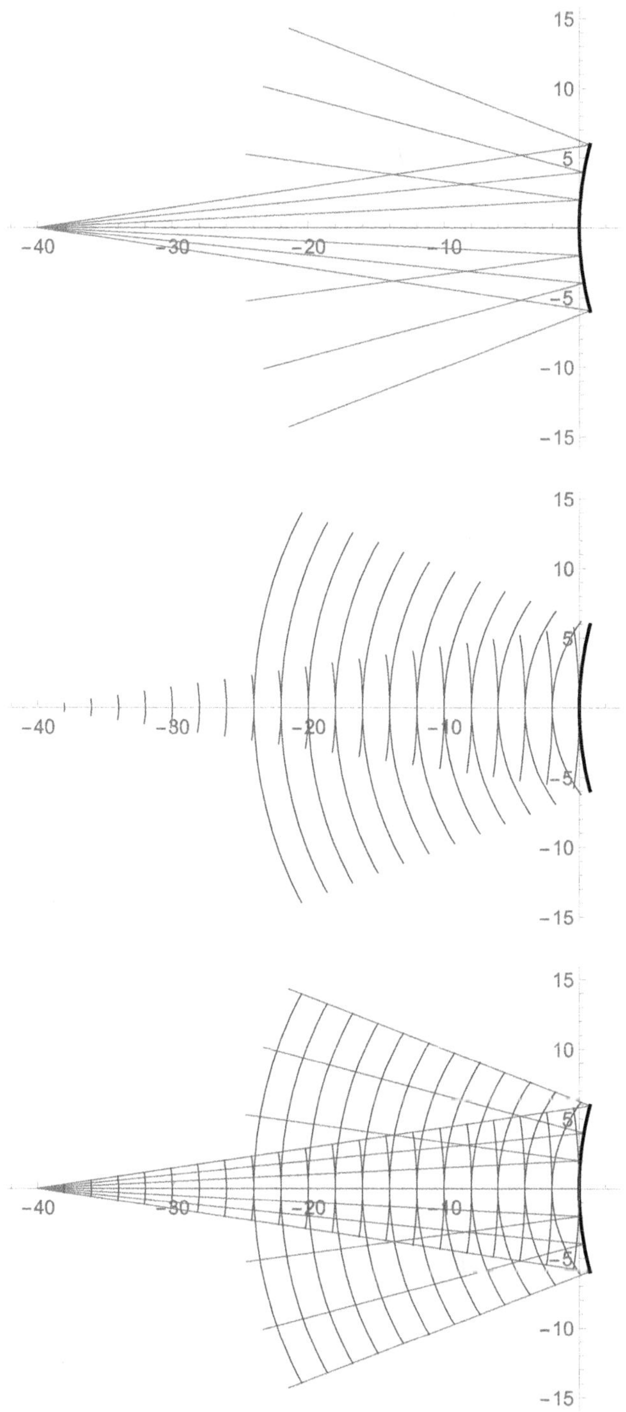

Figure 7.14. Specifications of the design: $n_o = -n_i = 1$, $z_o = 15$, $z_i = -40$ mm $z =$ equation (7.51) and $r =$ equation (7.52).

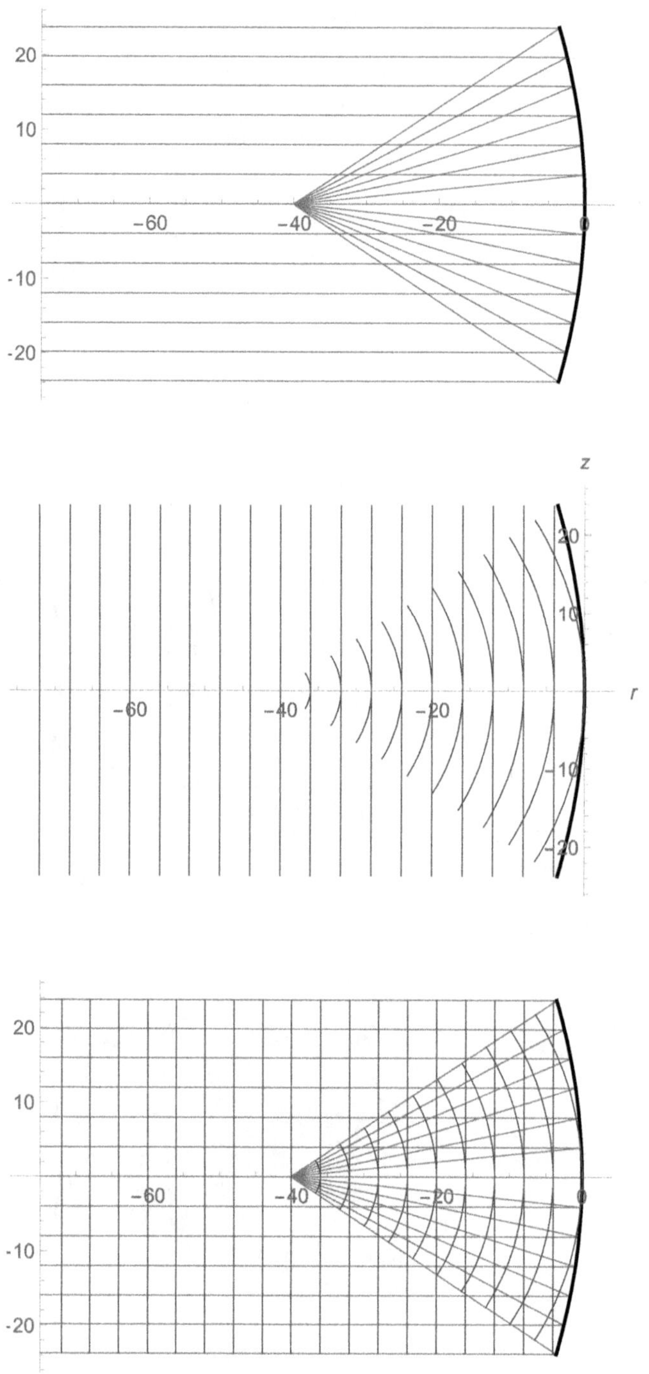

Figure 7.15. Specifications of the design: $n_o = -n_i = 1$, $z_o = -\infty$, $z_i = -40$ mm $z =$ equation (7.57) and $r =$ equation (7.58).

References

Avendaño-Alejo M, Román-Hernández E, Castañeda L and Moreno-Oliva V I 2017 Analytic conic constants to reduce the spherical aberration of a single lens used in collimated light *Appl. Opt.* **56** 6244–54

Bass M 1995 *Handbook of Optics, Volume I: Fundamentals Techniques and Design* (New York: McGraw-Hill)

Born M and Wolf E 2013 Principles of Optics: Electromagnetic Theory of Propagation *Interference and Diffraction of Light* (Amsterdam: Elsevier)

Braunecker B, Hentschel R and Tiziani H J 2008 *Advanced Optics Using Aspherical Elements* **vol 173** (Bellingham, WA: SPIE Press)

Castillo-Santiago G, Avendaño-Alejo M, Díaz-Uribe R and Castañeda L 2014 Analytic aspheric coefficients to reduce the spherical aberration of lens elements used in collimated light *Appl. Opt.* **53** 4939–46

Chaves J 2016 *Introduction to Nonimaging Optics* 2nd edn (Boca Raton, FL: CRC Press)

Estrada J C V, Calle Á H B and Hernández D M 2013 Explicit representations of all refractive optical interfaces without spherical aberration *J. Opt. Soc. Am.* **A30** 1814–24

Glassner A S 1989 *An Introduction to Ray Tracing* (Amsterdam: Elsevier)

González-Acuña R G and Chaparro-Romo H A 2018 General formula for bi-aspheric singlet lens design free of spherical aberration *App. Opt.* **57** 9341–5

González-Acuña R G and Guitiérrez-Vega J C 2018 Generalization of the axicon shape: the gaxicon *J. Opt. Soc. Am.* **A35** 1915–8

González-Acuña R G and Gutiérrez-Vega J C 2019 Analytic formulation of a refractive-reflective telescope free of spherical aberration *Opt. Eng.* **58** 085105

González-Acuña R G, Avendaño-Alejo M and Gutiérrez-Vega J C 2019a Singlet lens for generating aberration-free patterns on deformed surfaces *J. Opt. Soc. Am.* **A36** 925–9

González-Acuña R G, Chaparro-Romo H A and Gutiérrez-Vega J C 2019b General formula to design freeform singlet free of spherical aberration and astigmatism *Appl. Opt.* **58** 1010–5

González-Acuña R G, Chaparro-Romo H A and Gutiérrez-Vega J C 2020a Analytic aplanatic singlet lens: setting and design for three-point objects and images in the meridional plane *Opt. Eng.* **59** 055104

González-Acuña R G, Chaparro-Romo H A and Gutiérrez-Vega J C 2020b Analytic solution of the eikonal for a stigmatic singlet lens *Phys. Scr.*

González-Acuña R G, Chaparro-Romo H A and Gutiérrez-Vega J C 2020c *Analytical Lens Design* (Bristol: Institute of Physics Publishing)

González-Acuña R G, Chaparro-Romo H A and Gutíerrez-Vega J C 2019c *Single Lens Telescope* (arXiv:1903.11129)

González-Acuña R G and Gutiérrez-Vega J C 2019a Analytic formulation of a refractive-reflective telescope free of spherical aberration *Opt. Eng.* **58** 1–5

González-Acuña R G and Gutiérrez-Vega J C 2019b General formula to eliminate spherical aberration produced by an arbitrary number of lenses *Opt. Eng.* **58** 1–6

González-Acuña R G and Gutiérrez-Vega J C 2019c General formula for aspheric collimator lens design free of spherical aberration *Current Developments in Lens Design and Optical Engineering XX* ed R B Johnson, V N Mahajan and S Thibault (Bellingham, WA: International Society for Optics and Photonics, SPIE) 181–4 pp

González Acuña R G and Gutiérrez-Vega J C 2019 General formula to design freeform collimator lens free of spherical aberration and astigmatism *Novel Optical Systems,*

Methods, and Applications XXII **vol 11 105** (Bellingham, WA: International Society for Optics and Photonics) 111050A p

González Acuña R G and Gutiérrez-Vega J C 2019 General formula of the refractive telescope design free spherical aberration *Novel Optical Systems, Methods, and Applications XXII* **vol 11 105** ed C F Hahlweg and J R Mulley (Bellingham, WA: International Society for Optics and Photonics, SPIE) 162–6 p

Kingslake R and Johnson R B 2009 *Lens Design Fundamentals* (New York: Academic)

Lefaivre J 1951 A new approach in the analytical study of the spherical aberrations of any order *J. Opt. Soc. Am.* **41** 647

Luneburg R K 1964 *Mathematical Theory of Optics* (Berkeley, CA: University of California Press)

Malacara D 1965 Two lenses to collimate red laser light *Appl. Opt.* **4** 1652–4

Malacara-Hernández D and Malacara-Hernández Z 2016 *Handbook of Optical Design* (Boca Raton, FL: CRC Press)

González-Acuña R G and Gutiérrez-Vega J C 2020 Analytic design of a spherochromatic singlet *J. Opt. Soc. Am.* **A37** 149–53

Schulz G 1983 Achromatic and sharp real imaging of a point by a single aspheric lens *Appl. Opt.* **22** 3242–8

Silva-Lora A and Torres R 2020 Explicit Cartesian oval as a superconic surface for stigmatic imaging optical systems with real or virtual source or image *Proc. R. Soc.* A476

Stavroudis O 2012 *The Optics of Rays, Wavefronts, and Caustics* **vol 38** (Amsterdam: Elsevier)

Stavroudis O N 2006 *The Mathematics of Geometrical and Physical Optics: The K-function and Its Ramifications* (New York: Wiley)

Stavroudis O N and Feder D P 1954 Automatic computation of spot diagrams *J. Opt. Soc. Am.* **44** 163–70

Sun H 2016 *Lens Design: A Practical Guide* (Boca Raton, FL: CRC Press)

Valencia-Estrada J C and Malacara-Doblado D 2014 Parastigmatic corneal surfaces *Appl. Opt.* **53** 3438–47

Valencia-Estrada J C, Flores-Hernández R B and Malacara-Hernández D 2015 Singlet lenses free of all orders of spherical aberration *Proc. R. Soc.* **A471** 20140608

Vaskas E M 1957 Note on the Wassermann-Wolf method for designing aspheric surfaces *J. Opt. Soc. Am.* **47** 669–70

Wassermann G D and Wolf E 1949 On the theory of aplanatic aspheric systems *Proc. Phys. Soc.* **B62** 2

Winston R, Miñano J C and Benitez P G *et al* 2005 *Nonimaging Optics* (Amsterdam: Elsevier)

Wolf E 1948 On the designing of aspheric surfaces *Proc. Phys. Soc.* **61** 494

Wolf E and Preddy W S 1947 On the determination of aspheric profiles *Proc. Phys. Soc.* **59** 704

Yang T, Jin G-F and Zhu J 2017 Automated design of freeform imaging systems *Light Sci. Appl.* **6** e17081

IOP Publishing

Stigmatic Optics (Second Edition)

Rafael G González-Acuña and Héctor A Chaparro-Romo

Chapter 8

The stigmatic lens generated by Cartesian ovals

From two stigmatic surfaces, Cartesian ovals, a stigmatic lens can be generated. In this chapter, we will study the generation of stigmatic lenses from the Cartesian ovals model proposed by Silva–Torres.

8.1 Introduction

In the previous chapter, we presented the rigorous model of Silva–Torres Cartesian ovals, which is general enough to cover all cases of Cartesian ovals and conical mirrors.

In this chapter, we now address the design of a stigmatic lens using two Silva–Torres Cartesian ovals. The general idea is that the image formed by the first Cartesian oval of Silva–Torres is the object of the second Cartesian oval of Silva–Torres. The amazing thing about this strategy is that, as we mentioned, the model of Silva–Torres Cartesian ovals covers all cases of Cartesian ovals and conical mirrors. So, the image of the first Cartesian oval can be taken as a real or virtual object for the second Cartesian oval.

8.2 Mathematical model

The central idea behind this mathematical model is based on two Silva–Torres Cartesian ovals. The image produced by the first Silva–Torres Cartesian oval is taken as the object of the second Silva–Torres Cartesian oval. See the diagram of the model in figure 8.1. n_o is the refraction index of the medium that surrounds the object. z_o is the distance from the object to the vertex of the first Silva–Torres Cartesian oval, which is denoted by z_a. Each suffix with a concerns the first Silva–Torres Cartesian oval. We call the first Silva–Torres Cartesian oval the first surface. Notice that the coordinate system is placed in the vertex of the first surface. n is the refraction index of the lens generated by the two Silva–Torres Cartesian ovals. z_{in} is the distance from the first surface to the image generated by the first surface. The second surface takes this image as an object. That is the reason why z_{in} has *in* as a

doi:10.1088/978-0-7503-6423-2ch8

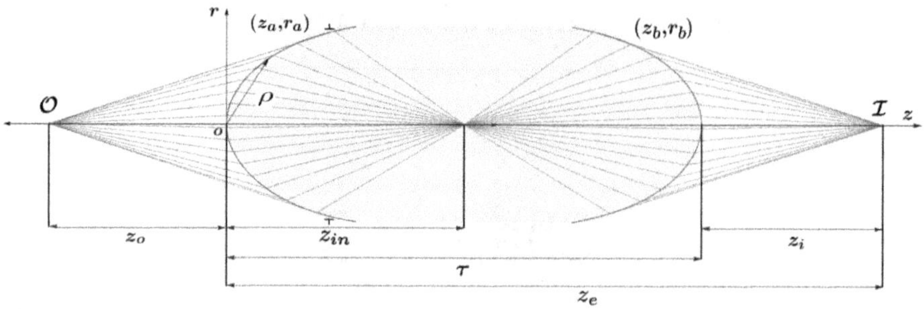

Figure 8.1. Sketch of a lens designed by two Cartesian ovals as refractive surfaces. The first Cartesian oval is (z_a, r_a) and the second Cartesian oval is (z_b, r_b). The refraction index in the object space is n_o, inside the lens, the refraction index is n, and in the image space is n_i. The origin is placed in the vertex of the first Cartesian oval. ρ is the distance from the origin to a point in the first surface. The length from the object to the first Cartesian oval is z_o; the distance from the first Cartesian oval to the image generated by the first surface is z_{in}. The center depth of the lens is τ. From the origin to the image the gap is given by z_e and for the gap from the second Cartesian oval to the image, the length is z_i.

suffix, *in* is from the inside. τ is the central thickness of the lens. z_e is the distance from the origin to the image. The Silva–Torres Cartesian oval is denoted by z_b, and it is also called the second surface in this chapter. The suffix b is the member mark that indicates that they are related to the second surface.

Taking all the considerations of the last paragraph into account, the mathematical model provided in the last chapter and figure 8.1; it is easy to see that the first surface is given by equation (8.1),

$$\boxed{z_a \equiv \frac{\rho^2(c_{1_a}\rho^2 + c_{0_a})}{\sqrt{\rho^2\left(2b_{1_a} - c_{0_a}^2 K_a\right) + 1} + b_{1_a}\rho^2 + 1}}, \tag{8.1}$$

where its radius, equation (8.2), is

$$\boxed{r_a \equiv \operatorname{sgn}(\rho)\sqrt{\rho^2 - z_a(\rho)^2}}, \tag{8.2}$$

and the inside parameters are

$$K_a \equiv \frac{(n^2 z_{in} - n_o^2 z_o)^2}{n n_o(n z_{in} - n_o z_o)(n z_o - n_o z_{in})}. \tag{8.3}$$

Then, c_{0_a} is expressed as

$$c_{0_a} \equiv \frac{n z_o - n_o z_{in}}{z_{in} z_o (n - n_o)}, \tag{8.4}$$

with c_{1_a}, we have

$$c_{1_a} \equiv \frac{(n - n_o)(n + n_o)^2}{4 n n_o z_{in} z_o (n z_{in} - n_o z_o)}, \tag{8.5}$$

and finally, b_{1_a},

$$b_{1_a} \equiv \frac{(n+n_o)(n^2 z_{in} - n_o^2 z_o)}{2nn_o z_{in} z_o (nz_{in} - n_o z_o)}. \tag{8.6}$$

Applying the same strategy as for obtaining the first surface, we can write the second surface as

$$\boxed{z_{b_0} \equiv \frac{\rho^2(c_{1_b}\rho^2 + c_{0_b})}{\sqrt{\rho^2\left(2b_{1_b} - c_{0_b}^2 K_b\right) + 1} + b_{1_b}\rho^2 + 1}}, \tag{8.7}$$

where

$$z_b \equiv \tau + z_{b_0}(\rho), \tag{8.8}$$

where the radius is

$$\boxed{r_b \equiv \text{sgn}(\rho)\sqrt{\rho^2 - z_{b_0}(\rho)^2}}. \tag{8.9}$$

If we take $n_o = n_i$, the parameters related to the second surfaces are

$$K_b \equiv \frac{(n_o^2(z_e - \tau) - n^2(z_{in} - \tau))^2}{nn_o[n(z_e - \tau) - n_o(z_{in} - \tau)][n(z_{in} - \tau) - n_o(z_e - \tau)]}. \tag{8.10}$$

c_{0_b} is given by

$$c_{0_b} \equiv \frac{n(z_e - \tau) - n_o(z_{in} - \tau)}{(n - n_o)(z_e - \tau)(z_{in} - \tau)}. \tag{8.11}$$

c_{1_b} is written as

$$c_{1_b} \equiv \frac{(n - n_o)(n + n_o)^2}{4nn_o(z_e - \tau)(z_{in} - \tau)[n(z_{in} - \tau) - n_o(z_e - \tau)]}, \tag{8.12}$$

and finally, b_{1_b},

$$b_{1_b} \equiv \frac{(n + n_o)(n^2(z_{in} - \tau) - n_o^2(z_e - \tau))}{2nn_o(z_e - \tau)(z_{in} - \tau)[n(z_{in} - \tau) - n_o(z_e - \tau)]}. \tag{8.13}$$

Notice that the second surface has the same structure as the first surface but it is out-placed by the central thickness of the lens, τ.

With equations (8.1)–(8.13) we can design stigmatic lenses for countless configurations, where z_o, z_{in}, z_e can be placed anywhere in the optical axis. Therefore, we can design stigmatic lenses for real and virtual object/images.

8.3 Examples

In this section, we present examples of stigmatic lenses designed with equations (8.1)–(8.13), where the objects and images are finite but they can be real or virtual.

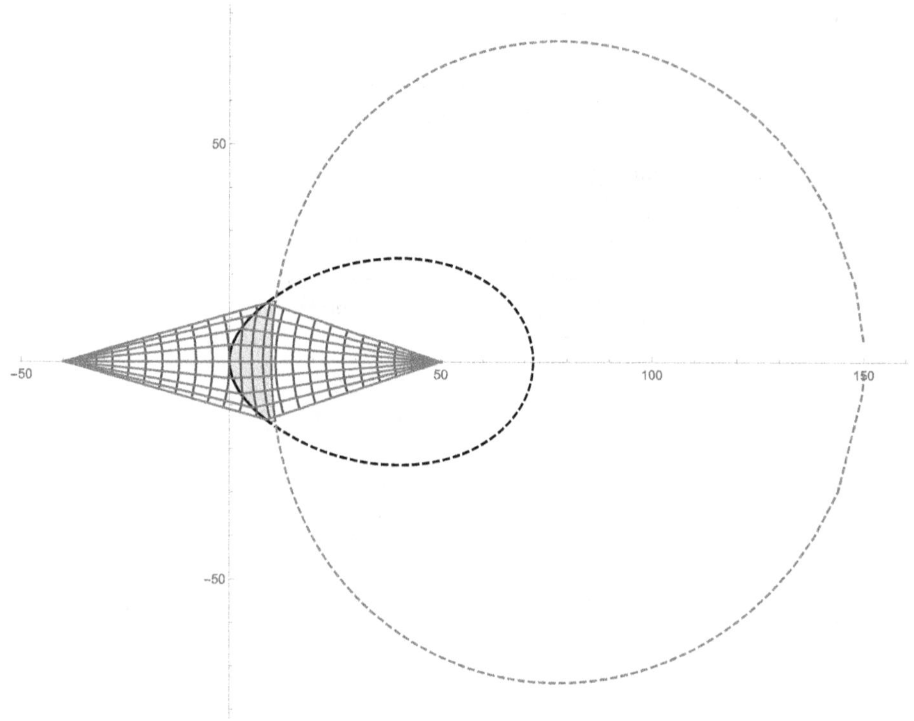

Figure 8.2. Specifications of the design: $n = 1.5$, $z_o = -40$, $\tau = 12$, $z_i = 40$, $z_{in} = n\, z_i$, z_a =equation (8.1), r_a = equation (8.2), z_b = equation (8.8), r_b = equation (8.9), using equations (8.3)–(8.7), (8.10)–(8.13).

The specification of each design is presented in the captions of the respective figure. The figures of this section are 8.2–8.7.

8.4 Collector

A collector lens is a lens that receives rays from minus infinity; this means that the image is placed at minus infinity. Therefore, if we are interested in designing a stigmatic lens with Cartesian ovals, we need to compute the limit when $z_o \to -\infty$ over the parameters of the first surface. Notice that we only need to apply the limit mentioned above in the first surface, because the second surface and its parameters do not depend on z_o.

We start by computing the limit when $z_o \to -\infty$ in equation (8.3),

$$\lim_{z_o \to -\infty} K_a = \lim_{z_o \to -\infty} \left[\frac{(n^2 z_{in} - n_o^2 z_o)^2}{n n_o (n z_{in} - n_o z_o)(n z_o - n_o z_{in})} \right] = -\frac{n_o^2}{n^2}. \tag{8.14}$$

Then we compute the same limit in equation (8.4),

$$\lim_{z_o \to -\infty} c_{0_a} = \lim_{z_o \to -\infty} \left[\frac{n z_o - n_o z_{in}}{z_{in} z_o (n - n_o)} \right] = \frac{n}{z_{in}(n - n_o)}, \tag{8.15}$$

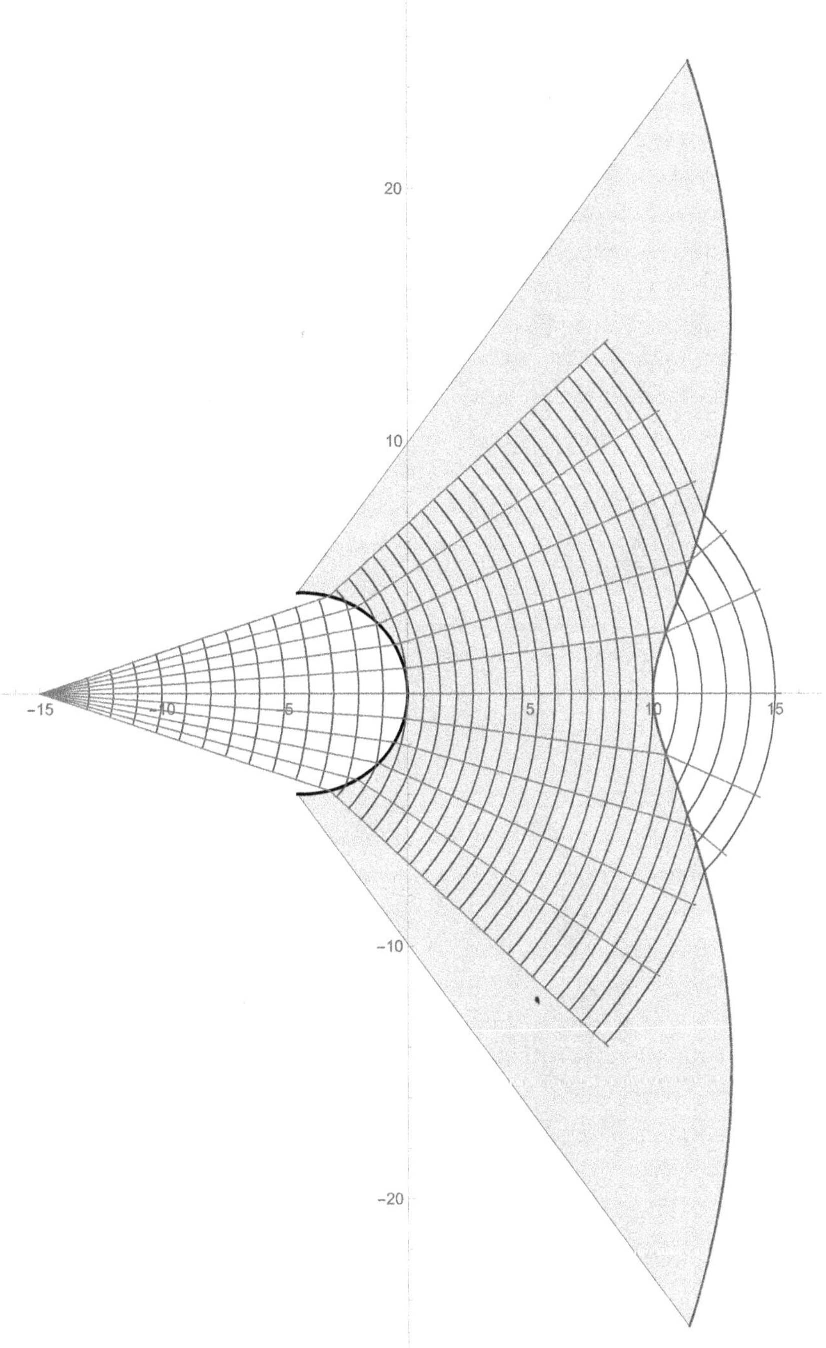

Figure 8.3. Specifications of the design: $n = 1.5$, $z_o = -15$, $\tau = 10$, $z_i = -5$, $z_{in} = n\, z_i$, z_a =equation (8.1), r_a = equation (8.2), z_b = equation (8.8), r_b = equation (8.9), using equations (8.3)–(8.7), (8.10)–(8.13).

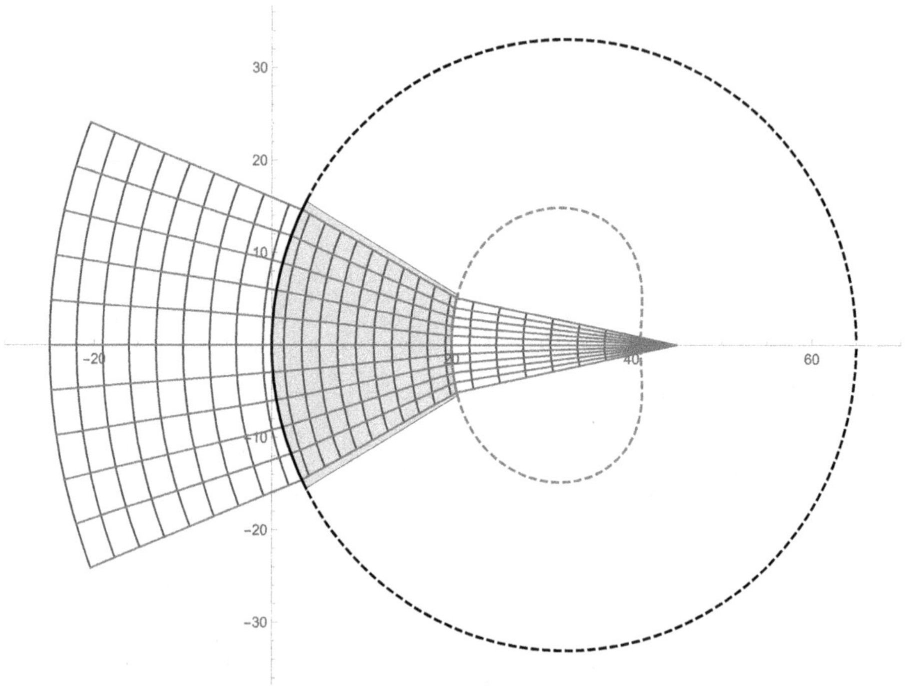

Figure 8.4. Specifications of the design: $n = 1.5$, $z_o = 40$, $\tau = 20$, $z_i = 25$, $z_{in} = n\ z_i$, $z_a =$ equation (8.1), $r_a =$ equation (8.2), $z_b =$ equation (8.8), $r_b =$ equation (8.9), using equations (8.3)–(8.7), (8.10)–(8.13).

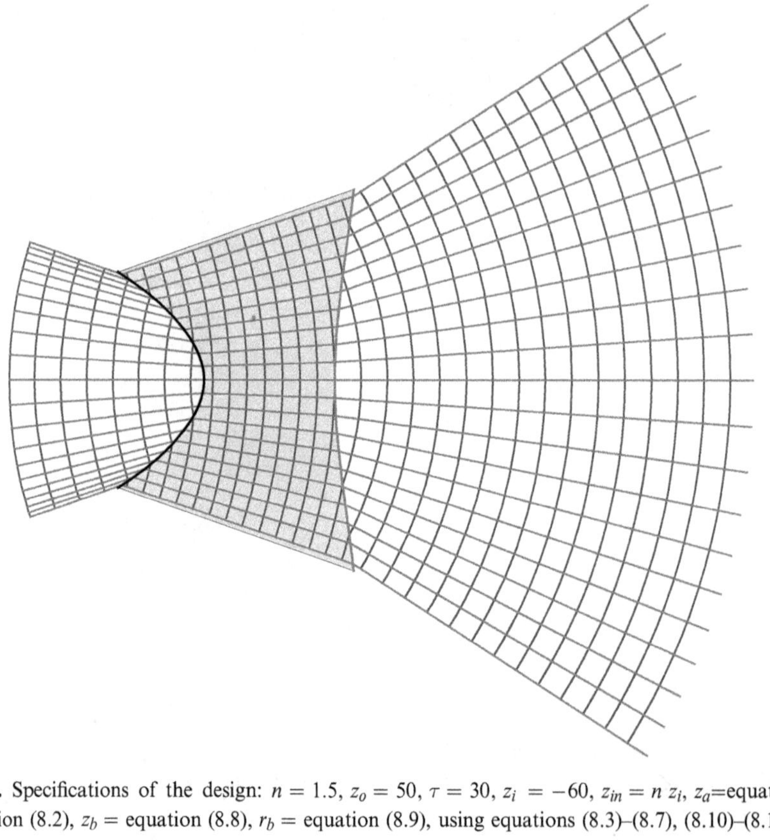

Figure 8.5. Specifications of the design: $n = 1.5$, $z_o = 50$, $\tau = 30$, $z_i = -60$, $z_{in} = n\ z_i$, z_a=equation (8.1), $r_a =$ equation (8.2), $z_b =$ equation (8.8), $r_b =$ equation (8.9), using equations (8.3)–(8.7), (8.10)–(8.13).

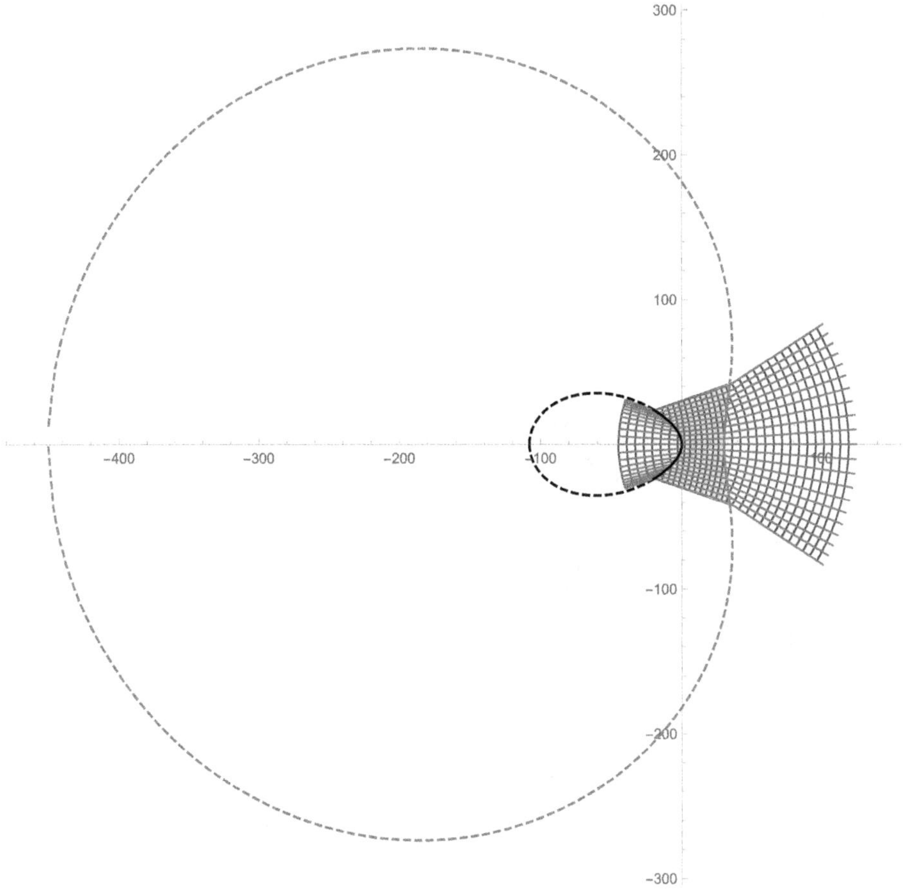

Figure 8.6. Specifications of the design: $n = 1.5$, $z_o = 50$, $\tau = 30$, $z_i = -60$, $z_{in} = n\, z_i$, z_a =equation (8.1), r_a = equation (8.2), z_b = equation (8.8), r_b = equation (8.9), using equations (8.3)–(8.7), (8.10)–(8.13).

followed by computing the limit when $z_o \to \infty$ in equation (8.5),

$$\lim_{z_o \to -\infty} c_{1_a} = \lim_{z_o \to -\infty} \left[\frac{(n - n_o)(n + n_o)^2}{4nn_o z_{in} z_o (nz_{in} - n_o z_o)} \right] = 0, \tag{8.16}$$

and finally, we apply the limit when $z_o \to -\infty$ in equation (8.6),

$$\lim_{z_o \to -\infty} b_{1_a} = \lim_{z_o \to -\infty} \left[\frac{(n + n_o)(n^2 z_{in} - n_o^2 z_o)}{2nn_o z_{in} z_o (nz_{in} - n_o z_o)} \right] = 0. \tag{8.17}$$

Therefore, we can write the first surface as

$$\boxed{\lim_{z_o \to -\infty} z_a = \frac{\left(\lim\limits_{z_o \to -\infty} c_{0_a}\right)\rho^2}{\sqrt{1 - \left(\lim\limits_{z_o \to -\infty} c_{0_a}\right)^2 \left(\lim\limits_{z_o \to -\infty} K_a\right)\rho^2 + 1}}}, \tag{8.18}$$

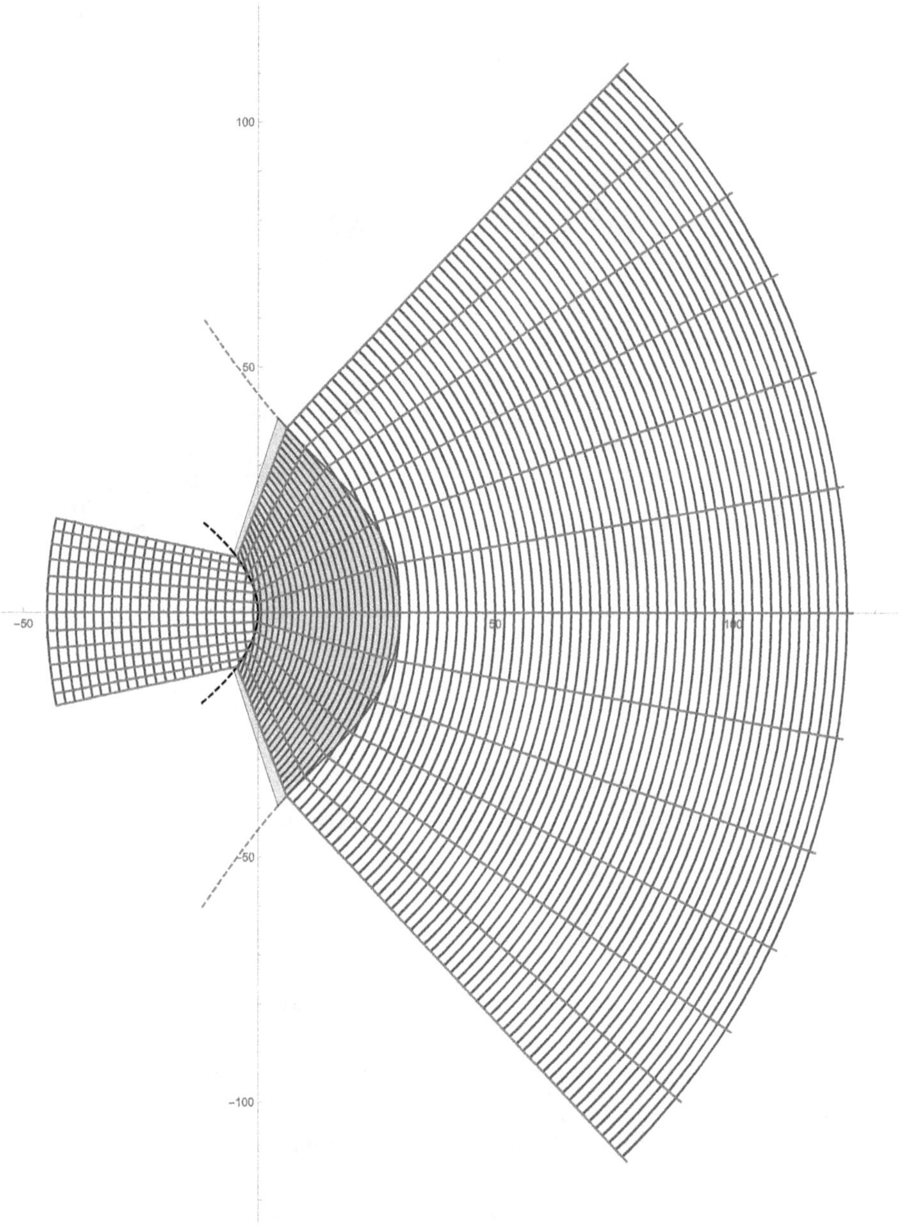

Figure 8.7. Example of a lens designed by two Cartesian ovals with the approach of Silva–Torres.

and its radius as

$$\boxed{\lim_{z_o \to -\infty} r_a = \text{sgn}(\rho)\sqrt{\rho^2 - \lim_{z_o \to -\infty}(z_a^2)}}. \tag{8.19}$$

Equations (8.7)–(8.19) give us the correct shape of a Cartesian oval when $z_o \to -\infty$. In the next section we will compute some examples with these equations.

8.5 Examples

In the following section, we start with some examples of lenses with real objects and real/virtual images. The parameters to design the lenses are in the captions of the figures. The presented designs are evaluated using equations (8.7)–(8.19) without any optimization or modification.

Figures 8.8–8.11 show lenses of the gallery. All the rays that come from minus infinity are a focus on the image point at z_e.

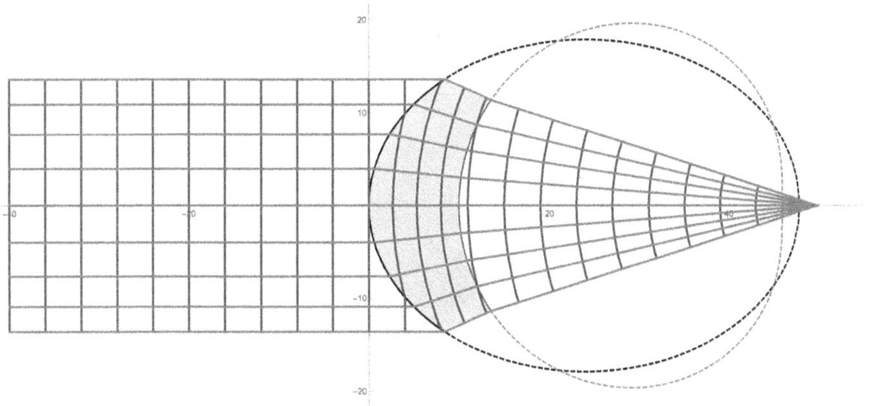

Figure 8.8. Specifications of the design: $n = 1.5$, $\tau = 10$, $z_i = 40$, $zin = n\ zi$ using $z_a =$ equation (8.18), $r_a =$ equation (8.19), $z_b =$ equation (8.8), $r_b =$ equation (8.9) with (8.10)–(8.13), (8.7), (8.14)–(8.19).

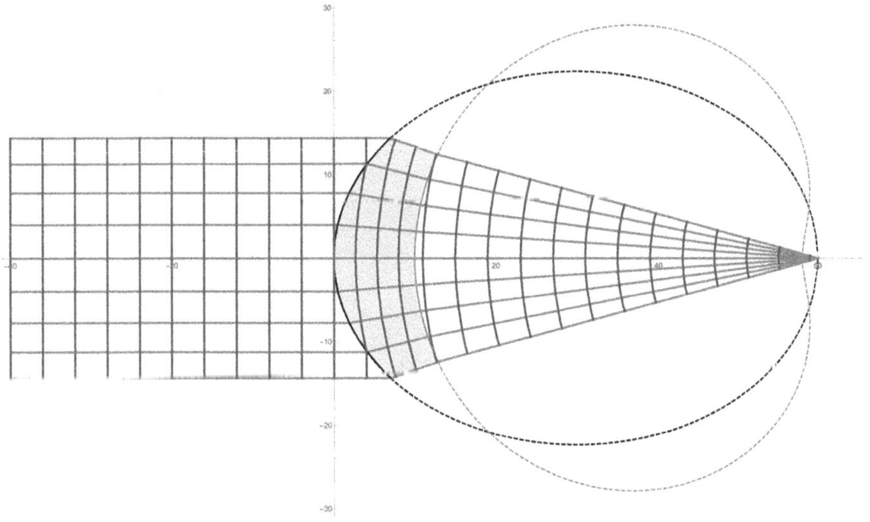

Figure 8.9. Specifications of the design: $n = 1.5$, $\tau = 10$, $z_i = 50$, $zin = n\ zi$ using $z_a =$ equation (8.18), $r_a =$ equation (8.19), $z_b =$ equation (8.8), $r_b =$ equation (8.9) with (8.10)–(8.13), (8.7), (8.14)–(8.19).

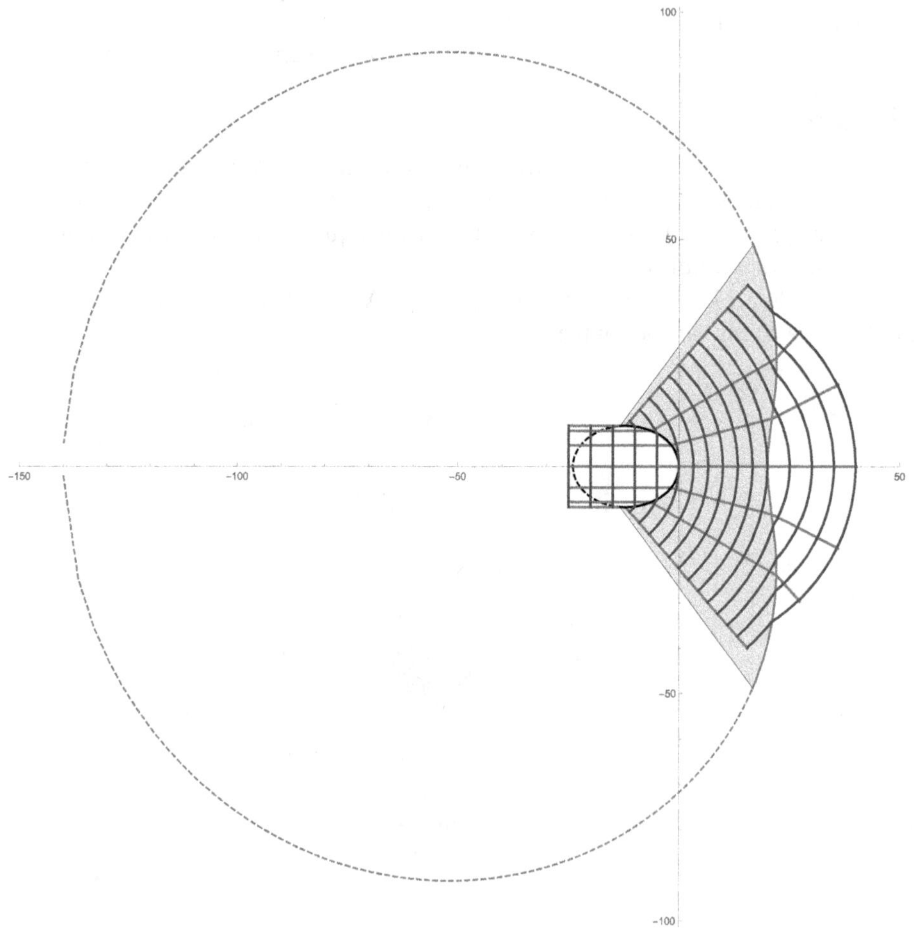

Figure 8.10. Specifications of the design: $n = 1.5$, $\tau = 28$, $z_i = -20$, $zin = n\ zi$ using $z_a =$ equation (8.18), $r_a =$ equation (8.19)$z_b =$ equation (8.8), $r_b =$ equation (8.9) with (8.10)–(8.13), (8.7), (8.14)–(8.19).

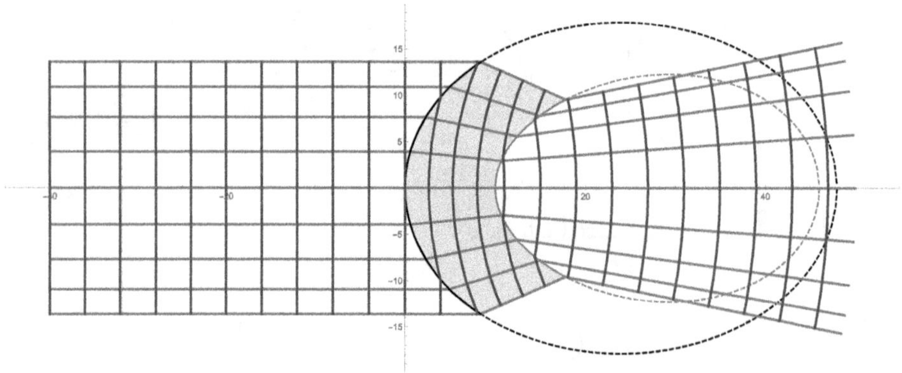

Figure 8.11. Specifications of the design: $n = 1.5$, $\tau = 10$, $z_i = 40$, $zin = n\ zi$ using $-z_a =$ equation (8.18), $r_a =$ equation (8.19)$z_b =$ equation (8.8), $r_b =$ equation (8.9) with (8.10)–(8.13), (8.7), (8.14)–(8.19).

8.6 Collimator

In this section, we are going to work with the inverse problem of the last part. Now the input rays come from a point source located at a finite distance with respect to the first surface, and the output rays are collimated. Therefore, we need to compute the limit when $z_e \to \infty$ for the parameter of the second surface.

We shall begin with equation (8.10). The limit gives us

$$\lim_{z_e\to\infty} K_b = \lim_{z_e\to\infty}\left[\frac{[n_o^2(z_e-\tau)-n^2(z_{in}-\tau)]^2}{nn_o[n(z_e-\tau)-n_o(z_{in}-\tau)][n(z_{in}-\tau)-n_o(z_e-\tau)]}\right] = -\frac{n_o^2}{n^2}. \quad (8.20)$$

Then, it is appropriated to compute the limit when $z_e \to \infty$ over equation (8.11),

$$\lim_{z_e\to\infty} c_{0_b} = \lim_{z_e\to\infty}\left[\frac{n(z_e-\tau)-n_o(z_{in}-\tau)}{(n-n_o)(z_e-\tau)(z_{in}-\tau)}\right] = \frac{n}{-n\tau+nz_{in}+n_o\tau-n_oz_{in}}. \quad (8.21)$$

Now, let's apply the limit when $z_e \to \infty$ in equation (8.12),

$$\lim_{z_e\to\infty} c_{1_b} = \lim_{z_e\to\infty}\left[\frac{(n-n_o)(n+n_o)^2}{4nn_o(z_e-\tau)(z_{in}-\tau)[n(z_{in}-\tau)-n_o(z_e-\tau)]}\right] = 0, \quad (8.22)$$

and finally, we compute the mentioned limit in equation (8.13),

$$\lim_{z_e\to\infty} b_{1_b} = \lim_{z_e\to\infty}\left[\frac{(n+n_o)[n^2(z_{in}-\tau)-n_o^2(z_e-\tau)]}{2nn_o(z_e-\tau)(z_{in}-\tau)[n(z_{in}-\tau)-n_o(z_e-\tau)]}\right] = 0. \quad (8.23)$$

As a result of the above computations we write the second surface as

$$\boxed{\lim_{z_e\to\infty} z_{b_0} = \frac{\left(\frac{n}{-n\tau+nz_{in}+n_o\tau-n_oz_{in}}\right)\rho^2}{\sqrt{1-\left(\frac{n}{-n\tau+nz_{in}+n_o\tau-n_oz_{in}}\right)^2\left(-\frac{n_o^2}{n^2}\right)\rho^2+1}}}, \quad (8.24)$$

where

$$\boxed{\lim_{z_e\to\infty} z_b = \tau + \lim_{z_e\to\infty} z_{b_0}} \quad (8.25)$$

and the radius becomes

$$\lim_{z_e\to\infty} r_b = \mathrm{sgn}(\rho)\sqrt{\rho^2 - \lim_{z_e\to\infty}(z_{b_0})^2}. \quad (8.26)$$

Equations (8.14)–(8.19), (8.20)–(8.26) are the ones we need if we want to design a collimator stigmatic singlet lens using Cartesian ovals. Notice that we do not compute any limit over the first surface and its parameters because z_e is not presented in them.

8.7 Examples

The following figures have the ray tracing and specification of several examples of collimator lenses. All the presented models are estimated using equations (8.1)–(8.6), (8.20)–(8.26). The figures are 8.12 and 8.13.

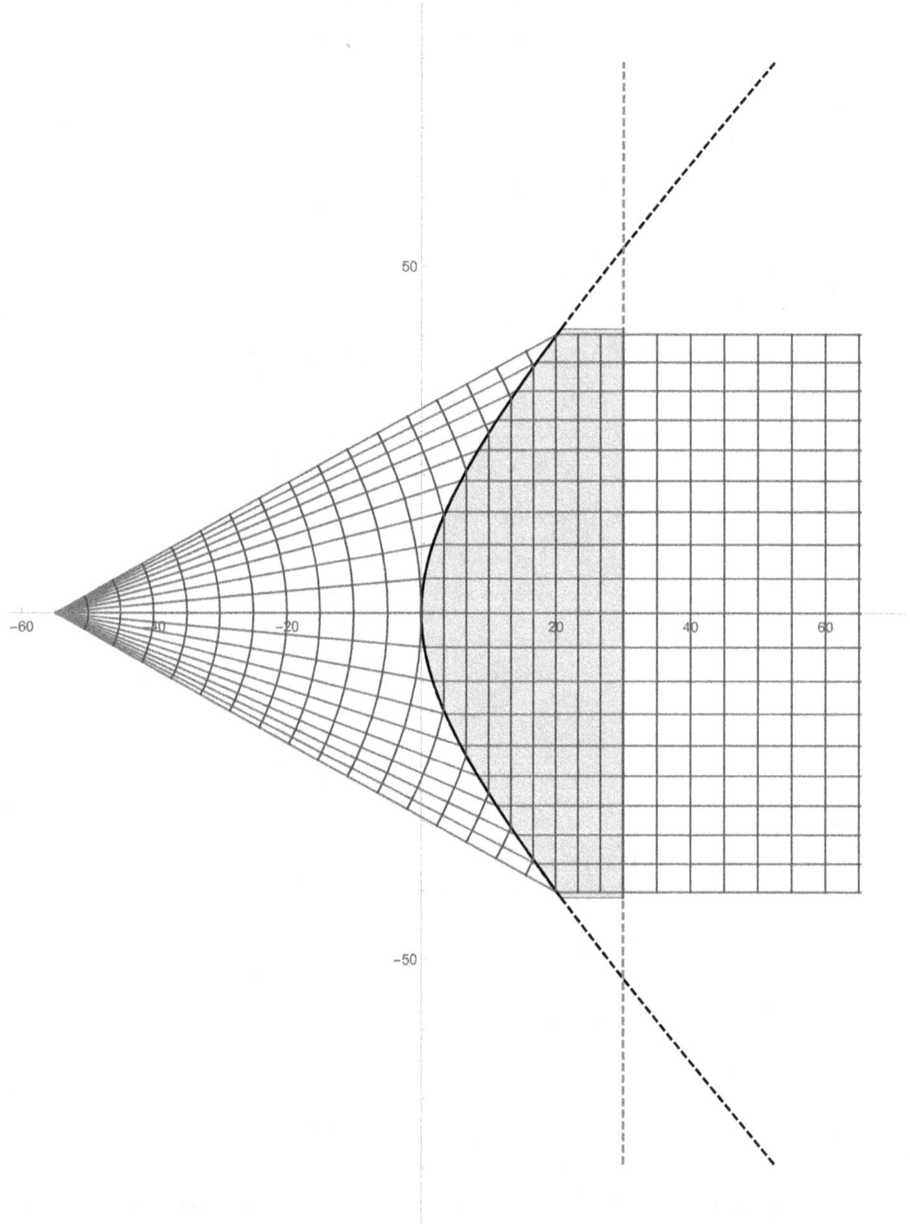

Figure 8.12. Specifications of the design: $n = 1.5$, $z_o = -55$, $\tau = 30$, $z_i = \infty$, $z_{in} = n\ z_i$, $z_a =$ equation (8.1), $r_a =$ equation (8.2), $z_b =$ equation (8.25), $r_b =$ equation (8.26), using equations (8.3)–(8.6), (8.20)–(8.23). This is a special case because the inner wavefront is flat (it happens that the second surface is flat too).

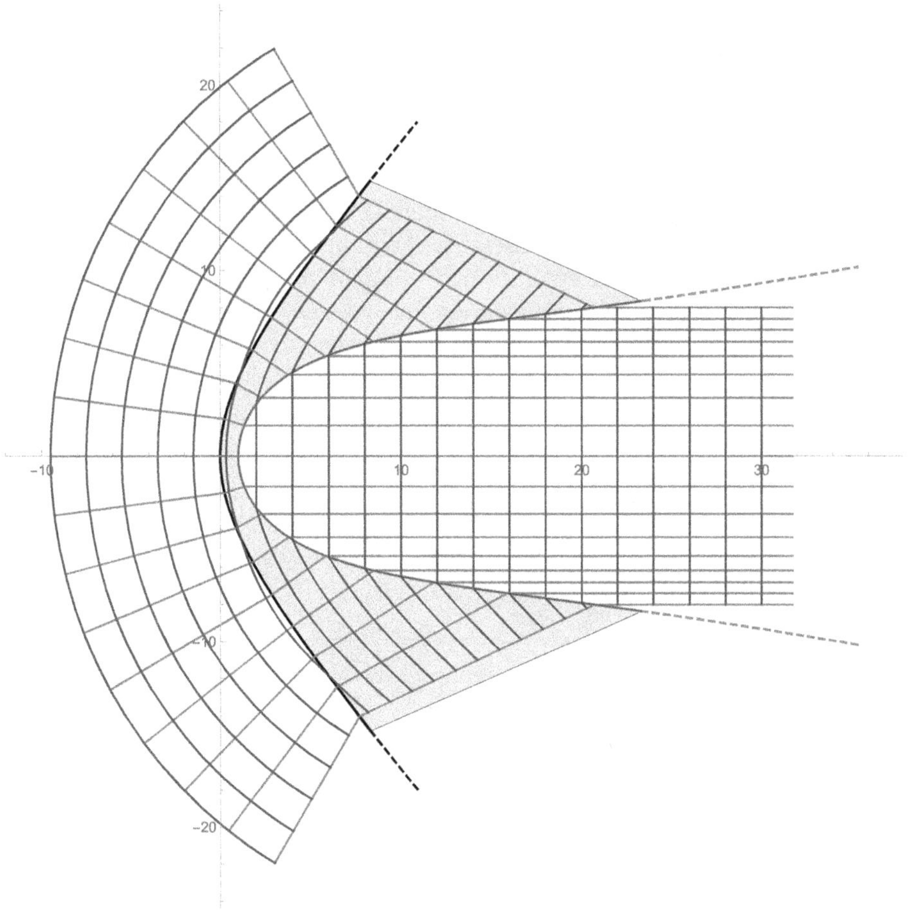

Figure 8.13. Specifications of the design: $n = 1.5$, $z_o = 16$, $\tau = 1$, $z_i = \infty$, $z_{in} = n\ z_i$, $z_a =$ equation (8.1), $r_a =$ equation (8.2), $z_b =$ equation (8.25), $r_b =$ equation (8.26), using equations (8.3)–(8.6), (8.20)–(8.24).

8.8 Single-lens telescope with Cartesian ovals

The single-lens telescope is more a myth than a practical lens used in engineering. The legend of the single-lens telescope comes from paraxial optics. In paraxial optics, all lenses can be represented by matrices. But there is not a matrix for the single-lens telescope; there is a matrix for flat glass. However, the rays that come collimated along the optical axis refracted by a flat glass at the output are collimated, but not amplified. The idea of the single-lens telescope is a lens such that the collimated rays entering the lens are refracted such as they suffer an amplification, and in the output they are collimated. Also, the myth comes from the first telescopes, which were made by at least two lenses.

In this section, we are going to partially demystify the single-lens telescope by obtaining the Cartesian ovals of the single-lens telescope. This procedure is to show the robustness of the method implemented during this chapter.

To design it we only use equations (8.1)–(8.6), (8.20)–(8.26).

8.9 Example

A possible application of the single-lens telescope is not as a telescope, but as a beam expander (typically an array of two lenses that expand a laser beam).

The single-beam expander lens and the single-lens telescope are the same because ray optics are invertible. The only difference is that in the singlet beam expander lens at the output, the amplification is positive and in the single-lens telescope it is negative. For us, a positive amplification means that the beam expands and a negative amplification means that the beam contracts (figures 8.14–8.16).

In figures 8.14–8.16 there is a single-lens telescope. At the input and output, the rays are collimated. The example is computed using equations (8.14)–(8.26). The design specifications are included in the captions of the aforementioned figures.

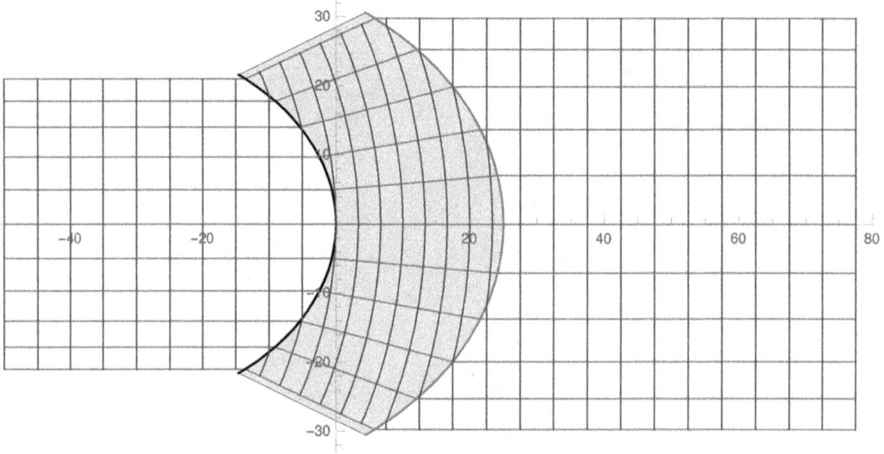

Figure 8.14. Specifications of the design: $n = 1.5$, $\tau = 25$, $z_{in} = -60$, $z_a =$ equation (8.18), $r_a =$ equation (8.19), $z_b =$ equation (8.25), $r_b =$ equation (8.26), using equations (8.14)–(8.17), (8.20)–(8.24).

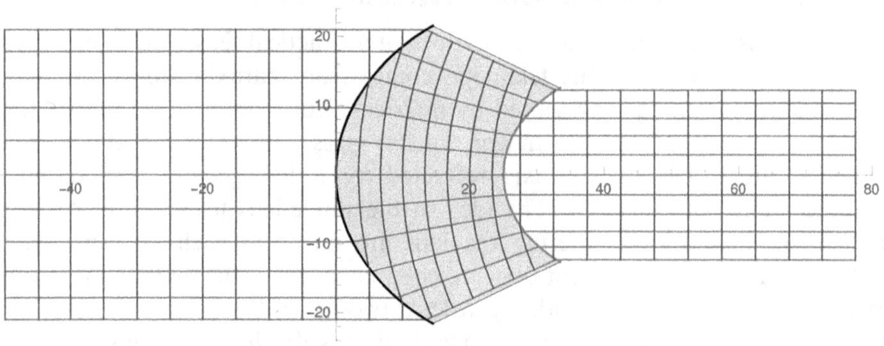

Figure 8.15. Specifications of the design: $n = 1.5$, $\tau = 25$, $z_{in} = 60$, $z_a =$ equation (8.18), $r_a =$ equation (8.19), $z_b =$ equation (8.25), $r_b =$ equation (8.26), using equations (8.14)–(8.17), (8.20)–(8.24).

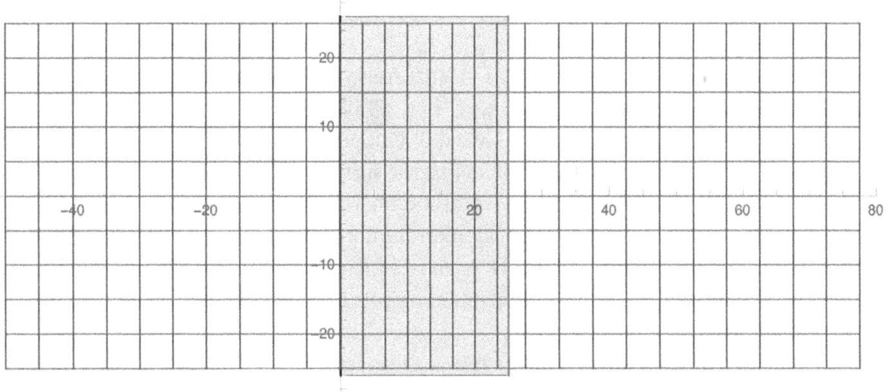

Figure 8.16. Specifications of the design: $n = 1.5$, $\tau = 25$, $z_{in} = \infty$, $z_a =$ equation (8.18), $r_a =$ equation (8.19), $z_b =$ equation (8.25), $r_b =$ equation (8.26), using equations (8.14)–(8.17), (8.20)–(8.24).

8.10 End notes

In this chapter, we have demonstrated that stigmatic singlet lenses exist for real/virtual objects and real/virtual images using Cartesian ovals. Also, we have tested several for several configurations; when the object is finite/infinite and the image is finite/infinite.

In the next chapter, we are going to generalize the concept of stigmatic lenses outside the Cartesian ovals and we are going to obtain the general equation of stigmatic lenses.

References

Avendaño-Alejo M, Román-Hernández E, Castañeda L and Moreno-Oliva V I 2017 Analytic conic constants to reduce the spherical aberration of a single lens used in collimated light *Appl. Opt.* **56** 6244–54

Bass M 1995 *Handbook of Optics, Volume I: Fundamentals Techniques and Design* (New York: McGraw-Hill)

Born M and Wolf E 2013 Principles of Optics: Electromagnetic Theory of Propagation *Interference and Diffraction of Light* (Amsterdam: Elsevier)

Braunecker B, Hentschel R and Tiziani H J 2008 *Advanced Optics Using Aspherical Elements* **vol 173** (Bellingham, WA: SPIE Press)

Castillo-Santiago G, Avendaño-Alejo M, Díaz-Uribe R and Castañeda L 2014 Analytic aspheric coefficients to reduce the spherical aberration of lens elements used in collimated light *Appl. Opt.* **53** 4939–46

Chaves J 2016 *Introduction to Nonimaging Optics* 2nd edn (Boca Raton, FL: CRC Press)

Estrada J C V, Calle Á H B and Hernández D M 2013 Explicit representations of all refractive optical interfaces without spherical aberration *J. Opt. Soc. Am.* **A30** 1814–24

Glassner A S 1989 *An Introduction to Ray Tracing* (Amsterdam: Elsevier)

González-Acuña R G and Chaparro-Romo H A 2018 General formula for bi-aspheric singlet lens design free of spherical aberration *App. Opt.* **57** 9341–5

González-Acuña R G and Guitiérrez-Vega J C 2018 Generalization of the axicon shape: the gaxicon *J. Opt. Soc. Am.* **A35** 1915–8

González-Acuña R G and Gutiérrez-Vega J C 2019 Analytic formulation of a refractive-reflective telescope free of spherical aberration *Opt. Eng.* **58** 085105

González-Acuña R G, Avendaño-Alejo M and Gutiérrez-Vega J C 2019a Singlet lens for generating aberration-free patterns on deformed surfaces *J. Opt. Soc. Am.* **A36** 925–9

González-Acuña R G, Chaparro-Romo H A and Gutiérrez-Vega J C 2019b General formula to design freeform singlet free of spherical aberration and astigmatism *Appl. Opt.* **58** 1010–5

González-Acuña R G, Chaparro-Romo H A and Gutiérrez-Vega J C 2020a Analytic aplanatic singlet lens: setting and design for three-point objects and images in the meridional plane *Opt. Eng.* **59** 055104

González-Acuña R G, Chaparro-Romo H A and Gutiérrez-Vega J C 2020b Analytic solution of the eikonal for a stigmatic singlet lens *Phys. Scr.*

González-Acuña R G, Chaparro-Romo H A and Gutiérrez-Vega J C 2020c *Analytical Lens Design* (Bristol: Institute of Physics Publishing)

González-Acuña R G, Chaparro-Romo H A and Gutíerrez-Vega J C 2019c *Single Lens Telescope* (arXiv:1903.11129)

González-Acuña R G and Gutiérrez-Vega J C 2019a Analytic formulation of a refractive-reflective telescope free of spherical aberration *Opt. Eng.* **58** 1–5

González-Acuña R G and Gutiérrez-Vega J C 2019b General formula to eliminate spherical aberration produced by an arbitrary number of lenses *Opt. Eng.* **58** 1–6

González-Acuña R G and Gutiérrez-Vega J C 2019c General formula for aspheric collimator lens design free of spherical aberration *Current Developments in Lens Design and Optical Engineering XX* ed R B Johnson, V N Mahajan and S Thibault (Bellingham, WA: International Society for Optics and Photonics, SPIE) 181–4 pp

González Acuña R G and Gutiérrez-Vega J C 2019 General formula to design freeform collimator lens free of spherical aberration and astigmatism *Novel Optical Systems, Methods, and Applications XXII* **vol 11 105** (Bellingham, WA: International Society for Optics and Photonics) 111050A p

González Acuña R G and Gutiérrez-Vega J C 2019 General formula of the refractive telescope design free spherical aberration *Novel Optical Systems, Methods, and Applications XXII* **vol 11 105** ed C F Hahlweg and J R Mulley (Bellingham, WA: International Society for Optics and Photonics, SPIE) 162–6 p

Kingslake R and Johnson R B 2009 *Lens Design Fundamentals* (New York: Academic)

Lefaivre J 1951 A new approach in the analytical study of the spherical aberrations of any order *J. Opt. Soc. Am.* **41** 647

Luneburg R K 1964 *Mathematical Theory of Optics* (Berkeley, CA: University of California Press)

Malacara D 1965 Two lenses to collimate red laser light *Appl. Opt.* **4** 1652–4

Malacara-Hernández D and Malacara-Hernández Z 2016 *Handbook of Optical Design* (Boca Raton, FL: CRC Press)

González-Acuña R G and Gutiérrez-Vega J C 2020 Analytic design of a spherochromatic singlet *J. Opt. Soc. Am.* **A37** 149–53

Schulz G 1983 Achromatic and sharp real imaging of a point by a single aspheric lens *Appl. Opt.* **22** 3242–8

Silva-Lora A and Torres R 2020 Explicit Cartesian oval as a superconic surface for stigmatic imaging optical systems with real or virtual source or image *Proc. R. Soc.* A476

Stavroudis O 2012 *The Optics of Rays, Wavefronts, and Caustics* **vol 38** (Amsterdam: Elsevier)

Stavroudis O N 2006 *The Mathematics of Geometrical and Physical Optics: The K-function and Its Ramifications* (New York: Wiley)

Stavroudis O N and Feder D P 1954 Automatic computation of spot diagrams *J. Opt. Soc. Am.* **44** 163–70

Sun H 2016 *Lens Design: A Practical Guide* (Boca Raton, FL: CRC Press)

Valencia-Estrada J C and Malacara-Doblado D 2014 Parastigmatic corneal surfaces *Appl. Opt.* **53** 3438–47

Valencia-Estrada J C, Flores-Hernández R B and Malacara-Hernández D 2015 Singlet lenses free of all orders of spherical aberration *Proc. R. Soc.* **A471** 20140608

Vaskas E M 1957 Note on the Wassermann-Wolf method for designing aspheric surfaces *J. Opt. Soc. Am.* **47** 669–70

Wassermann G D and Wolf E 1949 On the theory of aplanatic aspheric systems *Proc. Phys. Soc.* **B62** 2

Winston R, Miñano J C and Benitez P G *et al* 2005 *Nonimaging Optics* (Amsterdam: Elsevier)

Wolf E 1948 On the designing of aspheric surfaces *Proc. Phys. Soc.* **61** 494

Wolf E and Preddy W S 1947 On the determination of aspheric profiles *Proc. Phys. Soc.* **59** 704

Yang T, Jin G-F and Zhu J 2017 Automated design of freeform imaging systems *Light Sci. Appl.* **6** e17081

IOP Publishing

Stigmatic Optics (Second Edition)

Rafael G González-Acuña and Héctor A Chaparro-Romo

Chapter 9

The general equation of the stigmatic lenses

In the previous chapter, we studied an on-axis stigmatic singlet lens when the two refractive surfaces are Cartesian ovals. Here, we present the general formula of the on-axis stigmatic singlet lens. The input of the general formula is the first surface of the singlet lens. The first surface needs to be continuous. The output is the correcting second surface of the lens; the second surface is such that the singlet is stigmatic.

9.1 Introduction

In the previous chapter, we found a particular design of a lens-free spherical aberration. The lens is made up of two Silva–Torres Cartesian ovals. In this chapter, we introduce a more general expression for spherical aberration-free lenses. We will call this expression the general equation for stigmatic lenses. Said equation receives a first surface as the input parameter. The equation gives as output a second surface such that the pair of surfaces form a stigmatic lens.

We will first study the case when the image object distance is finite. Then we will study when the object is in the minus infinite, when the image is in the plus infinite. Finally, we will analyze when the object and image are at the minus and plus infinity, respectively.

9.2 Finite object finite image

The problem to solve in this chapter is announced as: given the first surface of the lens, the positions of the object and image, the refractive indexes in the object space, inside the glass and in the image space and central thickness of the singlet, how must the second surface be such that the singlet is free of spherical aberration?

The answer we are expecting is now the shape of the second surface (z_b, r_b), given the first surface (z_a, r_a), such that the singlet is free of spherical aberration, where r_a is the only independent variable. Thus, z_b, r_b and z_a are functions of r_a. z_b is the sagitta of the second surface and its radius is r_b. The sagitta of the first surface is z_a and the radius is r_a. As usual, the coordinate system is set at the vertex of z_a (see figure 9.1).

doi:10.1088/978-0-7503-6423-2ch9

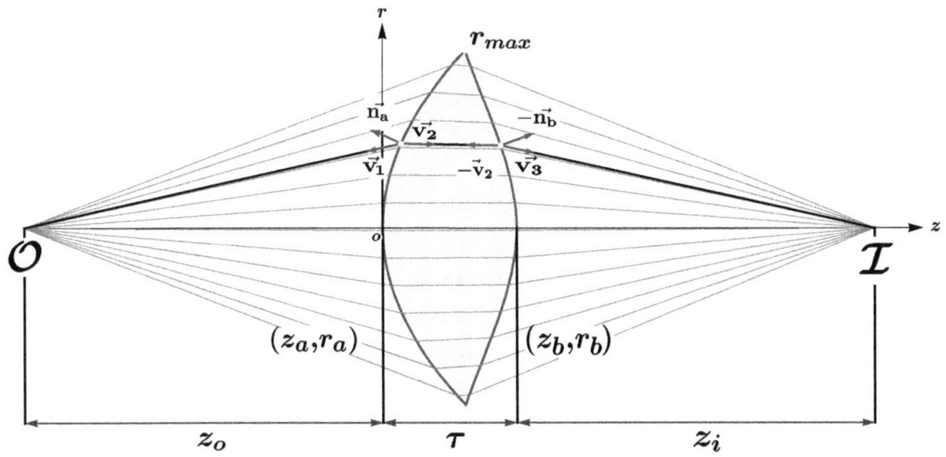

Figure 9.1. Diagram of an on-axis stigmatic singlet. The first and second surfaces are defined by (z_a, r_a) and (z_b, r_b), sequentially. The axial interval between the first surface and the object is z_o, the axial diameter of the lens is τ, and the axial length within the second surface and the image is z_i. The refraction index in the object space is n_o. Inside the lens, the refraction index is n and in the image space is n_i.

We pretend that the object is surrounded by a medium, which has a refractive index of n_o. The image is in a medium with a refractive index of n_i. The refraction index n of the lens is constant, and the singlet is radially symmetric. At the center, the singlet lens has a thickness of τ. The length from the object to the first surface is z_o. The range from the second surface to the image is z_i (see figure 9.1).

9.2.1 Fermat's principle

We have noticed before that for a stigmatic design, the optical path length of all rays that emanate from a point object on the axis and finish up on a point image on the axis requirement will be the equivalent. Thus, the optical path length of all rays is constant.

Also, from the Fermat principle, we know that the light travels the paths such as it spends the least time.

Consequently, combining both statements, we can compare the optical path length of the axial ray and the optical path length of the non-axial ray.

We start with the optical path length of the axial ray. It comes from $-z_o$, advances inside the lens and progresses a length τ and ultimately, from the other surface, it goes a length z_i to meet the point image (see figure 9.1). The optical path of the axial ray is given by

$$-n_o z_o + n\tau + n_i z_i = \text{constant}. \tag{9.1}$$

The optical path length of a non-axial ray is not simple, since it depends on if the object/image is real or virtual.

We start with the more natural case, when the object and the image are real, $z_o < 0$ and $z_i > 0$,

$$\begin{aligned} -n_o z_o + n\tau + n_i z_i = n_o\sqrt{r_a^2 + (z_a - z_o)^2} + n\sqrt{(r_b - r_a)^2 + (z_b - z_a)^2} \\ + n_i\sqrt{r_b^2 + (z_b - \tau - z_i)^2}, \end{aligned} \tag{9.2}$$

when $z_o > 0$ and $z_i > 0$, is when object is virtual and image is real,

$$\begin{aligned} -n_o z_o + nT + n_i z_i = & -n_o\sqrt{r_a^2 + (z_a - z_o)^2} + n\sqrt{(r_b - r_a)^2 + (z_b - z_a)^2} \\ & + n_i\sqrt{r_b^2 + (z_b - T - z_i)^2}. \end{aligned} \tag{9.3}$$

Real object and virtual image: $z_o < 0$ and $z_i < 0$,

$$\begin{aligned} -n_o z_o + nT + n_i z_i = & n_o\sqrt{r_a^2 + (z_a - z_o)^2} + n\sqrt{(r_b - r_a)^2 + (z_b - z_a)^2} \\ & - n_i\sqrt{r_b^2 + (z_b - T - z_i)^2}. \end{aligned} \tag{9.4}$$

Ultimately, if both are virtual, $z_o > 0$ and $z_i < 0$,

$$\begin{aligned} -n_o z_o + nT + n_i z_i = & -n_o\sqrt{r_a^2 + (z_a - z_o)^2} + n\sqrt{(r_b - r_a)^2 + (z_b - z_a)^2} \\ & - n_i\sqrt{r_b^2 + (z_b - T - z_i)^2}. \end{aligned} \tag{9.5}$$

The four previous equations can be combined in a single expression given by

$$\boxed{\begin{aligned} -n_o z_o + nT + n_i z_i\$ = & -n_o \operatorname{sgn}(z_o)\sqrt{r_a^2 + (z_a - z_o)^2} + n\sqrt{(r_b - r_a)^2 + (z_b - z_a)^2} \\ & + n_i \operatorname{sgn}(z_i)\sqrt{r_b^2 + (z_b - T - z_i)^2}. \end{aligned}} \tag{9.6}$$

Note that the sgn(·) is the sign of their argument, thus if the argument is zero, the function is not defined. Notice that the two unknowns are (z_b, r_b), and we have only one equation. Thus, we need another. We are going to get it from Snell's law.

9.2.2 Snell's law

Now we focus on Snell's law, which is the equation that contains information on when light crosses from one medium to another. The vector form of Snell's law at the first surface is

$$\boxed{\vec{\mathbf{v}_2} = \frac{n_o}{-\operatorname{sgn}(z_o)n}[\vec{\mathbf{v}_1} - (\vec{\mathbf{n}_a}\cdot\vec{\mathbf{v}_1})\vec{\mathbf{n}_a}] - \vec{\mathbf{n}_a}\sqrt{1 - \frac{n_o^2}{n^2}\left(\vec{\mathbf{n}_a}\times\vec{\mathbf{v}_1}\right)^2} \quad \text{for } \vec{\mathbf{v}_2}, \vec{\mathbf{v}_1}, \vec{\mathbf{n}_a} \in \mathbb{R}^2,} \tag{9.7}$$

where $\vec{\mathbf{v}_1}$ is the unit vector of the incident ray, $\vec{\mathbf{v}_2}$ is unit vector of the refracted ray and finally, $\vec{\mathbf{n}_a}$ is the normal vector of the first surface. Finally, $-\operatorname{sgn}(z_o)$ comes from the fact that the object is real/virtual (see figure 9.1).

The related unit vectors at the first surface are

$$\vec{\mathbf{v}_1} = \frac{r_a\vec{\mathbf{e}_1} + (z_a - z_o)\vec{\mathbf{e}_2}}{\sqrt{r_a^2 + (z_a - z_o)^2}}, \quad \vec{\mathbf{v}_2} = \frac{(r_b - r_a)\vec{\mathbf{e}_1} + (z_b - z_a)\vec{\mathbf{e}_2}}{\sqrt{(r_b - r_a)^2 + (z_b - z_a)^2}}, \quad \vec{\mathbf{n}_a} = \frac{z_a'\vec{\mathbf{e}_1} - \vec{\mathbf{e}_2}}{\sqrt{1 + z_a'^2}}, \tag{9.8}$$

where $\vec{\mathbf{e}_1}$ is for the r direction and $\vec{\mathbf{e}_2}$ is for the z direction. The idea is to replace the unit vectors in equation (9.7). The procedure is long, thus we first focus on $(\vec{\mathbf{n}_a} \times \vec{\mathbf{v}_1})^2$,

$$(\vec{\mathbf{n}_a} \times \vec{\mathbf{v}_1}) = \frac{[r_a + (z_a - z_o)z_a']}{\sqrt{r_a^2 + (z_a - z_o)^2}\sqrt{1 + z_a'2}}(\vec{\mathbf{e}_1} \times \vec{\mathbf{e}_2}). \tag{9.9}$$

Notice that $(\vec{\mathbf{e}}_1 \times \vec{\mathbf{e}}_2)^2 = \vec{\mathbf{e}}_3{}^2 = 1$, thus,

$$(\vec{\mathbf{n}}_a \times \vec{\mathbf{v}}_1)^2 = \frac{[r_a + (z_a - z_o)z_a{}']^2}{[r_a^2 + (z_a - z_o)^2](1 + z_a{}'^2)}. \tag{9.10}$$

Thus, the squared root in (9.7) is

$$-\vec{\mathbf{n}}_a\sqrt{1 + \frac{1}{n^2}(\vec{\mathbf{n}}_a \times \vec{\mathbf{v}}_1)^2} = -\frac{(z_a{}'\vec{\mathbf{e}}_1 - \vec{\mathbf{e}}_2)}{\sqrt{1 + z_a{}'^2}}\sqrt{1 - \frac{n_o^2[r_a + (z_a - z_o)z_a{}']^2}{n^2[r_a^2 + (z_a - z_o)^2](1 + z_a{}'^2)}}. \tag{9.11}$$

In equation (9.7), the term $\vec{\mathbf{v}}_1 - (\vec{\mathbf{n}}_a \cdot \vec{\mathbf{v}}_1)\vec{\mathbf{n}}_a$ is

$$\vec{\mathbf{v}}_1 - (\vec{\mathbf{n}}_a \cdot \vec{\mathbf{v}}_1)\vec{\mathbf{n}}_a = \frac{r_a\vec{\mathbf{e}}_1 + (z_a - z_o)\vec{\mathbf{e}}_2}{\sqrt{r_a^2 + (z_a - z_o)^2}} - \left[\frac{r_a z_a{}' - (z_a - z_o)}{\sqrt{r_a^2 + (z_a - z_o)^2}\sqrt{1 + z_a{}'^2}}\right]\frac{(z_a{}'\vec{\mathbf{e}}_1 - \vec{\mathbf{e}}_2)}{\sqrt{1 + z_a{}'^2}}. \tag{9.12}$$

Simplifying and multiplying by $\frac{n_o}{(-\mathrm{sgn}(z_o)n}$ the above expression,

$$\begin{aligned}\frac{n_o}{-\mathrm{sgn}(z_o)n}[\vec{\mathbf{v}}_1 - (\vec{\mathbf{n}}_a \cdot \vec{\mathbf{v}}_1)\vec{\mathbf{n}}_a] = {} & \frac{n_o[(z_a - z_o)z_a{}' + r_a]}{-\mathrm{sgn}(z_o)n\sqrt{r_a^2 + (z_o - z_a)^2}(1 + z_a{}'^2)}\vec{\mathbf{e}}_1 \\ & + \frac{n_o[r_a + (z_a - z_o)z_a{}']z_a{}'}{-\mathrm{sgn}(z_o)n\sqrt{r_a^2 + (z_o - z_a)^2}(1 + z_a{}'^2)}\vec{\mathbf{e}}_2.\end{aligned} \tag{9.13}$$

Summing (9.11) and (9.13), we have $\vec{\mathbf{v}}_2$ as the output of equation (9.7),

$$\begin{aligned}\vec{\mathbf{v}}_2 = {} & \frac{n_o[(z_a - z_o)z_a{}' + r_a]}{-\mathrm{sgn}(z_o)n\sqrt{r_a^2 + (z_o - z_a)^2}(1 + z_a{}'^2)}\vec{\mathbf{e}}_1 + \frac{n_o[r_a + (z_a - z_o)z_a{}']z_a{}'}{-\mathrm{sgn}(z_o)n\sqrt{r_a^2 + (z_o - z_a)^2}(1 + z_a{}'^2)}\vec{\mathbf{e}}_2 \\ & - \frac{(z_a{}'\vec{\mathbf{e}}_1 - \vec{\mathbf{e}}_2)}{\sqrt{1 + z_a{}'^2}}\sqrt{1 - \frac{n_o^2[r_a + (z_a - z_o)z_a{}']^2}{n^2[r_a^2 + (z_a - z_o)^2](1 + z_a{}'^2)}}.\end{aligned} \tag{9.14}$$

Separating coordinates $\vec{\mathbf{e}}_1$ and $\vec{\mathbf{e}}_2$,

$$\begin{aligned}\frac{r_b - r_a}{\sqrt{(z_b - z_a)^2 + (r_b - r_a)^2}} = {} & \frac{n_o[(z_a - z_o)z_a{}' + r_a]}{-\mathrm{sgn}(z_o)n\sqrt{r_a^2 + (z_o - z_a)^2}(1 + z_a{}'^2)} \\ & - z_a{}'\frac{\sqrt{1 - \frac{n_o^2[r_a + (z_a - z_o)z_a{}']^2}{n^2[r_a^2 + (z_o - z_a)^2](1 + z_a{}'^2)}}}{\sqrt{1 + z_a{}'^2}},\end{aligned} \tag{9.15}$$

and

$$\begin{aligned}\frac{z_b - z_a}{\sqrt{(z_b - z_a)^2 + (r_b - r_a)^2}} = {} & \frac{n_o[r_a + (z_a - z_o)z_a{}']z_a{}'}{-\mathrm{sgn}(z_o)n\sqrt{r_a^2 + (z_o - z_a)^2}(1 + z_a{}'^2)} \\ & + \frac{\sqrt{1 - \frac{n_o^2[r_a + (z_a - z_o)z_a{}']^2}{n^2[r_a^2 + (z_o - z_a)^2](1 + z_a{}'^2)}}}{\sqrt{1 + z_a{}'^2}}.\end{aligned} \tag{9.16}$$

The unknowns z_b and r_b are only in the left side of equations (9.15) and (9.16). Notice that the right side of equations (9.15) and (9.16) are the cosine directors inside the lens. Thus, $\wp_r^2 + \wp_z^2 = 1$. $\wp_r$ is the director in the direction of $\vec{\mathbf{e}}_1$. $\wp_z$ is the director along $\vec{\mathbf{e}}_2$. Therefore, we have the following expression if we assign $\wp_r$ and $\wp_z$ in the right side of equations (9.15) and (9.16),

$$\wp_r = \frac{n_o[(z_a - z_o)z_a' + r_a]}{-\text{sgn}(z_o)n\sqrt{r_a^2 + (z_o - z_a)^2}(1 + z_a'^2)} - z_a' \frac{\sqrt{1 - \frac{n_o^2[r_a + (z_a - z_o)z_a']^2}{n^2[r_a^2 + (z_o - z_a)^2](1 + z_a'^2)}}}{\sqrt{1 + z_a'^2}}, \tag{9.17}$$

and

$$\wp_z = \frac{n_o[r_a + (z_a - z_o)z_a']z_a'}{-\text{sgn}(z_o)n\sqrt{r_a^2 + (z_o - z_a)^2}(1 + z_a'^2)} + \frac{\sqrt{1 - \frac{n_o^2[r_a + (z_a - z_o)z_a']^2}{n^2[r_a^2 + (z_o - z_a)^2](1 + z_a'^2)}}}{\sqrt{1 + z_a'^2}}. \tag{9.18}$$

Now we can simplify the left side of equations (9.15) and (9.16) with the following parameter,

$$\vartheta \equiv \sqrt{(z_b - z_a)^2 + (r_b - r_a)^2}. \tag{9.19}$$

ϑ is the distance traveled by the ray inside the lens. Replacing ϑ in (9.15) and (9.16),

$$\frac{z_b - z_a}{\vartheta} = \wp_z, \tag{9.20}$$

and

$$\frac{r_b - r_a}{\vartheta} = \wp_r. \tag{9.21}$$

From the previous equation we can simply solve for z_b and r_b,

$$z_b = z_a + \vartheta \wp_z, \tag{9.22}$$

and

$$r_b = r_a + \vartheta \wp_r. \tag{9.23}$$

The above expression is the solution that we want to get. It describes, point by point, how the second surface of the lens must be such that it is stigmatic. The only problem in this equation is that we do not know the length of ϑ. But we can find the solution by mixing this result with the Fermat principle. In the next section, we are going to do it!

9.2.3 Solution

In the last section, we found how z_b and r_b must be, but in terms of ϑ. Therefore, we need an extra equation, and that equation is the Fermat principle, which relates the optical path of an axial ray with the optical path of a non-axial ray. It is essential to remark that we have just two unknowns, z_b and r_b, but we have a system with three equations. Two equations are given by Snell's law at the first surface and another one is provided by the Fermat principle. There is nothing wrong here since the equations granted by Snell's law are not independent. We recall equation (9.6),

$$\begin{aligned} & -n_o z_o + n\tau + n_i z_i = -n_o \operatorname{sgn}(z_o)\sqrt{r_a^2 + (z_a - z_o)^2} \\ & + n\sqrt{(r_b - r_a)^2 + (z_b - z_a)^2} + n_i \operatorname{sgn}(z_i)\sqrt{r_b^2 + (z_b - \tau - z_i)^2}. \end{aligned} \tag{9.6}$$

We assign the following parameter to simplify the last expression,

$$z_f \equiv -n_o z_o + n\tau + n_i z_i + \operatorname{sgn}(z_o) n_0 \sqrt{r_a^2 + (z_o - z_a)^2}, \tag{9.24}$$

replacing equation (9.24) in equation (9.6),

$$z_f = n\sqrt{(r_b - r_a)^2 + (z_b - z_a)^2} + n_i\sqrt{r_b^2 + (-\tau + z_b - z_i)^2}, \tag{9.25}$$

replacing equations (9.22) and (9.23) in equation (9.25),

$$z_f = n\vartheta + n_i\sqrt{(r_a + \vartheta\wp_r^2 + (-\tau + z_a - z_i + \vartheta\wp_z)^2}. \tag{9.26}$$

Also, we assign another parameter to clean equation (9.26),

$$z_\tau \equiv -\tau + z_a - z_i. \tag{9.27}$$

Then, we replace equation (9.27) in equation (9.26),

$$z_f = n\vartheta + n_i\sqrt{(r_a + \vartheta\wp_r)^2 + (z_\tau + \vartheta\wp_z)^2}, \tag{9.28}$$

squaring both sides of the last equation and manipulating,

$$\left(z_f - n\vartheta\right)^2 = n_i^2[(r_a + \vartheta\wp_r)^2 + (z_\tau + \vartheta\wp_z)^2], \tag{9.29}$$

expanding,

$$z_f^2 - 2z_f n\vartheta + n^2\vartheta^2 = n_i^2[r_a^2 + 2r_a\vartheta\wp_r + \vartheta^2\wp_r^2 + z_\tau^2 + 2z_\tau\vartheta\wp_z + \vartheta^2\wp_z^2]. \tag{9.30}$$

Remember that $\vartheta^2\wp_r^2 + \vartheta^2\wp_z^2 = \vartheta^2$, thus equation (9.30) is reduced to

$$z_f^2 - 2z_f n\vartheta + n^2\vartheta^2 = n_i^2[r_a^2 + 2r_a\vartheta\wp_r + \vartheta^2 + z_\tau^2 + 2z_\tau\vartheta\wp_z], \tag{9.31}$$

collecting for ϑ,

$$\vartheta^2(n_i^2 - n^2) + \vartheta\left(2z_f n + 2n_i^2 r_a\wp_r + 2n_i^2 z_\tau\wp_z\right) + n_i^2\left(r_a^2 + z_\tau^2\right) - z_f^2 = 0, \tag{9.32}$$

multiplying equation (9.32) by $4(n_i^2 - n^2)$,

$$\begin{aligned} & 4(n_i^2 - n^2)^2\vartheta^2 + 4\left(n_i^2 - n^2\right)\left(2z_f n + 2n_i^2 r_a\wp_r + 2n_i^2 z_\tau\wp_z\right)\vartheta \\ & = -4\left(n_i^2 - n^2\right)\left[n_i^2\left(r_a^2 + z_\tau^2\right) - z_f^2\right], \end{aligned} \tag{9.33}$$

summing $(2z_f n + 2n_i^2 r_a \wp_r + 2n_i^2 z_T \wp_z)^2$ on both sides of equation (9.33),

$$\begin{aligned}&4\left(n_i^2 - n^2\right)^2 \vartheta^2 + 4(n_i^2 - n^2)(2z_f n + 2n_i^2 r_a \wp_r + 2n_i^2 z_T \wp_z)\vartheta + \left(2z_f n + 2n_i^2 r_a \wp_r + 2n_i^2 z_T \wp_z\right)^2 \\ &= \left(2z_f n + 2n_i^2 r_a \wp_r + 2n_i^2 z_T \wp_z\right)^2 - 4(n_i^2 - n^2)\left[n_i^2(r_a^2 + z_T^2) - z_f^2\right],\end{aligned} \tag{9.34}$$

factoring the left side of equation (9.34),

$$\begin{aligned}&\left[2\left(n_i^2 - n^2\right)\vartheta + \left(2z_f n + 2n_i^2 r_a \wp_r + 2n_i^2 z_T \wp_z\right)\right]^2 \\ &= \left(2z_f n + 2n_i^2 r_a \wp_r + 2n_i^2 z_T \wp_z\right)^2 - 4(n_i^2 - n^2)\left[n_i^2(r_a^2 + z_T^2) - z_f^2\right],\end{aligned} \tag{9.35}$$

applying the square root on both sides,

$$\begin{aligned}&2\left(n_i^2 - n^2\right)\vartheta + \left(2z_f n + 2n_i^2 r_a \wp_r + 2n_i^2 z_T \wp_z\right) \\ &= \pm\sqrt{\left(2z_f n + 2n_i^2 r_a \wp_r + 2n_i^2 z_T \wp_z\right)^2 - 4(n_i^2 - n^2)\left[n_i^2(r_a^2 + z_T^2) - z_f^2\right]},\end{aligned} \tag{9.36}$$

solving for ϑ,

$$\vartheta = \frac{-2[z_f n + n_i^2(r_a \wp_r + z_T \wp_z)] - \sqrt{4\left(n_i^2 - n^2\right)\left[n_i^2\left(r_a^2 + z_T^2\right) - z_f^2\right] + 4\left[z_f n + n_i^2\left(r_a \wp_r + z_T \wp_z\right)\right]^2}}{2n_i^2 - 2n^2}, \tag{9.37}$$

eliminating the number 2,

$$\vartheta = \frac{-[z_f n + n_i^2(r_a \wp_r + z_T \wp_z)] \pm \sqrt{\left(n_i^2 - n^2\right)\left[n_i^2\left(r_a^2 + z_T^2\right) - z_f^2\right] + \left[z_f n + n_i^2\left(r_a \wp_r + z_T \wp_z\right)\right]^2}}{n_i^2 - n^2}. \tag{9.38}$$

With equations (9.22), (9.23) and (9.38) we can write the second surface as

$$\begin{cases} z_b = z_a + \vartheta \wp_z, \\ r_b = r_a + \vartheta \wp_r. \end{cases} \tag{9.39}$$

Equation (9.39) is the most important equation in the chapter. It tells us how the second surface must be, point by point, such that we get a stigmatic lens. It is significant to observe that equation (9.39) only works for singlet lenses such as the rays inside them do not cross each other. In other words, equations (9.22) and (9.23) tell us that for a point of the first surface, there is a unique point in the second surface for it to achieve stigmatism.

Also, it is necessary to remark that in the expression of ϑ, equation (9.38), there is no expression for $\mathrm{sgn}(t_b)$. In other words, the second surface adapts its shape according to where the image is located; it does not matter if the image is real or virtual.

Another essential remark about equation (9.38) is that it has a plus-minus sign $\pm$ multiplying the square root; the plus-minus sign led us to two solutions. The problem is, which is the solution that works for a given case? Well, it depends on the sign of the refraction index n and if the object/image is virtual or real. In the following sections, we are going to show several illustrative examples. But first we show the expanded version of equation (9.38),

$$
\begin{aligned}
z_b(r_a) = {} & z_a(r_a) \\
& + \frac{1}{n^2 - n_i^2}\left(\frac{n_o[z_a'(r_a)(r_a + (z_a(r_a) - z_o)z_a'(r_a))]}{n\sqrt{r_a + (z_o - z_a(r_a))^2}\left(r_a^2 + z_a'(r_a)^2\right)} + \frac{\sqrt{1 - \frac{n_o^2[r_a + (z_a(r_a) - z_o)z_a'(r_a)]^2}{n^2\left[r_a + (z_o - z_a(r_a))^2\right]\left(r_a^2 + z_a'(r_a)^2\right)}}}{\sqrt{r_a^2 + z_a'(r_a)^2}}\right) \\
& \left(
\begin{aligned}
& n\left(n_i z_i - n_o z_o + n\tau + n_o \,\mathrm{Sign}[z_o]\sqrt{r_a + (z_o - z_a(r_a))^2}\right) \\
& + n_i^2\left\{
\begin{aligned}
& \left[r_a \times \left(\frac{n_o[z_a'(r_a)(r_a + (z_a(r_a) - z_o)z_a'(r_a))]}{n\sqrt{r_a + (z_o - z_a(r_a))^2}\left(r_a^2 + z_a'(r_a)^2\right)} - \frac{z_a'(r_a)\sqrt{1 - \frac{n_o^2[r_a + (z_a(r_a) - z_o)z_a'(r_a)]^2}{n^2\left[r_a + (z_o - z_a(r_a))^2\right]\left(r_a^2 + z_a'(r_a)^2\right)}}}{\sqrt{r_a^2 + z_a'(r_a)^2}}\right)\right] \\
& + \\
& \left[\left(z_a(r_a) - z_i - \tau\right) \times \left(\frac{n_o[z_a'(r_a)(r_a + (z_a(r_a) - z_o)z_a'(r_a))]}{n\sqrt{r_a + (z_o - z_a(r_a))^2}\left(r_a^2 + z_a'(r_a)^2\right)} + \frac{\sqrt{1 - \frac{n_o^2[r_a + (z_a(r_a) - z_o)z_a'(r_a)]^2}{n^2\left[r_a + (z_o - z_a(r_a))^2\right]\left(r_a^2 + z_a'(r_a)^2\right)}}}{\sqrt{r_a^2 + z_a'(r_a)^2}}\right)\right]
\end{aligned}
\right\} \\
& - n_i \,\mathrm{Sign}[\eta_1] \times \\
& \sqrt{
\begin{aligned}
& \left(n_i z_i - n_o z_o + n\tau + n_o \,\mathrm{Sign}[z_o]\sqrt{r_a + (z_o - z_a(r_a))^2}\right)^2 + n^2\left(r_a + \left(z_a(r_a) - z_i - \tau\right)^2\right) \\
& \times \left(
\begin{aligned}
& -n_i^2\left\{
\begin{aligned}
& \left[\left(z_a(r_a) - z_i - \tau\right) \times \left(\frac{n_o[z_a'(r_a)(r_a + (z_a(r_a) - z_o)z_a'(r_a))]}{n\sqrt{r_a + (z_o - z_a(r_a))^2}\left(r_a^2 + z_a'(r_a)^2\right)} - \frac{z_a'(r_a)\sqrt{1 - \frac{n_o^2[r_a + (z_a(r_a) - z_o)z_a'(r_a)]^2}{n^2\left[r_a + (z_o - z_a(r_a))^2\right]\left(r_a^2 + z_a'(r_a)^2\right)}}}{\sqrt{r_a^2 + z_a'(r_a)^2}}\right)\right] \\
& + \\
& \left[-r_a \times \left(\frac{n_o[z_a'(r_a)(r_a + (z_a(r_a) - z_o)z_a'(r_a))]}{n\sqrt{r_a + (z_o - z_a(r_a))^2}\left(r_a^2 + z_a'(r_a)^2\right)} + \frac{\sqrt{1 - \frac{n_o^2[r_a + (z_a(r_a) - z_o)z_a'(r_a)]^2}{n^2\left[r_a + (z_o - z_a(r_a))^2\right]\left(r_a^2 + z_a'(r_a)^2\right)}}}{\sqrt{r_a^2 + z_a'(r_a)^2}}\right)\right]
\end{aligned}
\right\}^2 \\
& + 2n\left(n_i z_i - n_o z_o + n\tau + n_o \,\mathrm{Sign}[z_o]\sqrt{r_a + (z_o - z_a(r_a))^2}\right) \\
& \times \left\{
\begin{aligned}
& \left[r_a \times \left(\frac{n_o[z_a'(r_a)(r_a + (z_a(r_a) - z_o)z_a'(r_a))]}{n\sqrt{r_a + (z_o - z_a(r_a))^2}\left(r_a^2 + z_a'(r_a)^2\right)} - \frac{z_a'(r_a)\sqrt{1 - \frac{n_o^2[r_a + (z_a(r_a) - z_o)z_a'(r_a)]^2}{n^2\left[r_a + (z_o - z_a(r_a))^2\right]\left(r_a^2 + z_a'(r_a)^2\right)}}}{\sqrt{r_a^2 + z_a'(r_a)^2}}\right)\right] \\
& + \\
& \left[\left(z_a(r_a) - z_i - \tau\right) \times \left(\frac{n_o[z_a'(r_a)(r_a + (z_a(r_a) - z_o)z_a'(r_a))]}{n\sqrt{r_a + (z_o - z_a(r_a))^2}\left(r_a^2 + z_a'(r_a)^2\right)} + \frac{\sqrt{1 - \frac{n_o^2[r_a + (z_a(r_a) - z_o)z_a'(r_a)]^2}{n^2\left[r_a + (z_o - z_a(r_a))^2\right]\left(r_a^2 + z_a'(r_a)^2\right)}}}{\sqrt{r_a^2 + z_a'(r_a)^2}}\right)\right]
\end{aligned}
\right\}
\end{aligned}
\right)
\end{aligned}
}
\end{aligned}
\right)
\end{aligned}
$$

$$
\begin{aligned}
r_b(r_a) = {} & r_a \\
& + \frac{1}{n^2 - n_i^2}\left(\frac{n_o[z_a'(r_a)(r_a + (z_a(r_a) - z_o)z_a'(r_a))]}{n\sqrt{r_a + (z_o - z_a(r_a))^2}\left(r_a^2 + z_a'(r_a)^2\right)} - \frac{z_a'(r_a)\sqrt{1 - \dfrac{n_o^2[r_a + (z_a(r_a) - z_o)z_a'(r_a)]^2}{n^2\left[r_a + (z_o - z_a(r_a))^2\right]\left(r_a^2 + z_a'(r_a)^2\right)}}}{\sqrt{r_a^2 + z_a'(r_a)^2}}\right) \\
& \Bigg(n\left(n_i z_i - n_o z_o + n\tau + n_o \,\mathrm{Sign}[z_o]\sqrt{r_a + (z_o - z_a(r_a))^2}\right) \\
& + n_i^2 \left\{
\begin{aligned}
& \left[r_a \times \left(\frac{n_o[z_a'(r_a)(r_a + (z_a(r_a) - z_o)z_a'(r_a))]}{n\sqrt{r_a + (z_o - z_a(r_a))^2}\left(r_a^2 + z_a'(r_a)^2\right)} - \frac{z_a'(r_a)\sqrt{1 - \dfrac{n_o^2[r_a + (z_a(r_a) - z_o)z_a'(r_a)]^2}{n^2\left[r_a + (z_o - z_a(r_a))^2\right]\left(r_a^2 + z_a'(r_a)^2\right)}}}{\sqrt{r_a^2 + z_a'(r_a)^2}}\right)\right] \\
& + \left[\left(z_a(r_a) - z_i - \tau\right) \times \left(\frac{n_o[z_a'(r_a)(r_a + (z_a(r_a) - z_o)z_a'(r_a))]}{n\sqrt{r_a + (z_o - z_a(r_a))^2}\left(r_a^2 + z_a'(r_a)^2\right)} + \frac{\sqrt{1 - \dfrac{n_o^2[r_a + (z_a(r_a) - z_o)z_a'(r_a)]^2}{n^2\left[r_a + (z_o - z_a(r_a))^2\right]\left(r_a^2 + z_a'(r_a)^2\right)}}}{\sqrt{r_a^2 + z_a'(r_a)^2}}\right)\right]
\end{aligned}
\right\} \\
& - n_i \,\mathrm{Sign}[n] \times \\
& \times \Bigg[\left(n_i z_i - n_o z_o + n\tau + n_o \,\mathrm{Sign}[z_o]\sqrt{r_a + (z_o - z_a(r_a))^2}\right)^2 + n^2\left(r_a + \left(z_a(r_a) - z_i - \tau\right)^2\right) \\
& - n_i^2 \left\{
\begin{aligned}
& \left[\left(z_a(r_a) - z_i - \tau\right) \times \left(\frac{n_o[z_a'(r_a)(r_a + (z_a(r_a) - z_o)z_a'(r_a))]}{n\sqrt{r_a + (z_o - z_a(r_a))^2}\left(r_a^2 + z_a'(r_a)^2\right)} - \frac{z_a'(r_a)\sqrt{1 - \dfrac{n_o^2[r_a + (z_a(r_a) - z_o)z_a'(r_a)]^2}{n^2\left[r_a + (z_o - z_a(r_a))^2\right]\left(r_a^2 + z_a'(r_a)^2\right)}}}{\sqrt{r_a^2 + z_a'(r_a)^2}}\right)\right] \\
& + \left[-r_a \times \left(\frac{n_o[z_a'(r_a)(r_a + (z_a(r_a) - z_o)z_a'(r_a))]}{n\sqrt{r_a + (z_o - z_a(r_a))^2}\left(r_a^2 + z_a'(r_a)^2\right)} + \frac{\sqrt{1 - \dfrac{n_o^2[r_a + (z_a(r_a) - z_o)z_a'(r_a)]^2}{n^2\left[r_a + (z_o - z_a(r_a))^2\right]\left(r_a^2 + z_a'(r_a)^2\right)}}}{\sqrt{r_a^2 + z_a'(r_a)^2}}\right)\right]
\end{aligned}
\right\}^2 \\
& + 2n\left(n_i z_i - n_o z_o + n\tau + n_o \,\mathrm{Sign}[z_o]\sqrt{r_a + (z_o - z_a(r_a))^2}\right) \\
& \times \left\{
\begin{aligned}
& \left[r_a \times \left(\frac{n_o[z_a'(r_a)(r_a + (z_a(r_a) - z_o)z_a'(r_a))]}{n\sqrt{r_a + (z_o - z_a(r_a))^2}\left(r_a^2 + z_a'(r_a)^2\right)} - \frac{z_a'(r_a)\sqrt{1 - \dfrac{n_o^2[r_a + (z_a(r_a) - z_o)z_a'(r_a)]^2}{n^2\left[r_a + (z_o - z_a(r_a))^2\right]\left(r_a^2 + z_a'(r_a)^2\right)}}}{\sqrt{r_a^2 + z_a'(r_a)^2}}\right)\right] \\
& + \left[\left(z_a(r_a) - z_i - \tau\right) \times \left(\frac{n_o[z_a'(r_a)(r_a + (z_a(r_a) - z_o)z_a'(r_a))]}{n\sqrt{r_a + (z_o - z_a(r_a))^2}\left(r_a^2 + z_a'(r_a)^2\right)} + \frac{\sqrt{1 - \dfrac{n_o^2[r_a + (z_a(r_a) - z_o)z_a'(r_a)]^2}{n^2\left[r_a + (z_o - z_a(r_a))^2\right]\left(r_a^2 + z_a'(r_a)^2\right)}}}{\sqrt{r_a^2 + z_a'(r_a)^2}}\right)\right]
\end{aligned}
\right\} \Bigg]^{1/2} \Bigg)
\end{aligned}
$$

The process to get equation (9.39) looks very easy if you read this chapter from the beginning to here. The process can be easy, but the equation is not; it is gigantic if you expand it. However, for centuries, people have tried to get the general analytical closed-form solution, but they failed. The secret is that in the whole process, we do not use any angle. The usual form of Snell's law complicates everything since there is no clear relation between the angles and the optical paths. With Snell's law in its vector form, it is very easy to see the association between the optical paths and the director cosines.

Note that we simply did not obtain the general analytical solution in a closed way for the problem, we also discovered that the solution is unique. We are going to study this statement in the following chapter.

9.2.4 The eikonal of the stigmatic lens

The eikonal of the stigmatic lens proposed in the last section is a two-dimensional piecewise function divided in three regions. Thus, we can express the eikonal $H(t, r_a)$ with the order pairs (t, r_a) as its input; and its output is the order pair given by $(k_x(t, r_a), k_y(t, r_a))$. Notice that t is the time. Thus, $H(t, r_a) = (k_x(t, r_a), k_y(t, r_a))$. The three regions are: from the object to the first surface, inside the lens, and from the second surface to the image.

There is a one to one relationship between the first surface z_a and the second surface z_b. Thus, any two rays inside the lens do not cross each other, and they meet only at the object and image points, i.e. the stigmatic pair. For that reason, we can generate the eikonal of the stigmatic lens using the angles θ_1, θ_2, and θ_3 shown in figure 9.2.

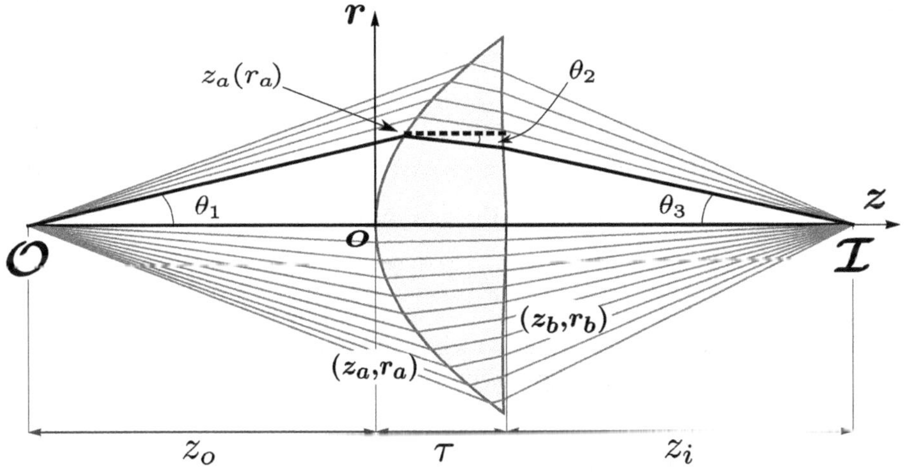

Figure 9.2. Ray tracing of an on-axis stigmatic singlet. The first and second surfaces are described by (z_a, r_a) and (z_b, r_b), respectively. The axial distance between the first surface and the object is z_o, the axial thickness of the lens is τ, and the axial distance between the second surface and the image is z_i. θ_1 is the angle subtended between an input ray that strikes the first surface at $z_a(r_a)$ and the optical axis. θ_2 is the angle subtended between the optical axis and the ray traveling inside the lens. θ_3 is the angle subtended by the output ray and optical axis.

The angle θ_1 is subtended between an input ray that strikes the first surface at $z_a(r_a)$ and the optical axis, thus θ_1 is given by

$$\theta_1 = \arctan\left[\frac{r_a}{z_a(r_a) - z_o}\right]. \tag{9.40}$$

θ_2 is the angle subtended by the optical axis and the ray traveling inside the lens, thus,

$$\theta_2 = \arctan\left[\frac{r_b(r_a) - r_a}{z_b(r_a) - z_a(r_a)}\right]. \tag{9.41}$$

Finally, θ_3 is the angle subtended by the output ray and optical axis, thus we have

$$\theta_3 = \arctan\left[\frac{r_b(r_a)}{z_b(r_a) - \tau - z_o}\right]. \tag{9.42}$$

With these angles in mind it is easy to construct the eikonal $H(t, r_a)$ that follows the Fermat principle,

$$H(t, r_a) = \left(k_x(t, r_a), k_y(t, r_a)\right), \tag{9.43}$$

taking $n_o = n_i = 1$,

$$\boxed{H(t, r_a) = \begin{cases} (z_o + t\cos\theta_1, t\sin\theta_1), & t < c_1, \\ (v(t - c_1)\cos\theta_2, v(t - c_1)\sin\theta_2), & c_1 < t < c_2, \\ (z_b(r_a) + (t - c_2)\cos\theta_3, r_b(r_a) + (t - c_2)\sin\theta_3), & t > c_2. \end{cases}} \tag{9.44}$$

where $v = c/n$ is the speed of light inside the lens, and c is the speed of light in vacuum (outside the lens $n_o = n_i = 1$). c_1 is the condition that divides the eikonal in the first and second region. c_2 is the condition that divides the eikonal in the second and third region.

From figure 9.2, it is easy to see where c_1 and c_2 come from, outside the lens $t < c_1$

$$z_a(r_a) = \tau\cos\theta_1 + z_o \Rightarrow c_1 = [z_a(r_a) - z_o]\sec\theta_1. \tag{9.45}$$

Thus, if $c_1 > t > c_2$, we are inside the lens. Now for c_2,

$$z_b(r_a) = v(t - c_1)\cos\theta_2 \Rightarrow c_2 = c_1 + \sec\theta_2[z_b(r_a) - z_a(r_a)]/v. \tag{9.46}$$

If $t > c_2$, we are in the third region of the system. Equation (9.44) is an analytical solution of (2.16), the eikonal equation when the object is real.

To plot the eikonals when the object is virtual, we slightly modify $H(t, r_a)$ as follows,

$$\boxed{H(t, r_a) = \begin{cases} (z_o + (t - R)\cos\theta_1, (t - R)\sin\theta_1), & t < c_1, \\ (v(t - c_1)\cos\theta_2, v(t - c_1)\sin\theta_2), & c_1 < t < c_2, \\ (z_b(r_a) + (t - c_2)\cos\theta_3, r_b(r_a) + (t - c_2)\sin\theta_3), & t > c_2. \end{cases}} \tag{9.47}$$

where $R = z_o + n\tau$ and the conditions become

$$c_1 = (n\tau + z_o) + \sec\theta_1[z_a(r_a) - z_o], \tag{9.48}$$

and

$$c_2 = c_1 + \frac{\sec \theta_2[z_b(r_a) - z_a(r_a)]}{v}. \tag{9.49}$$

The angles are the same, thus θ_1, θ_2 and θ_3 are described by equations (9.40), (9.41) and (9.42), respectively. Equation (9.47) is the analytical solution of (2.16), the eikonal equation when the object is virtual.

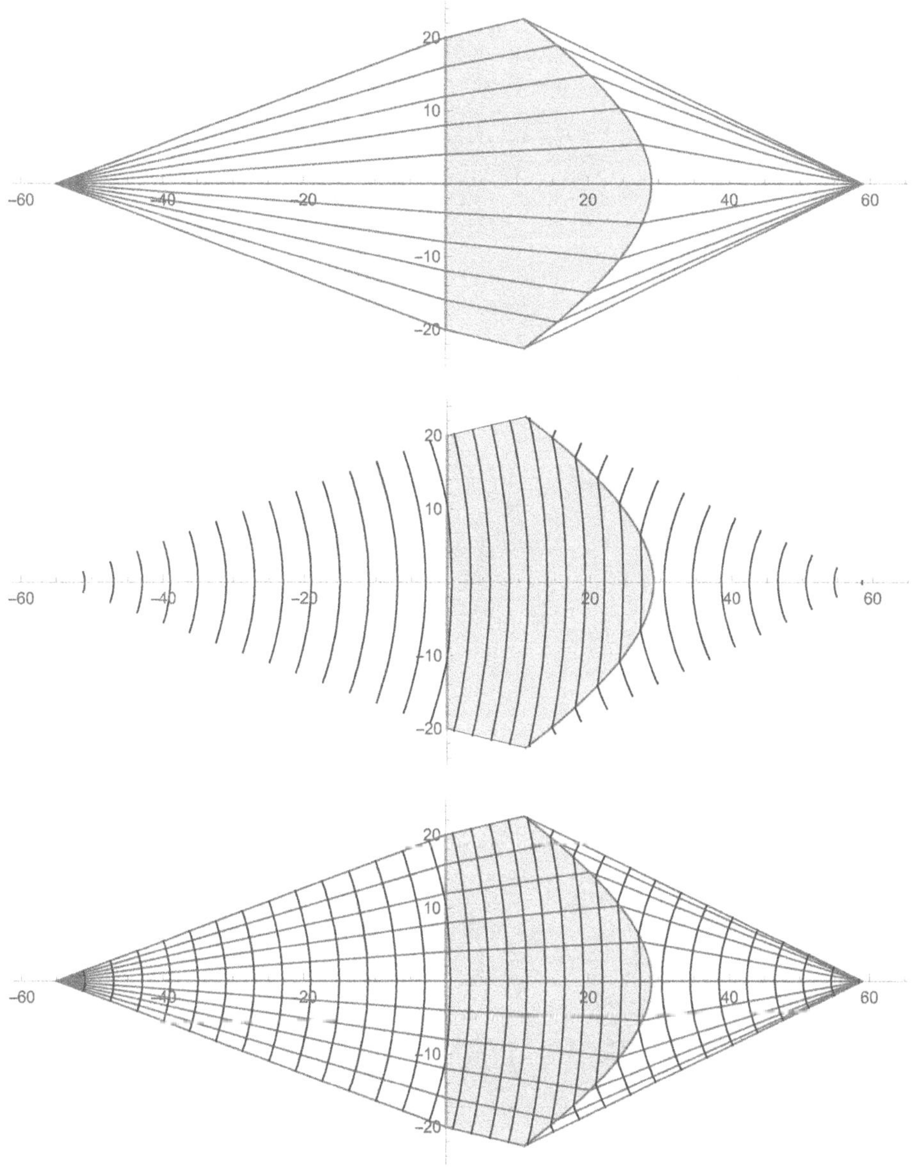

Figure 9.3. *Top:* ray tracing, *center:* waves, *bottom:* rays and waves. Design specifications: $n = 1.5$, $z_o = -55$ mm, $\tau = 29$ mm, $z_i = 30$ mm, $z_a = 0$, $z_b =$ equation (9.39) and $H(t, r_a) =$ equation (9.44).

9.2.5 Gallery

In this section, we present several examples of lenses free of spherical aberration when the object/image is real/virtual.

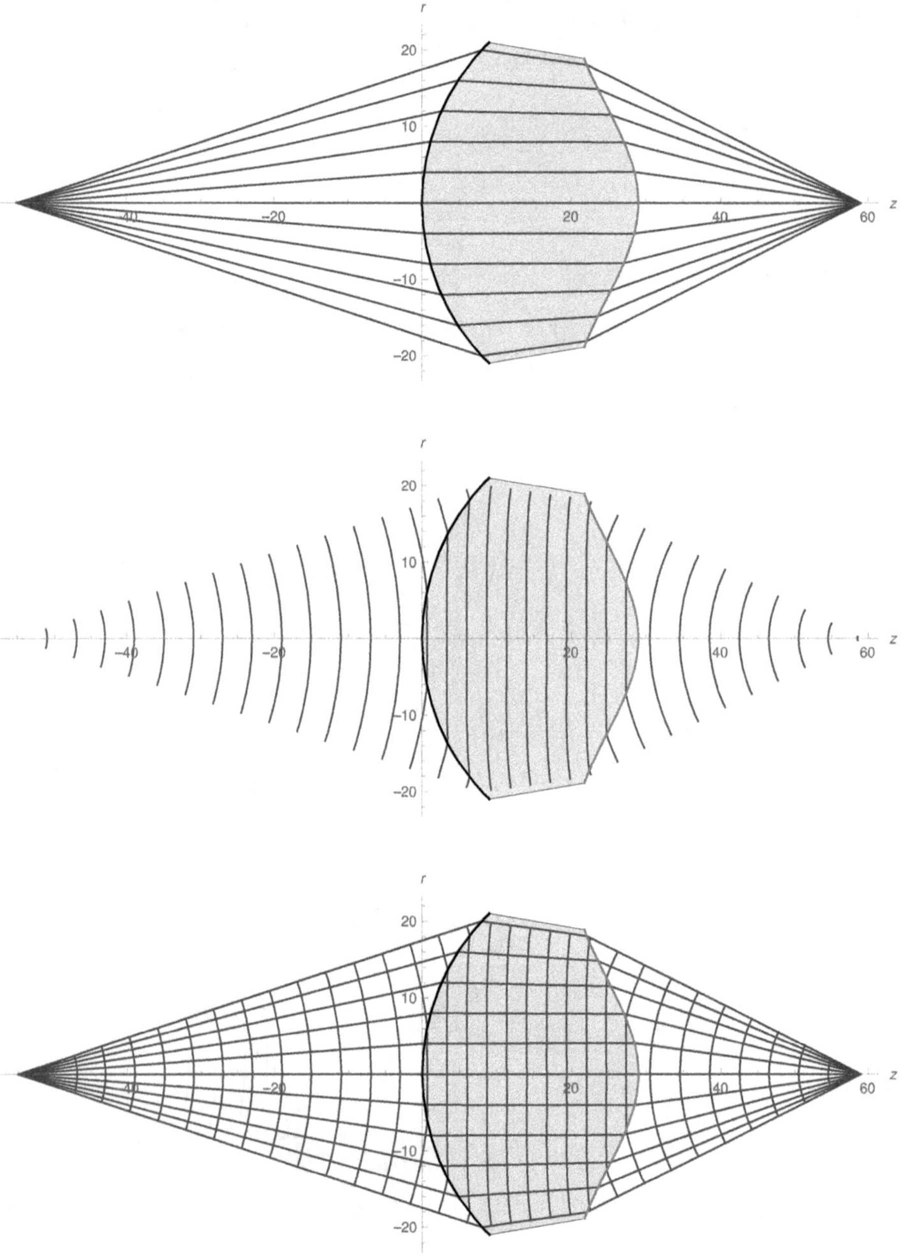

Figure 9.4. *Top:* rays, *center:* waves, *bottom:* rays and waves. Design specifications: $n = 1.5$, $z_o = -55$ mm, $\tau = 29$ mm, $z_i = 30$ mm, $z_a = (29 - \sqrt{29^2 - r_a^2})$, $z_b =$ equation (9.39) and $H(t, r_a) =$ equation (9.44).

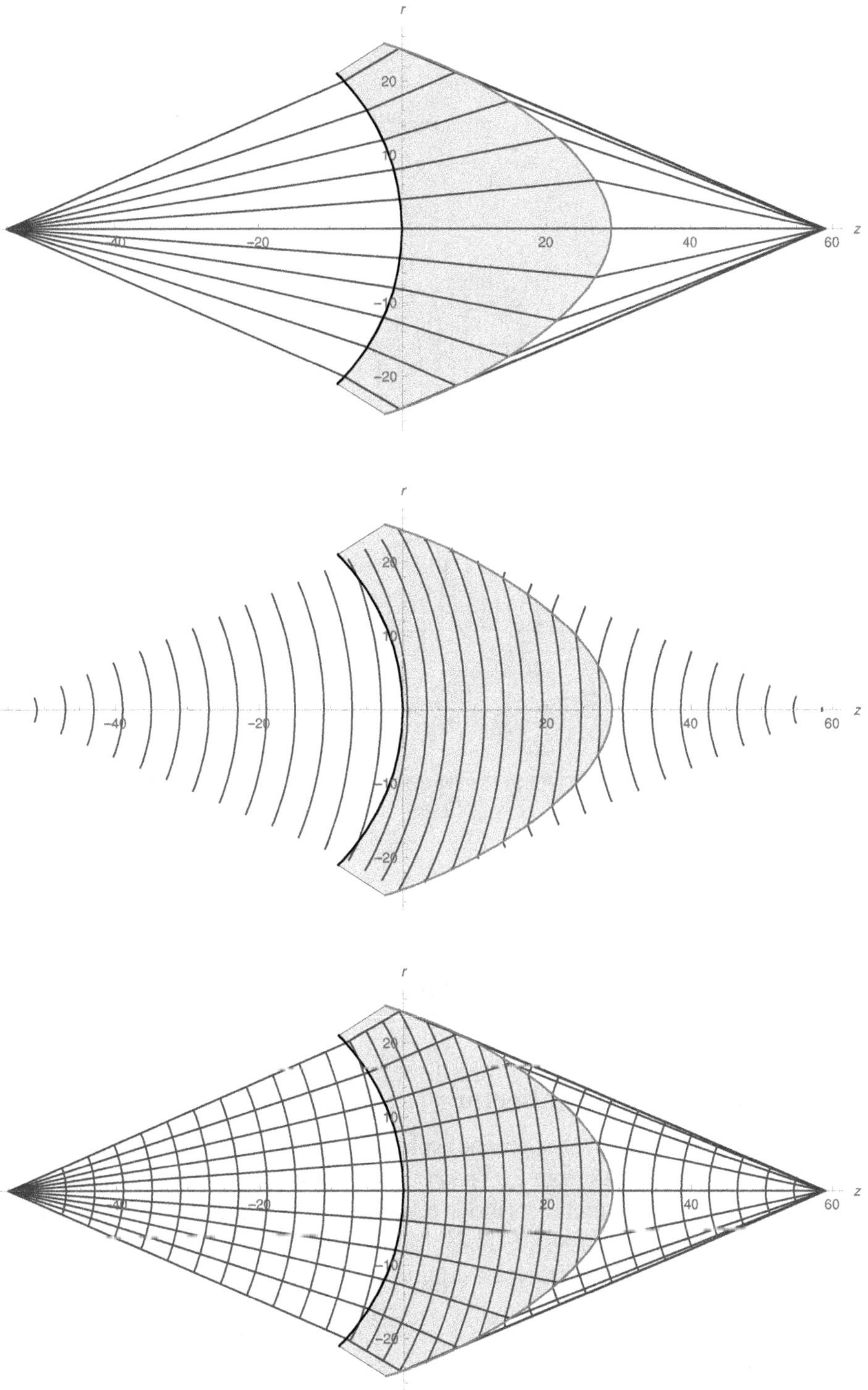

Figure 9.5. *Top:* rays in purple, *center:* waves in blue, *bottom:* rays and waves. Design specifications: $n = 1.5$, $z_o = -55$ mm, $\tau = 29$ mm, $z_i = 30$ mm, $z_a = -(29 - \sqrt{29^2 - r_a^2})$, $z_b =$ equation (9.39) and $H(t, r_a) =$ equation (9.44).

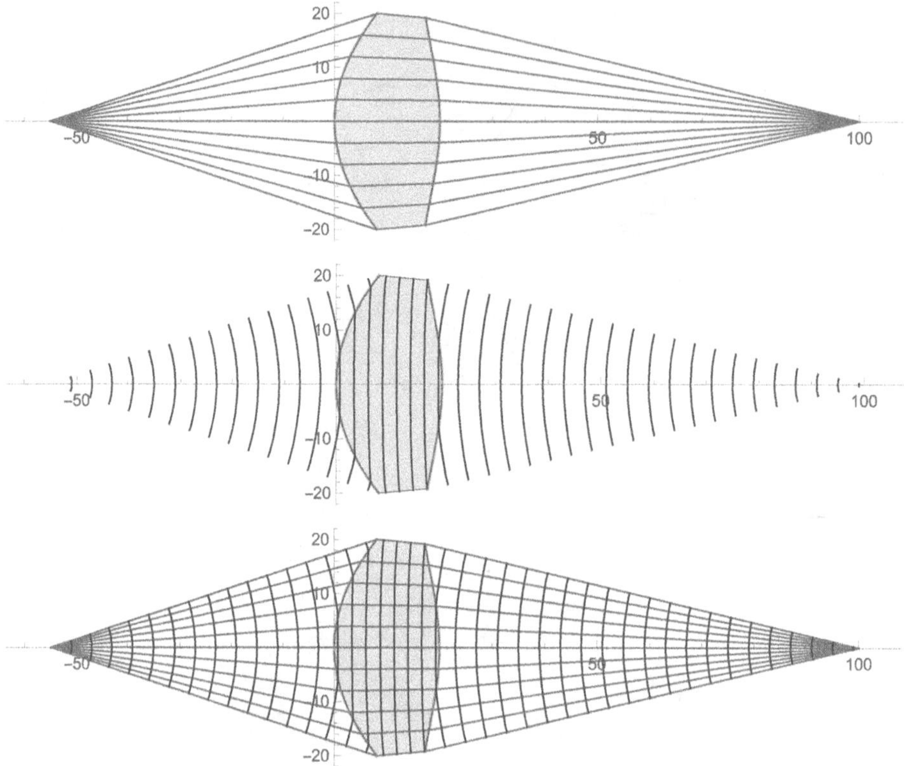

Figure 9.6. *Top:* rays in purple, *center:* waves in blue, *bottom:* rays and waves. Design specifications: $n = 1.5$, $z_o = -55$ mm, $\tau = 20$ mm, $z_i = 80$ mm, $z_a = r_a^2/50$, $z_b =$ equation (9.39) and $H(t, r_a) =$ equation (9.44).

The specification values of each design are in the captions of the figures. All the lenses presented in this section have been designed directly computing equation (9.39). No process of optimization has been applied.

The figures of this section are 9.3–9.16.

9.3 Stigmatic aspheric collector

The next example to study is the stigmatic aspheric collector. This case is when $z_o \to -\infty$, thus the lens collects the rays from z_o. To get the collector lens; we need to apply the limit when $z_o \to -\infty$ to some parameters inside equation (9.39). The parameters are the cosine directors $\wp_r$, $\wp_z$ and z_f, equations (9.17), (9.18) and (9.24). Let's recall them.

We start with the cosine directors, equation (9.17),

$$\wp_r = \frac{n_o[(z_a - z_o)z_a' + r_a]}{-\text{sgn}(z_o)n\sqrt{r_a^2 + (z_o - z_a)^2}(1 + z_a'^2)} - z_a'\frac{\sqrt{1 - \frac{n_o^2[r_a + (z_a - z_o)z_a']^2}{n^2[r_a^2 + (z_o - z_a)^2](1 + z_a'^2)}}}{\sqrt{1 + z_a'^2}}, \quad (9.17)$$

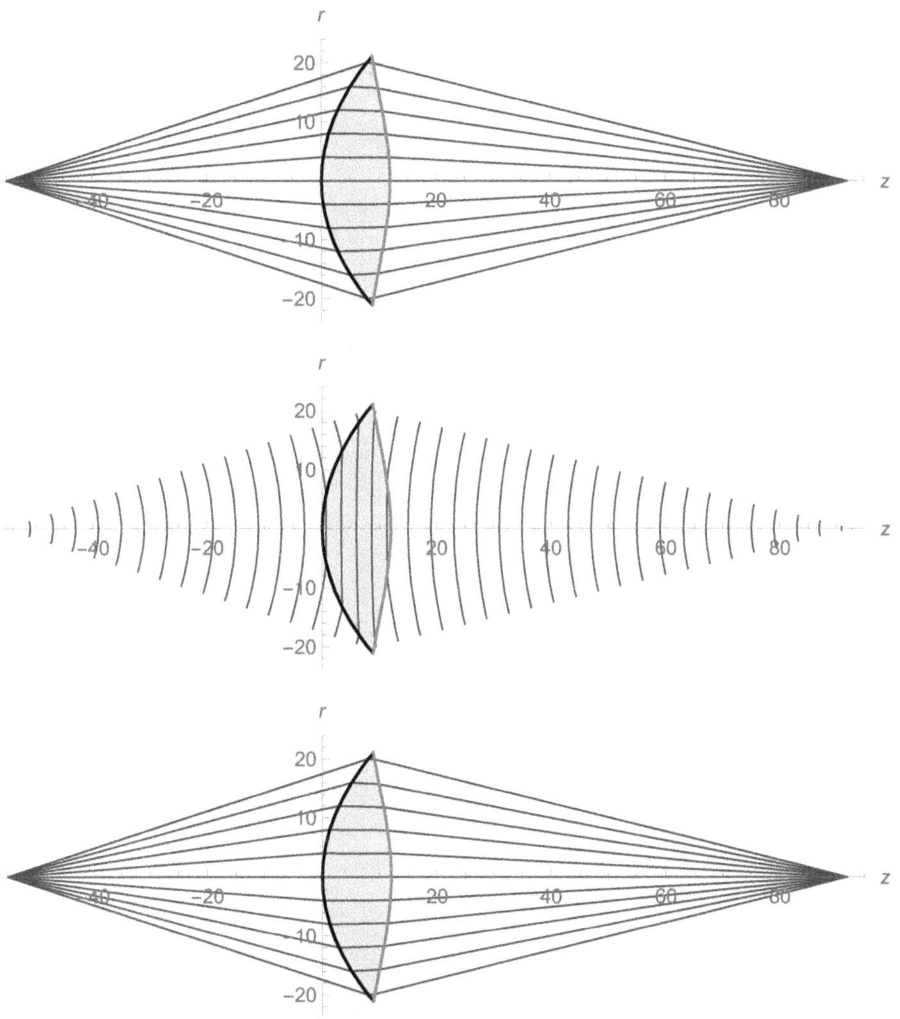

Figure 9.7. *Top:* rays, *center:* waves, *bottom:* rays and waves. Design specifications: $n = 1.5$, $z_o = -55$ mm, $\tau = 12$ mm, $z_i = 80$ mm, $z_a = r_a^2/50$, $z_b =$ equation (9.39) and $H(t, r_a) =$ equation (9.44).

and equation (9.18),

$$\wp_z = \frac{n_o\left[r_a + (z_a - z_o)z_a'\right]z_a'}{-\text{sgn}(z_o)n\sqrt{r_a^2 + (z_o - z_a)^2}\left(1 + z_a'^2\right)} + \frac{\sqrt{1 - \dfrac{n_o^2\left[r_a + (z_a - z_o)z_a'\right]^2}{n^2[r_a^2 + (z_o - z_a)^2]\left(1 + z_a'^2\right)}}}{\sqrt{1 + z_a'^2}}, \tag{9.18}$$

and the parameter z_f, equation (9.24),

$$z_f \equiv n_o\text{sgn}(z_o)\sqrt{(z_a - z_o)^2 + r_a^2} + n_i z_i - n_o z_o + n\tau. \tag{9.24}$$

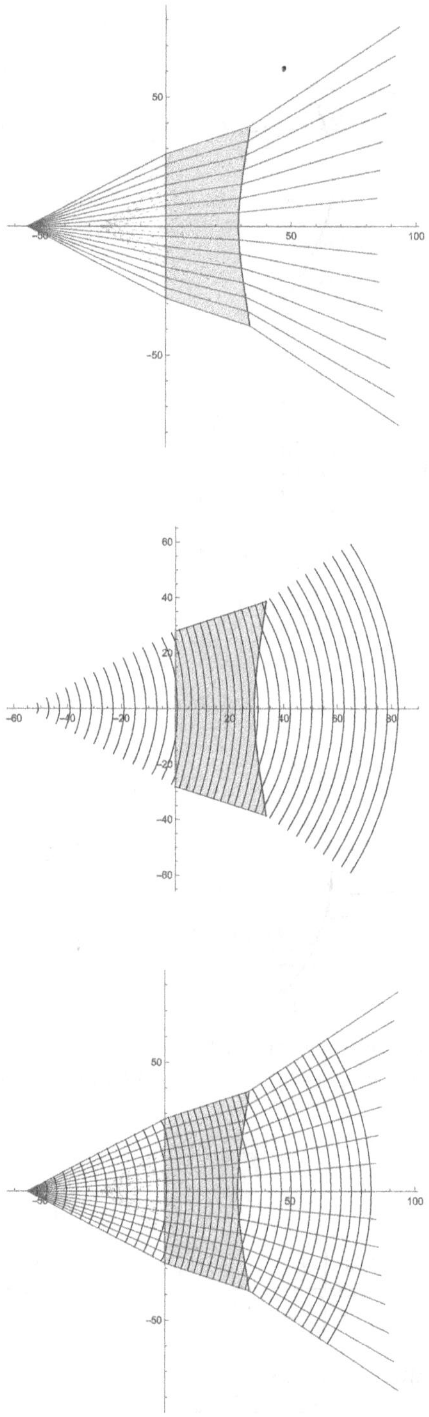

Figure 9.8. *Top:* rays, *center:* waves, *bottom:* rays and waves. Design specifications: $n = 1.5$, $z_o = -55$ mm, $\tau = 29$ mm, $z_i = -55$ mm, $z_a = 0$, $z_b =$ equation (9.39) and $H(t, r_a) =$ equation (9.44).

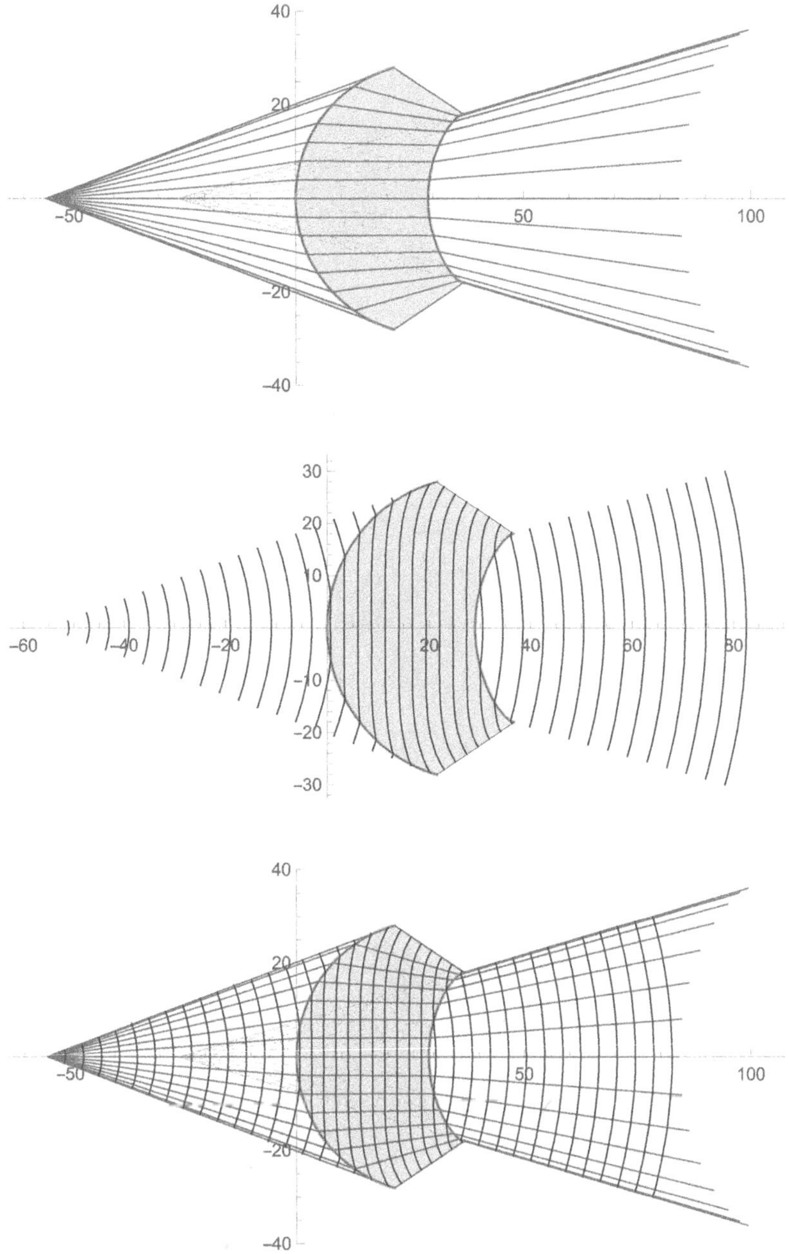

Figure 9.9. *Top:* rays, *center:* waves, *bottom:* rays and waves. Design specifications: $n = 1.5$, $z_o = -55$ mm, $\tau = 29$ mm, $z_i = -55$ mm, $z_a = 29 - \sqrt{29^2 - r_a^2}$, $z_b =$ equation (9.39) and $H(t, r_a) =$ equation (9.44).

Applying the $z_o \to -\infty$ in equation (9.17),

$$\lim_{z_o \to -\infty} \wp_z = \frac{z_a' \left(n_o - n\sqrt{(z_a')^2 + 1}\sqrt{\dfrac{(n^2 - n_o^2)(z_a')^2 + n^2}{n^2[(z_a')^2 + 1]}} \right)}{n[(z_a')^2 + 1]}, \tag{9.50}$$

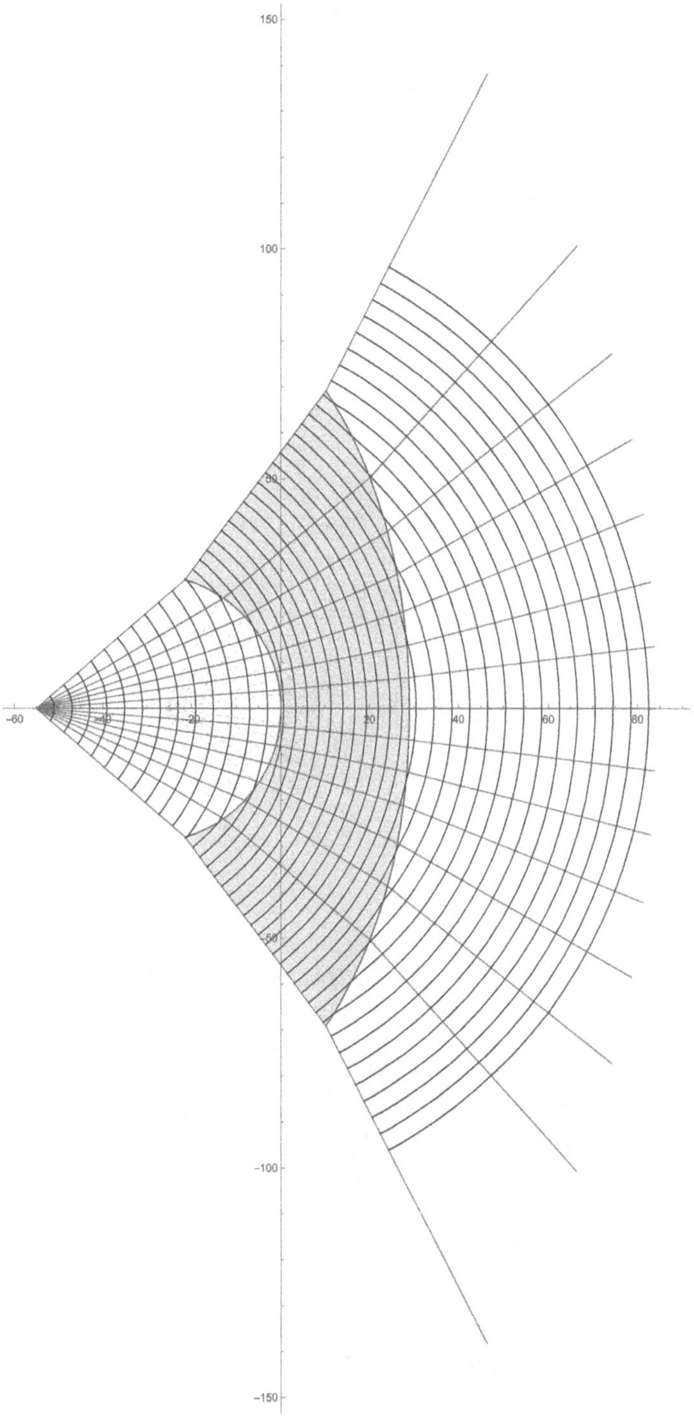

Figure 9.10. Design specifications: $n = 1.5$, $z_o = -55$ mm, $\tau = 29$ mm, $z_i = -55$ mm, $z_a = -(29 - \sqrt{29^2 - r_a^2})$, $z_b =$ equation (9.39) and $H(t, r_a) =$ equation (9.44).

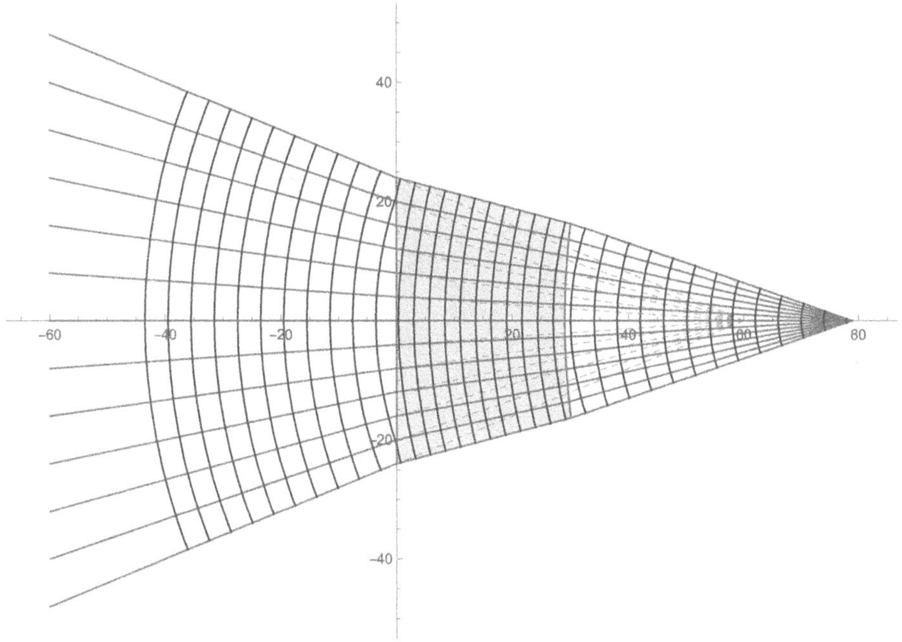

Figure 9.11. *Top:* rays, *center:* waves, *bottom:* rays and waves. Design specifications: $n = 1.5$, $z_o = 60$ mm, $\tau = 29$ mm, $z_i = 50$ mm, $z_a = 0$, $z_b =$ equation (9.39) and $H(t, r_a) =$ equation (9.47).

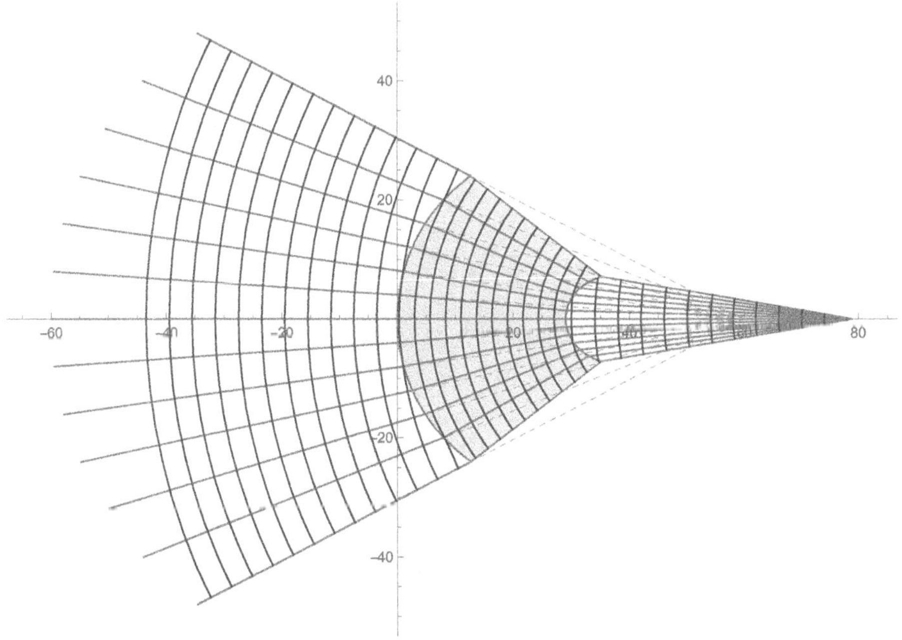

Figure 9.12. *Top:* rays, *center:* waves, *bottom:* rays and waves. Design specifications: $n = 1.5$, $z_o = 60$ mm, $\tau = 29$ mm, $z_i = 50$ mm, $z_a = (29 - \sqrt{29^2 - r_a^2})$, $z_b =$ equation (9.39) and $H(t, r_a) =$ equation (9.47).

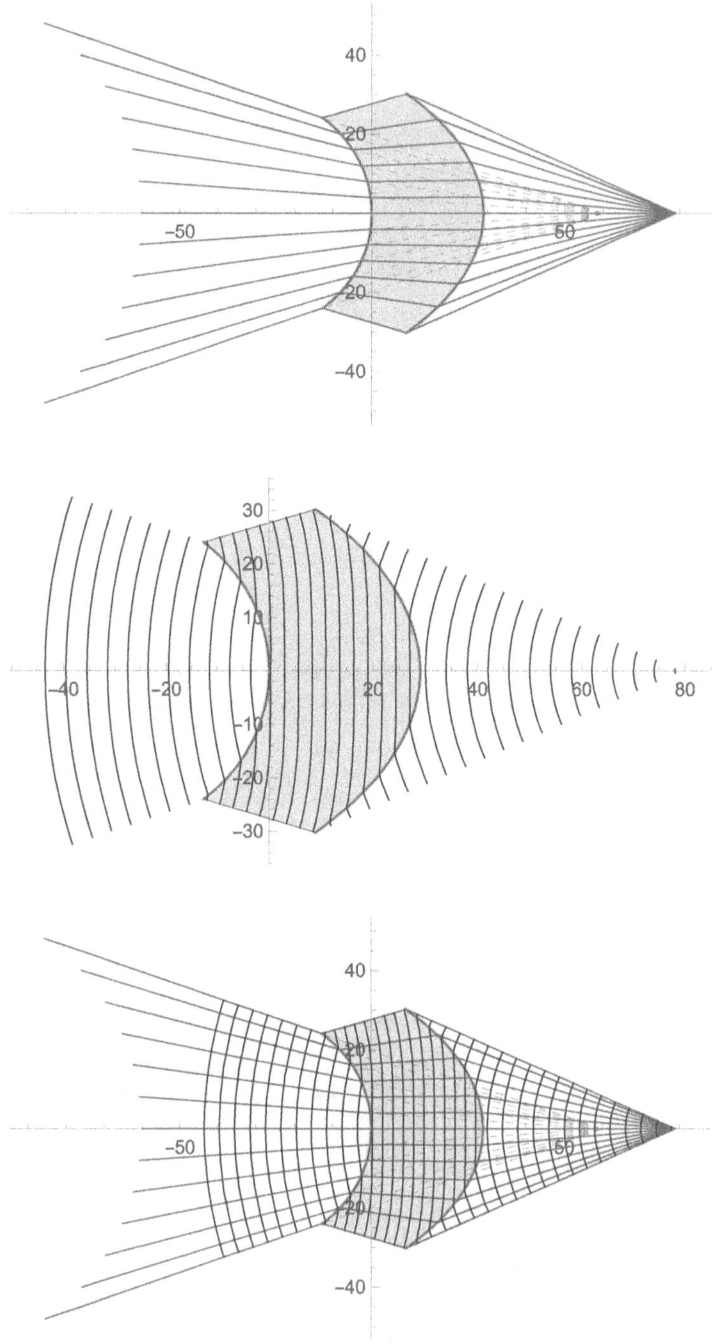

Figure 9.13. *Top:* rays, *center:* waves, *bottom:* rays and waves. Design specifications: $n = 1.5$, $z_o = 60$ mm, $\tau = 29$ mm, $z_i = 50$ mm, $z_a = -(29 - \sqrt{29^2 - r_a^2})$, $z_b =$ equation (9.39) and $H(t, r_a) =$ equation (9.47).

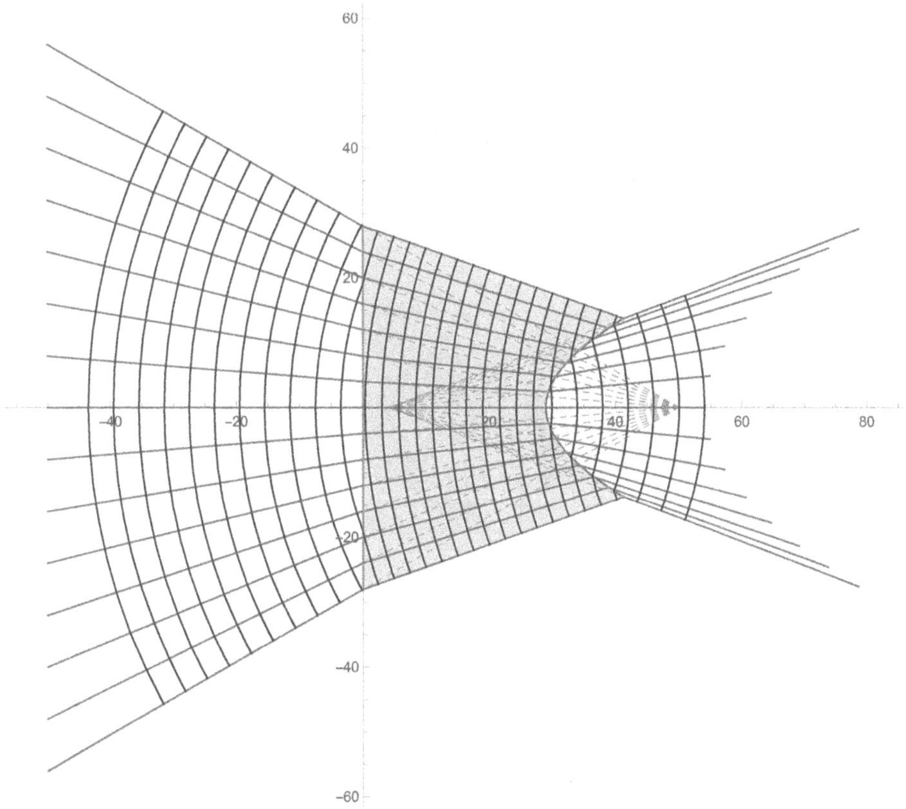

Figure 9.14. Design specifications: $n_o = n_i = 1$, $n = 1.5$, $z_o = 50$ mm, $\tau = 29$ mm, $z_i = -25$ mm, $z_a = 0$, $z_b =$ equation (9.39) and $H(t, r_a) =$ equation (9.47).

and computing the aforementioned limit in equation (9.18),

$$\lim_{z_o \to -\infty} \wp_z = \frac{\sqrt{\dfrac{(n^2 - n_o^2)(z_a')^2 + n^2}{n^2((z_a')^2 + 1)}}}{\sqrt{(z_a')^2 + 1}} + \frac{n_o (z_a')^2}{n(z_a')^2 + n}, \tag{9.51}$$

we compute the mentioned limit on parameter z_f with equation (9.24),

$$\lim_{z_o \to -\infty} z_f = -z_a n_o + n_i z_i + n\tau. \tag{9.52}$$

Then, when $z_o \to -\infty$ becomes ϑ, equation (9.38) becomes

$$\lim_{z_o \to -\infty} (\vartheta) = \frac{-\beta \pm \sqrt{\beta^2 + (n^2 - n_i^2)\left\{n_i^2 r_a^2 + n_i^2 z_\tau^2 - \left[\lim_{z_o \to -\infty} (z_f)\right]^2\right\}}}{(n_i^2 - n^2)}, \tag{9.53}$$

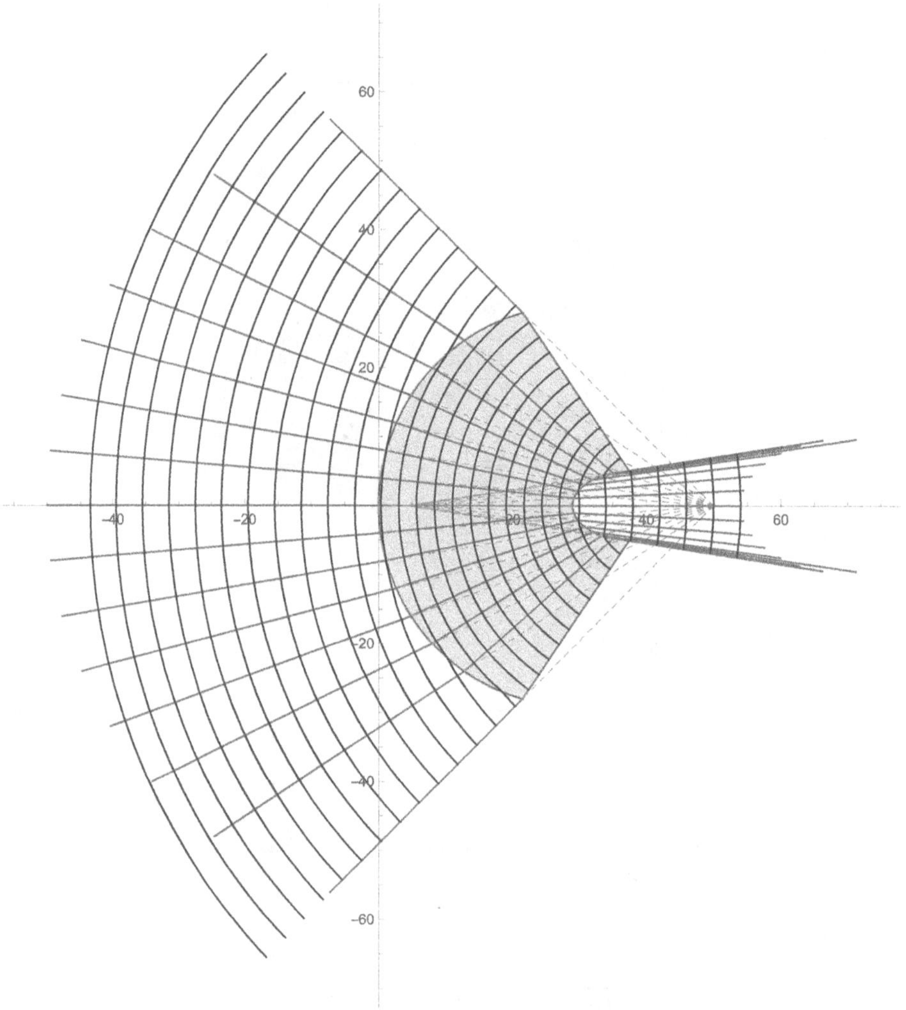

Figure 9.15. Design specifications: $n_o = n_i = 1$, $n = 1.5$, $z_o = 50$ mm, $\tau = 29$ mm, $z_i = -25$ mm, $z_a = (29 - \sqrt{29^2 - r_a^2})$, $z_b =$ equation (9.39) and $H(t, r_a) =$ equation (9.47).

where,

$$\beta \equiv \left\{\left[\lim_{z_o\to-\infty}(z_f)\right]n + n_i^2 r_a\left[\lim_{z_o\to-\infty}(\wp_r)\right] + n_i^2 z_\tau\left[\lim_{z_o\to-\infty}(\wp_z)\right]\right\}. \tag{9.54}$$

Therefore, the second surface is given by

$$\boxed{\begin{aligned}\lim_{z_o\to-\infty}(z_b) &= z_a + \left[\lim_{z_o\to-\infty}(\vartheta)\right]\left[\lim_{z_o\to-\infty}(\wp_z)\right],\\ \lim_{z_o\to-\infty}(r_b) &= r_a + \left[\lim_{z_o\to-\infty}(\vartheta)\right]\left[\lim_{z_o\to-\infty}(\wp_r)\right].\end{aligned}} \tag{9.55}$$

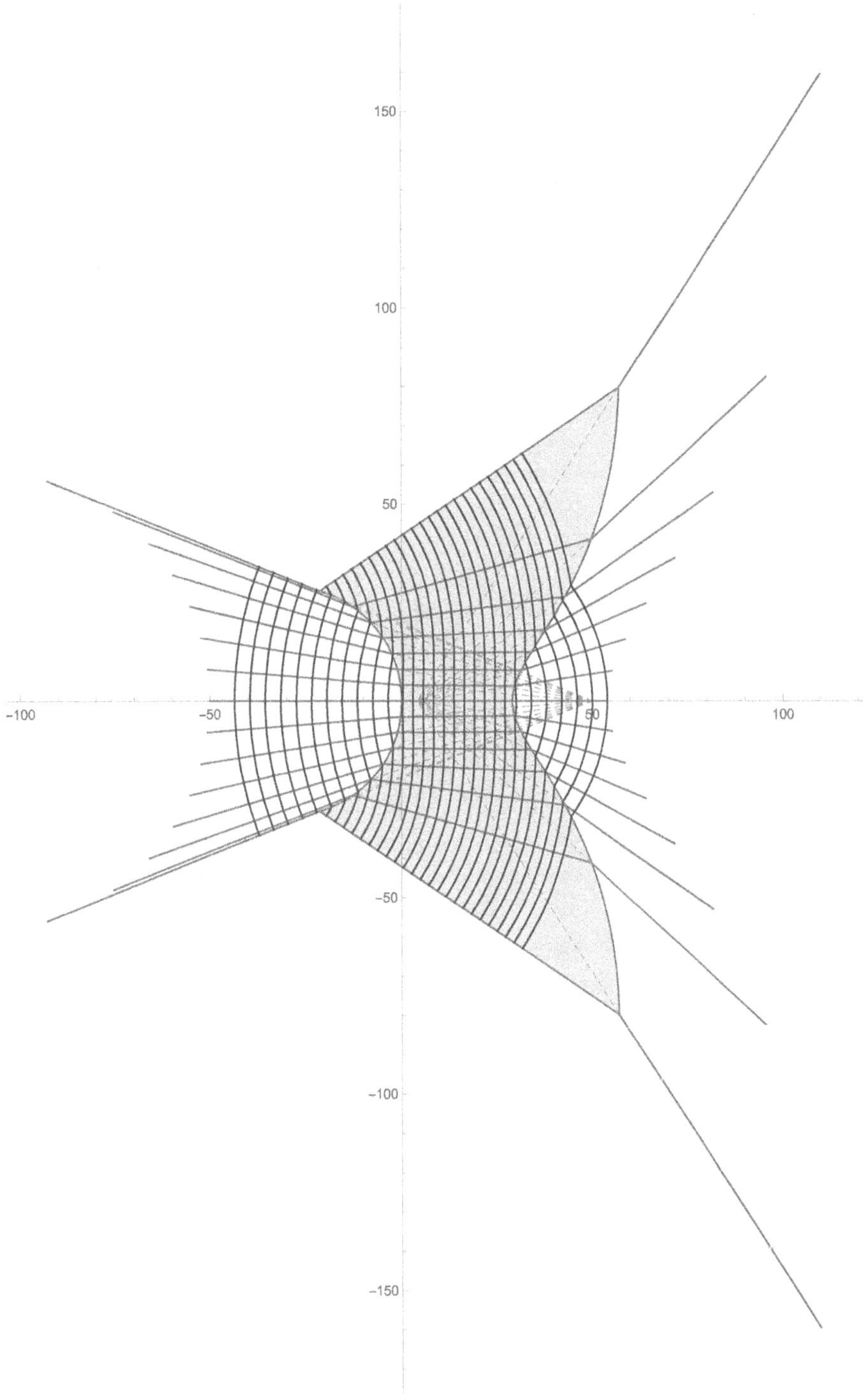

Figure 9.16. Design specifications: $n_o = n_i = 1$, $n = 1.5$, $z_o = 50\,\text{mm}$, $\tau = 29\,\text{mm}$, $z_i = -25\,\text{mm}$, $z_a = -(29 - \sqrt{29^2 - r_a^2})$, $z_b =$ equation (9.39) and $H(t, r_a) =$ equation (9.47).

To properly use (9.55) we need to use equations (9.50)–(9.54).

It is important to observe the novelty of equation (9.55). We have not made any changes in the procedure to get it; we apply a limit on equation (9.39) and it is perfectly supported.

In this case, equation (9.55) is only valid for real objects and virtual/real images. Finally, we want to note that the evaluation of the limit when $z_o \to -\infty$ is possible with the vector notation proposed. In another case, using angles with the usual form of Snell's law will lead to several difficulties.

9.3.1 The eikonal of the stigmatic collector

In the case of the eikonal of the stigmatic collector, we need to compute some limits, over equation (9.44) and equations (9.40), (9.41) and (9.42), respectively.

We start with θ_1, equation (9.40),

$$\lim_{z_o \to -\infty} (\theta_1) = \lim_{z_o \to -\infty} \arctan\left[\frac{r_a}{z_a(r_a) - z_o}\right] = 0. \tag{9.56}$$

θ_2 is when $z_o \to -\infty$, equation (9.41),

$$\lim_{z_o \to -\infty} (\theta_2) = \lim_{z_o \to -\infty} \arctan\left[\frac{r_b(r_a) - r_a}{z_b(r_a) - z_a(r_a)}\right] = \arctan\left[\frac{\lim_{z_o \to -\infty} r_b(r_a) - r_a}{\lim_{z_o \to -\infty} z_b(r_a) - z_a(r_a)}\right]. \tag{9.57}$$

Finally, θ_3, is when $z_o \to -\infty$, equation (9.42),

$$\lim_{z_o \to -\infty} (\theta_3) = \lim_{z_o \to -\infty} \arctan\left[\frac{r_b(r_a)}{z_b(r_a) - \tau - z_o}\right] = \arctan\left[\frac{\lim_{z_o \to -\infty} r_b(r_a)}{\lim_{z_o \to -\infty} z_b(r_a) - \tau - z_o}\right]. \tag{9.58}$$

Therefore, $H(t, r_a)$ when $z_o \to -\infty$, equation (9.44), becomes

$$\lim_{z_o \to -\infty} H(t, r_a) = \begin{cases} [Z_o + t, r_a], & t < c_1, \\ \left[v(t - c_1)\cos\left(\lim_{z_o \to -\infty} \theta_2\right), \right. \\ \left. v(t - c_1)\sin\left(\lim_{z_o \to -\infty} \theta_2\right)\right], & c_1 < t < c_2, \\ \left[\lim_{z_o \to -\infty} z_b(r_a) + (t - c_2)\cos\left(\lim_{z_o \to -\infty} \theta_3\right), \right. \\ \left. \lim_{z_o \to -\infty} r_b(r_a) + (t - c_2)\cos\left(\lim_{z_o \to -\infty} \theta_3\right)\right], & t > c_2, \end{cases} \tag{9.59}$$

where Z_o is just a negative constant where the plot starts. The condition c_1 becomes equation (9.45),

$$z_a(r_a) = \tau \cos\left(\lim_{z_o \to -\infty} \theta_1\right) + Z_o \Rightarrow c_1 = [z_a(r_a) - Z_o], \tag{9.60}$$

and c_2, equation (9.46),

$$z_b(r_a) = v(t - c_1)\cos\left(\lim_{z_o \to -\infty} \theta_2\right) \Rightarrow c_2 = c_1 + \sec\left(\lim_{z_o \to -\infty} \theta_2\right)\frac{[z_b(r_a) - z_a(r_a)]}{v}. \tag{9.61}$$

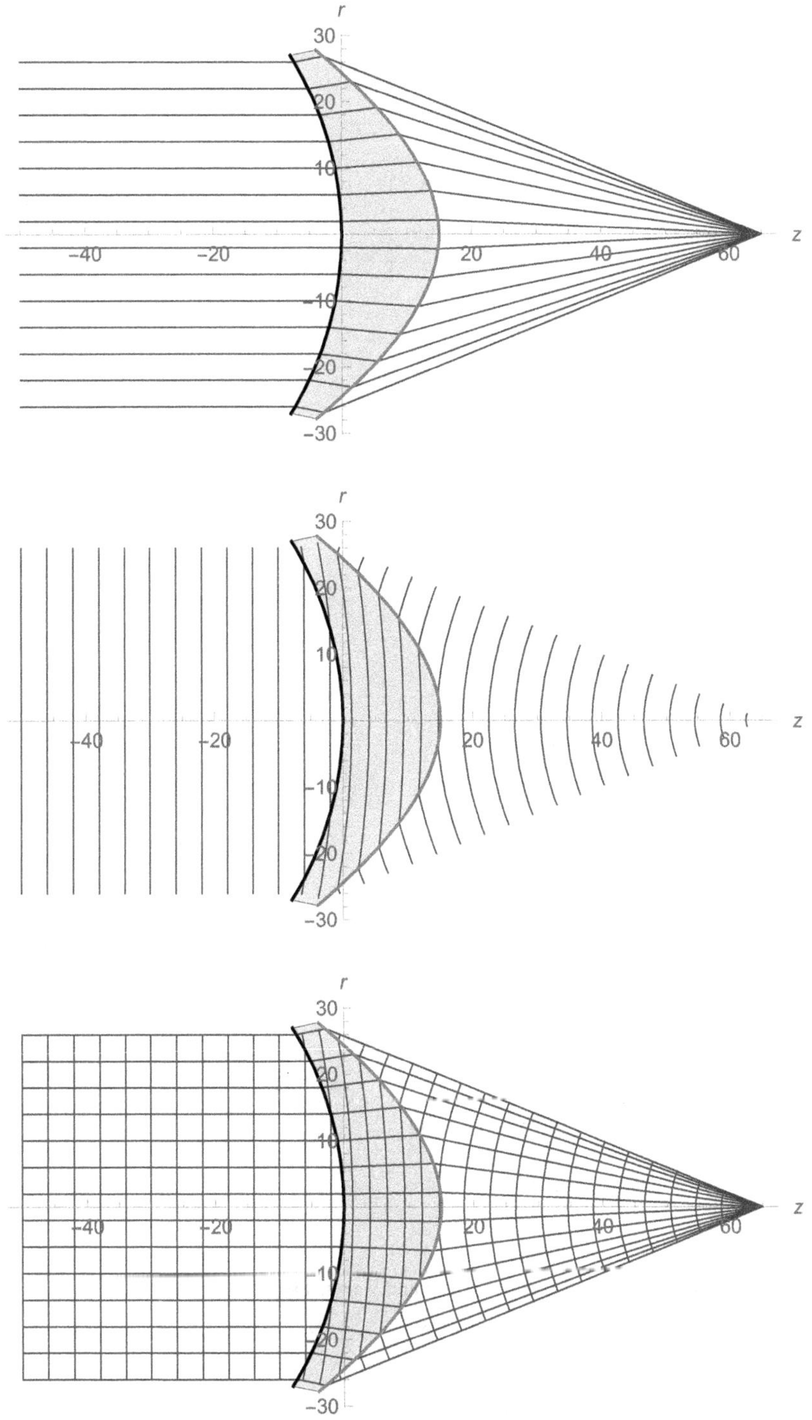

Figure 9.17. *Top:* rays, *center:* waves, *bottom:* rays and waves. Design specifications: $n_o = n_i = 1$, $n = 1.5$, $z_o \to -\infty$, $\tau = 15$ mm, $z_i = 50$ mm, $z_a = -\left(50 - \sqrt{50^2 - r_a^2}\right)$, $z_b =$ equation (9.55) and $H(t, r_a) =$ equation (9.59).

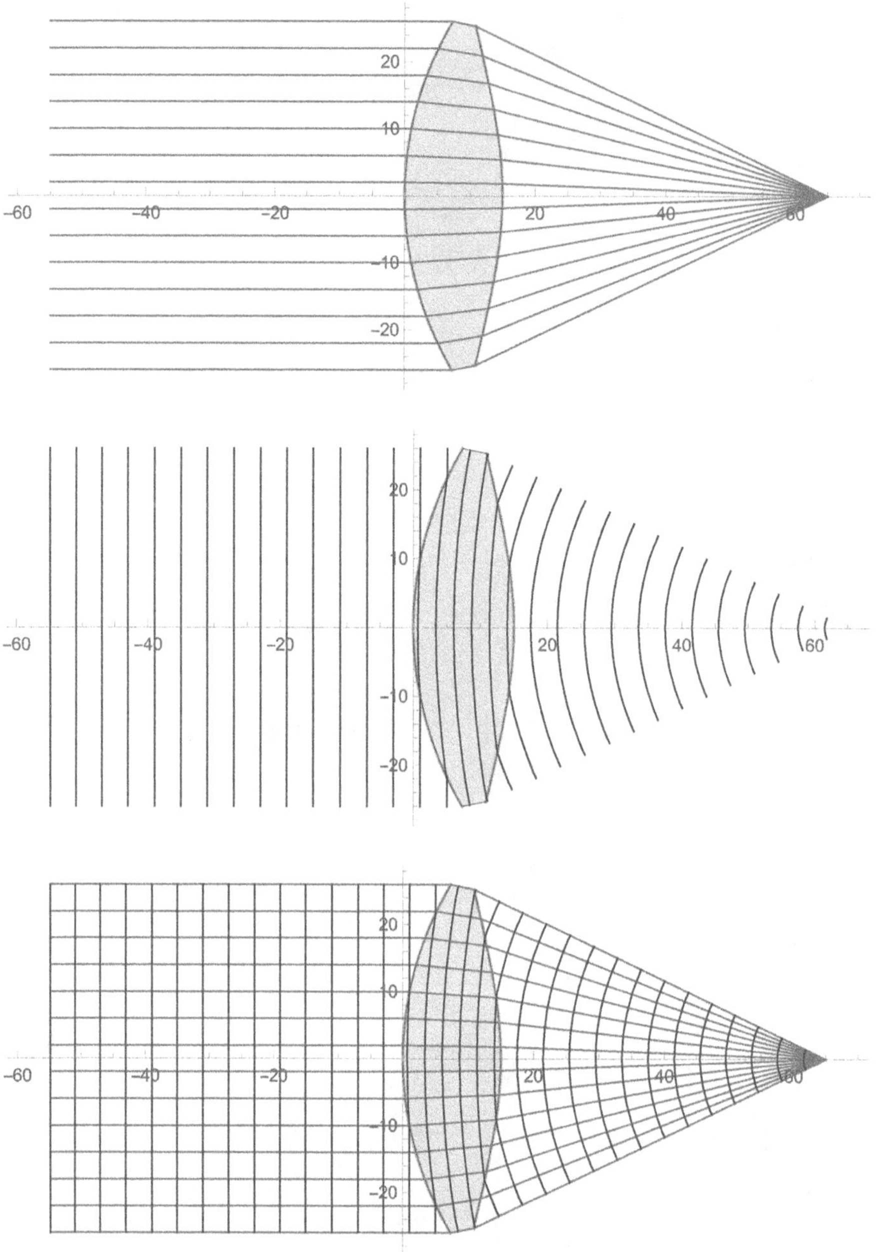

Figure 9.18. *Top:* rays, *center:* waves, *bottom:* rays and waves. Design specifications: $n_o = n_i = 1$, $n = 1.5$, $z_o \to -\infty$, $\tau = 15$ mm, $z_i = 50$ mm, $z_a = \left(50 - \sqrt{50^2 - r_a^2}\right)$, $z_b =$ equation (9.55) and $H(t, r_a) =$ equation (9.59).

In the following section, we are going to show several lenses analogous to the parabolic mirror. Equation (9.59) is the analytical solution of the eikonal equation, equation (2.16), for the collector lens.

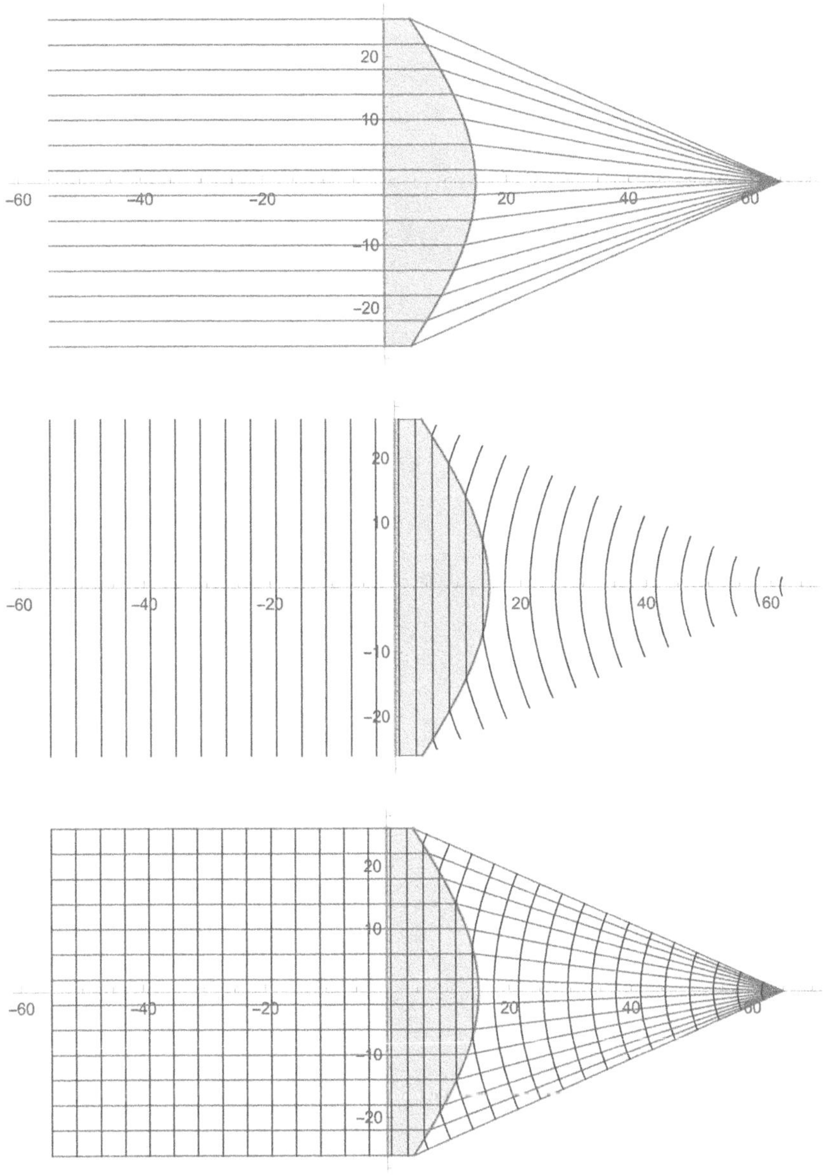

Figure 9.19. *Top:* rays, *center:* waves, *bottom:* rays and waves. Design specifications: $n_o = n_i = 1$, $n = 1.5$, $z_o \to -\infty$, $\tau = 15$ mm, $z_i = 50$ mm, $z_a = 0$, $z_b =$ equation (9.55) and $H(t, r_a) =$ equation (9.59).

9.3.2 Gallery

In this section we show a gallery of stigmatic collectors. As usual, the captions contain all the information needed to reproduce the respective examples. The figures of the gallery are 9.17–9.22.

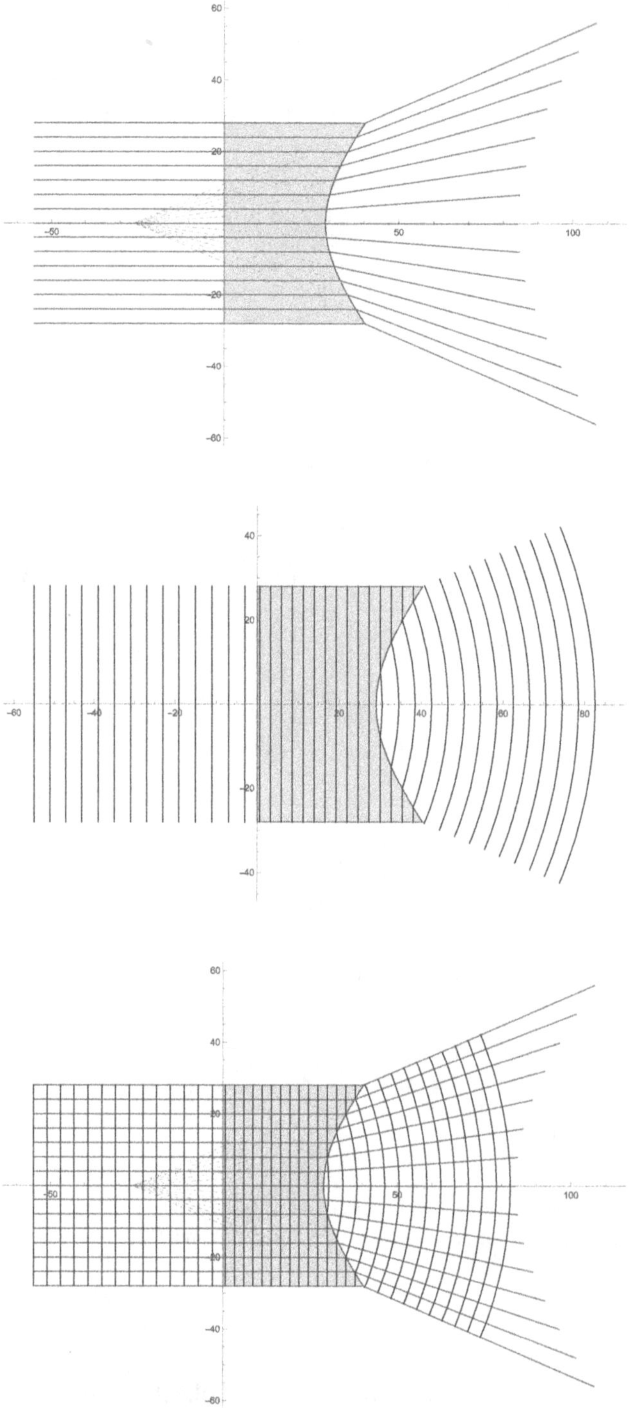

Figure 9.20. *Top:* rays, *center:* waves, *bottom:* rays and waves. Design specifications: $n_o = n_i = 1$, $n = 1.5$, $z_o \to -\infty$, $\tau = 29$ mm, $z_i = -55$ mm, $z_a = 0$, $z_b =$ equation (9.55) and $H(t, r_a) =$ equation (9.59).

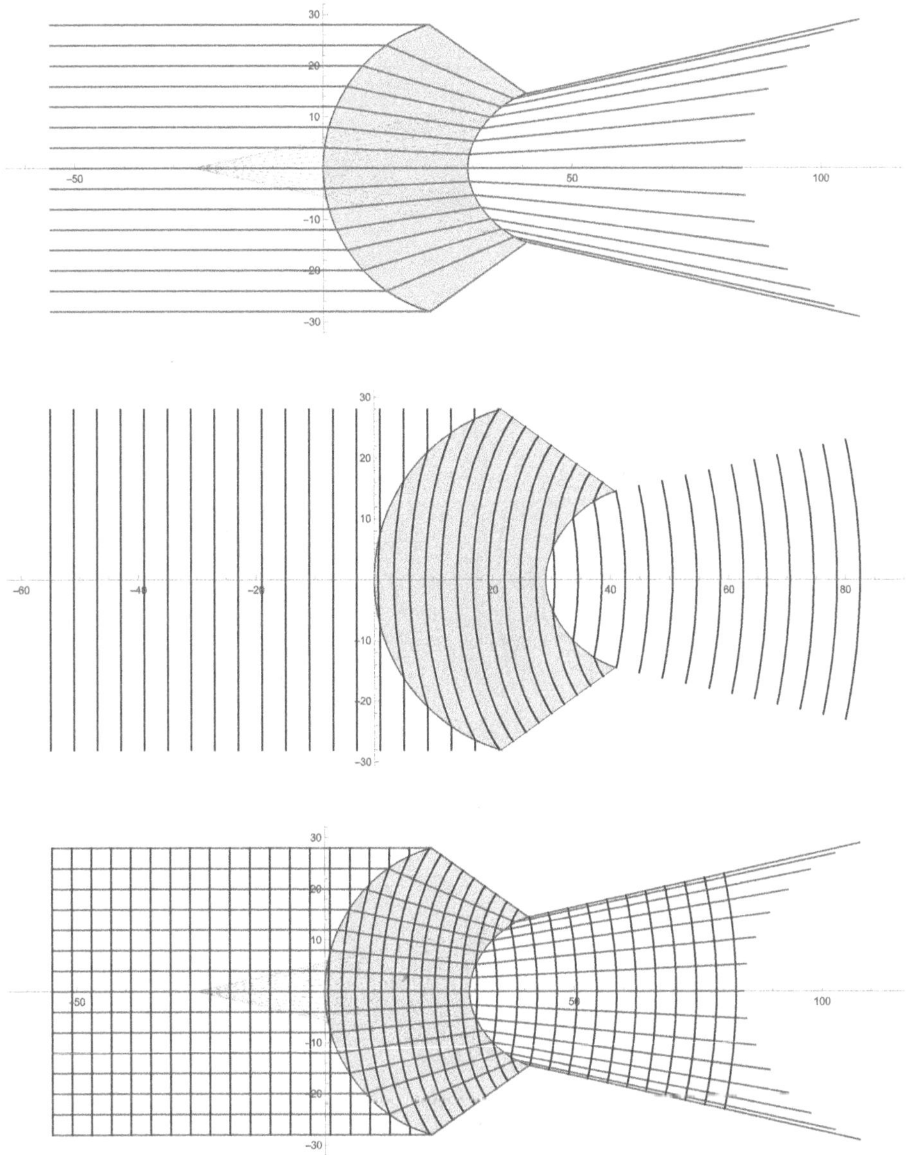

Figure 9.21. *Top:* rays, *center:* waves, *bottom:* rays and waves. Design specifications: $n_o = n_i = 1$, $n = 1.5$, $z_o \to -\infty$, $\tau = 29\,\text{mm}$, $z_i = -55\,\text{mm}$, $z_a = 29 - \sqrt{29^2 - r_a^2}$, $z_b =$ equation (9.55) and $H(t, r_a) =$ equation (9.59).

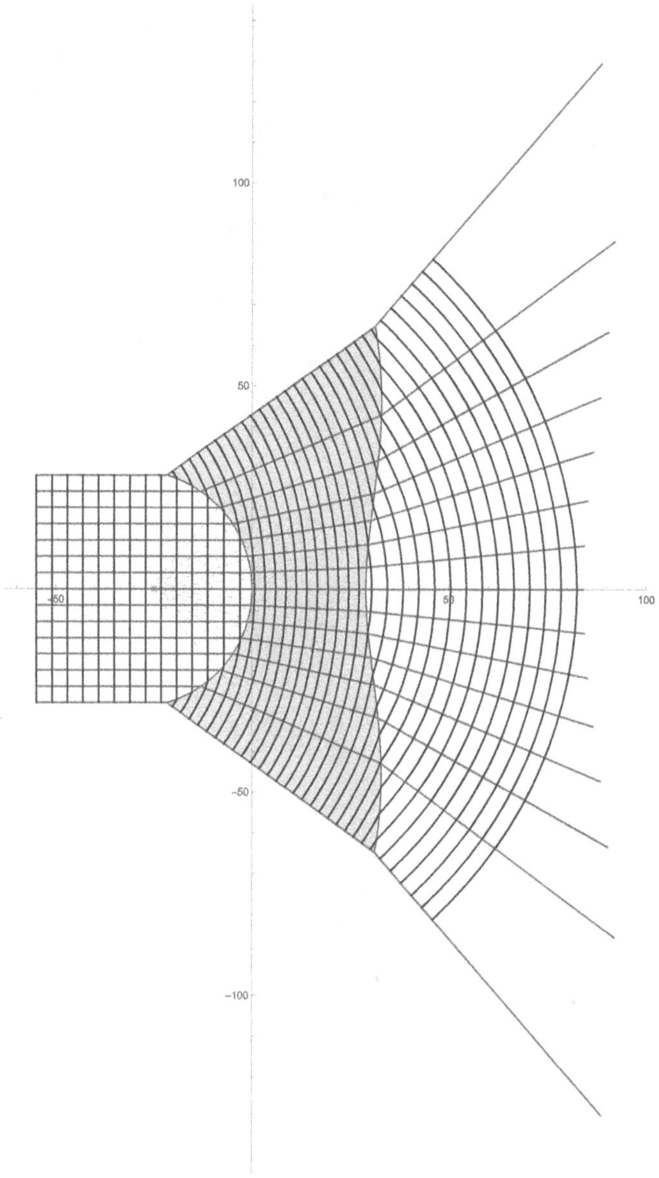

Figure 9.22. Design specifications: $n_o = n_i = 1$, $n = 1.5$, $z_o \to -\infty$, $\tau = 29$ mm, $z_i = -55$ mm, $z_a - (29 - \sqrt{29^2 - r_a^2})$, $z_b =$ equation (9.55) and $H(t, r_a) =$ equation (9.59).

9.4 Stigmatic aspheric collimator

If the collectors at the entrance of the ray are collimated along the optical axis, the collimator is the opposite. In the collimator, the output rays are those collimated along the axis.

To get the collimator we need to compute $z_i \to \infty$ in the optical path and then solve for the unknowns z_b and r_b. Notice that the cosine directors $\wp_r$ and $\wp_z$ are not affected by the aforementioned limit. Thus, $\wp_r$ and $\wp_z$ give equations (9.17) and (9.18).

Here, we need to implement the limit when $z_i \to \infty$, this time with the Fermat principle. We recall the Fermat principle of equation (9.6),

$$\begin{aligned} &- n_o z_o + n\tau + n_i z_i = -n_o \mathrm{sgn}(z_o)\sqrt{r_a^2 + (z_a - z_o)^2} \\ &+ n\sqrt{(r_b - r_a)^2 + (z_b - z_a)^2} + n_i \mathrm{sgn}(z_i)\sqrt{r_b^2 + (z_b - \tau - z_i)^2}. \end{aligned} \tag{9.6}$$

Replacing $\vartheta \equiv \sqrt{(r_b - r_a)^2 + (z_b - z_a)^2}$ in equation (9.6),

$$-z_o n_o + n\tau + z_i n_i - n_i \mathrm{sgn}(z_i)\sqrt{r_b^2 + (z_b - \tau - z_i)^2} - n\vartheta = -n_o \mathrm{sgn}(z_o)\sqrt{r_a^2 + (z_a - z_o)^2}. \tag{9.62}$$

Notice that on the right side of the above equation nothing depends on z_i. Then, it does not matter what the value of z_i is, the right side stays equal. On the other hand, on the left side, we have two times z_i. We can evaluate the limit when $z_i \to \infty$ only in the left side, thus,

$$\begin{aligned} &\lim_{z_i \to \infty}\left[-z_o n_o + n\tau + z_i n_i - n_i \mathrm{sgn}(z_i)\sqrt{r_b^2 + (z_b - \tau - z_i)^2} - n\vartheta\right] \\ &= (-n_i + n)\tau - n_o z_o + n_i\left[\lim_{z_i \to \infty}(z_b)\right] - n\left[\lim_{z_i \to \infty}(\vartheta)\right]. \end{aligned} \tag{9.63}$$

Computing when $z_i \to \infty$ in equation (9.63),

$$(-n_i + n)\tau - n_o z_o + n_i\left[\lim_{z_i \to \infty}(z_b)\right] - n\left[\lim_{z_i \to \infty}(\vartheta)\right] = -n_o \mathrm{sgn}(z_o)\sqrt{r_a^2 + (z_a - z_o)^2}. \tag{9.64}$$

Let's replace $z_b = z_a + \vartheta\wp_z$ in the previous equation,

$$(-n_i + n)\tau - n_o z_o + n_i\left[\lim_{z_i \to \infty}\left(z_a + \vartheta\wp_z\right)\right] - n\left[\lim_{z_i \to \infty}(\vartheta)\right] = n_o \mathrm{sgn}(z_o)\sqrt{r_a^2 + (z_a - z_o)^2}. \tag{9.65}$$

Since z_a and $\wp_z$ are not affected by the limit when $z_i \to \infty$, we have

$$(-n_i + n)\tau - n_o z_o + n_i z_a + n_i\left[\lim_{z_i \to \infty}(\vartheta)\right]\wp_z = -n_o \mathrm{sgn}(z_o)\sqrt{r_a^2 + (z_a - z_o)^2} + n\left[\lim_{z_i \to \infty}(\vartheta)\right]. \tag{9.66}$$

Solving for $\left[\lim_{z_i \to \infty}(\vartheta)\right]$,

$$\left[\lim_{z_i \to \infty}(\vartheta)\right] = \frac{\left(-n_i + n\right)\tau - n_o z_o + n_i z_a + n_o \mathrm{sgn}(z_o)\sqrt{r_a^2 + (z_a - z_o)^2}}{n - n_i\wp_z}. \tag{9.67}$$

Finally, the solution is given by equation (9.68), where $\left[\lim_{t_b\to\infty}(\vartheta)\right]$ is taken from equation (9.67) and $\wp_r$ and $\wp_z$ are obtained from equations (9.17) and (9.18),

$$\begin{cases} \lim_{z_i\to\infty}(z_b) = z_a + \left[\lim_{z_i\to\infty}(\vartheta)\right]\wp_z, \\ \lim_{z_i\to\infty}(r_b) = r_a + \left[\lim_{z_i\to\infty}(\vartheta)\right]\wp_r. \end{cases} \tag{9.68}$$

Equation (9.68) describes the second surface of the lens collimator.

To properly use equation (9.68),we need equation (9.67) for $\left[\lim_{z_i\to\infty}(\vartheta)\right]$ equation (9.17) and equation (9.18), for $\wp_r$ and $\wp_z$, respectively.

9.4.1 The eikonal of the stigmatic collimator

We now focus on the eikonal of the stigmatic collimator. The first thing to notice is that θ_1 does not depend on z_i, thus,

$$\lim_{z_o\to\infty}(\theta_1) = \theta_1. \tag{9.40}$$

θ_2 is when $z_i \to \infty$, equation (9.41) becomes

$$\lim_{z_i\to\infty}(\theta_2) = \lim_{z_i\to-\infty}\arctan\left[\frac{r_b(r_a) - r_a}{z_b(r_a) - z_a(r_a)}\right] = \arctan\left[\frac{\lim_{z_i\to\infty} r_b(r_a) - r_a}{\lim_{z_i\to\infty} z_b(r_a) - z_a(r_a)}\right]. \tag{9.69}$$

Now we look for θ_3, when $z_i \to \infty$, equation (9.42) becomes

$$\lim_{z_i\to\infty}(\theta_3) = \lim_{z_i\to\infty}\arctan\left[\frac{r_b(r_a)}{z_b(r_a) - \tau - z_o}\right] = \arctan\left[\frac{\lim_{z_i\to\infty} r_b(r_a)}{\lim_{z_i\to\infty} z_b(r_a) - \tau - z_o}\right]. \tag{9.70}$$

Finally, the eikonal $H(t, r_a)$ when $z_o \to \infty$ is given,

$$\lim_{z_i\to\infty} H(t, r_a) = \begin{cases} [z_o + t\cos\theta_1,\ t\sin\theta_1], & t < c_1, \\ \left[v(t - c_1)\cos\left(\lim_{z_i\to\infty}\theta_2\right),\right. \\ \left. v(t - c_1)\sin\left(\lim_{z_i\to\infty}\theta_2\right)\right], & c_1 < t < c_2, \\ \left[\lim_{z_i\to\infty} z_b(r_a) + (t - c_2)\cos\left(\lim_{z_i\to\infty}\theta_3\right),\right. \\ \left. \lim_{z_i\to\infty} r_b(r_a) + (t - c_2)\cos\left(\lim_{z_i\to\infty}\theta_3\right)\right], & t > c_2. \end{cases} \tag{9.71}$$

The condition c_1 stays as equation (9.45),

$$z_a(r_a) = t\cos(\theta_1) + z_o \Rightarrow c_1 = [z_a(r_a) - z_o], \tag{9.45}$$

but c_2 is modified too, equation (9.46) becomes

$$\begin{gathered} \lim_{z_o\to-\infty} z_b(r_a) = v(t - c_1)\cos\left(\lim_{z_o\to-\infty}\theta_2\right) \\ \Rightarrow c_2 = c_1 + \sec\left(\lim_{z_o\to-\infty}\theta_2\right)\frac{\left[\lim_{z_o\to-\infty} z_b(r_a) - z_a(r_a)\right]}{v}. \end{gathered} \tag{9.72}$$

To plot the eikonals when the object is virtual we use $H(t, r_a)$ as

$$\boxed{\lim_{z_i\to\infty} H(t, r_a) = \begin{cases} [z_o + (t - R)\cos\theta_1, (t - R)\sin\theta_1], & t < c_1, \\ \left[v(t - c_1)\cos\left(\lim_{z_i\to\infty}\theta_2\right),\right. & \\ \left. v(t - c_1)\sin\left(\lim_{z_i\to\infty}\theta_2\right)\right], & c_1 < t < c_2, \\ \left[\lim_{z_i\to\infty} z_b(r_a) + (t - c_2)\cos\left(\lim_{z_i\to\infty}\theta_3\right),\right. & \\ \left.\lim_{z_i\to\infty} r_b(r_a) + (t - c_2)\cos\left(\lim_{z_i\to\infty}\theta_3\right)\right], & t > c_2, \end{cases}} \tag{9.73}$$

where $R = z_o + n\tau$. The c_1 is given by

$$c_1 = (n\tau + z_o) + \sec\theta_1[z_a(r_a) - z_o], \tag{9.74}$$

and c_2 is given by

$$c_2 = c_1 + \frac{\sec\theta_2[z_b(r_a) - z_a(r_a)]}{v}. \tag{9.75}$$

The angles θ_1, θ_2 and θ_3 are given by equations (9.40), (9.69) and (9.70), respectively.

9.4.2 Gallery

This gallery has several collimator lenses free of spherical aberration. All the examples have been computed using (9.68). The details of each design are in the caption of the figure that corresponds to it. The figures of the gallery are 9.23–9.28.

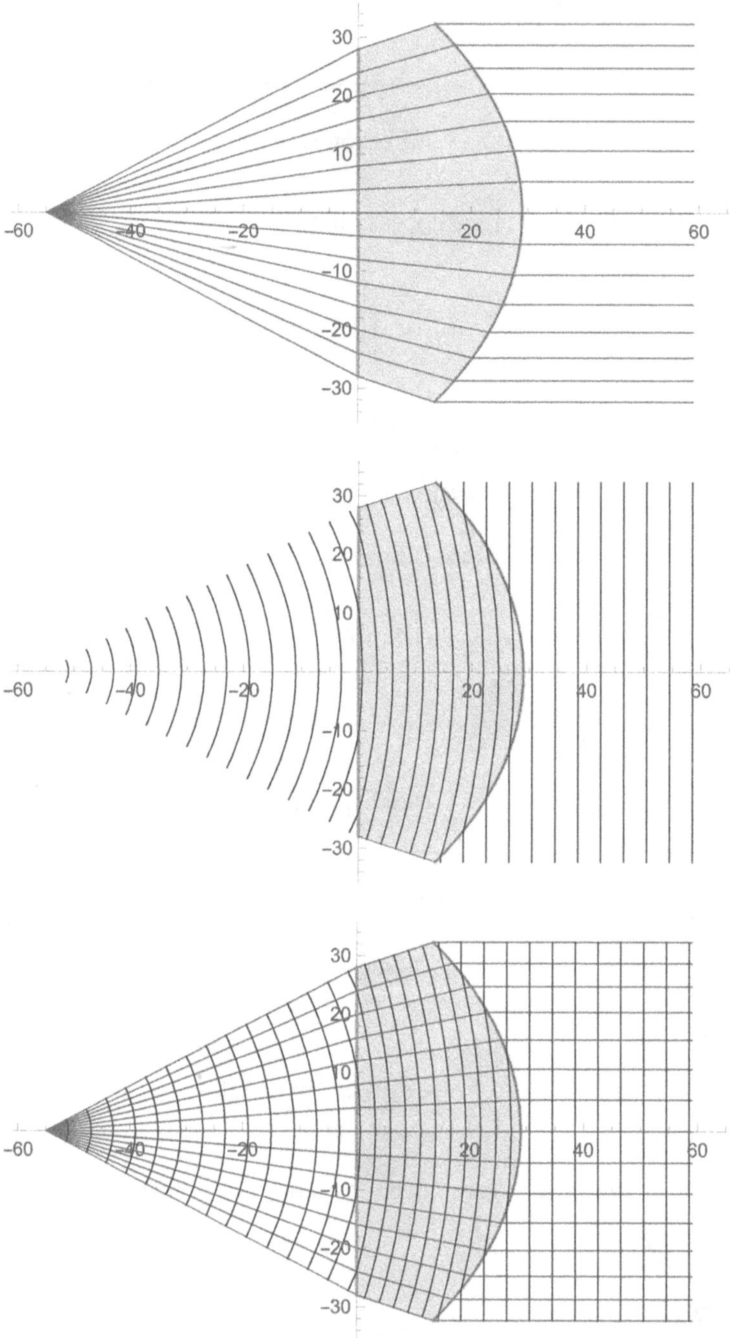

Figure 9.23. *Top:* rays, *center:* waves, *bottom:* rays and waves. Design specifications: $n_o = n_i = 1$, $n = 1.5$, $z_o = -55$ mm, $\tau = 29$ mm, $z_i \to \infty$, $z_a = 0$, $z_b =$ equation (9.68) and $H(t, r_a) =$ equation (9.71).

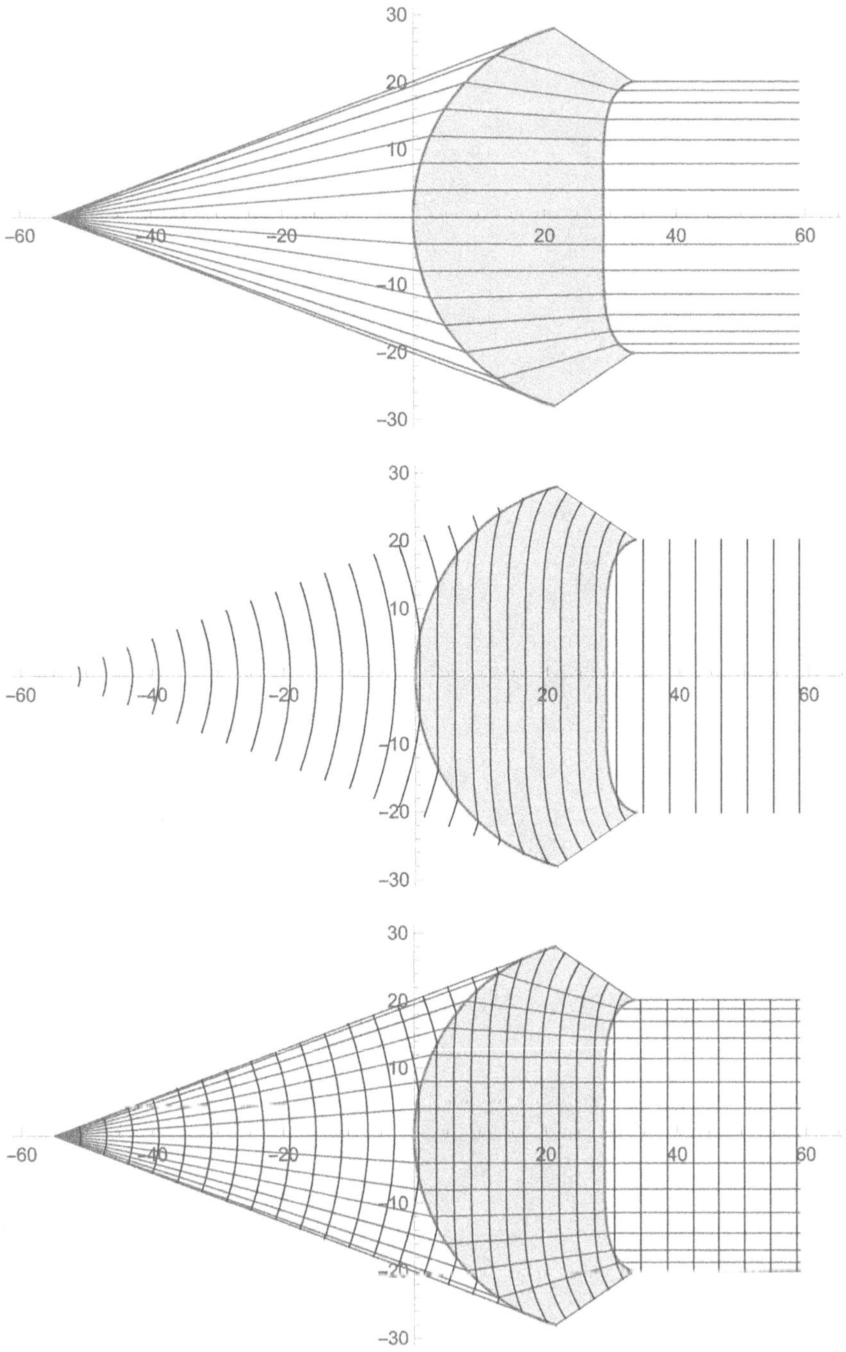

Figure 9.24. *Top:* rays, *center:* waves, *bottom:* rays and waves. Design specifications: $n_o = n_i = 1$, $n = 1.5$, $z_o = -55$ mm, $\tau = 29$ mm, $z_i \to \infty$, $z_a = 29 - \sqrt{29^2 - r_a^2}$, $z_b =$ equation (9.68) and $H(t, r_a) =$ equation (9.71).

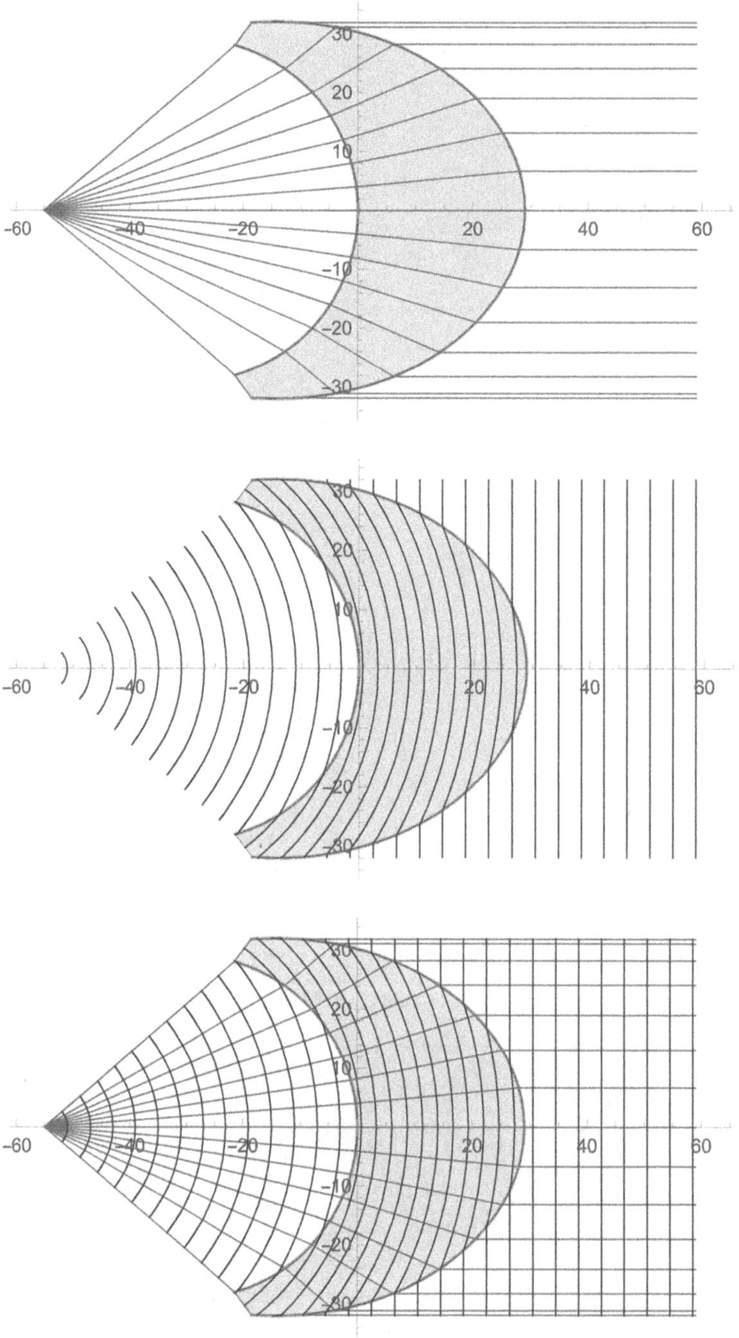

Figure 9.25. *Top:* rays, *center:* waves, *bottom:* rays and waves. Design specifications: $n_o = n_i = 1$, $n = 1.5$, $z_o = -55$ mm, $\tau = 29$ mm, $z_i \to \infty$, $z_a = -(29 - \sqrt{29^2 - r_a^2})$, $z_b =$ equation (9.68) and $H(t, r_a) =$ equation (9.71).

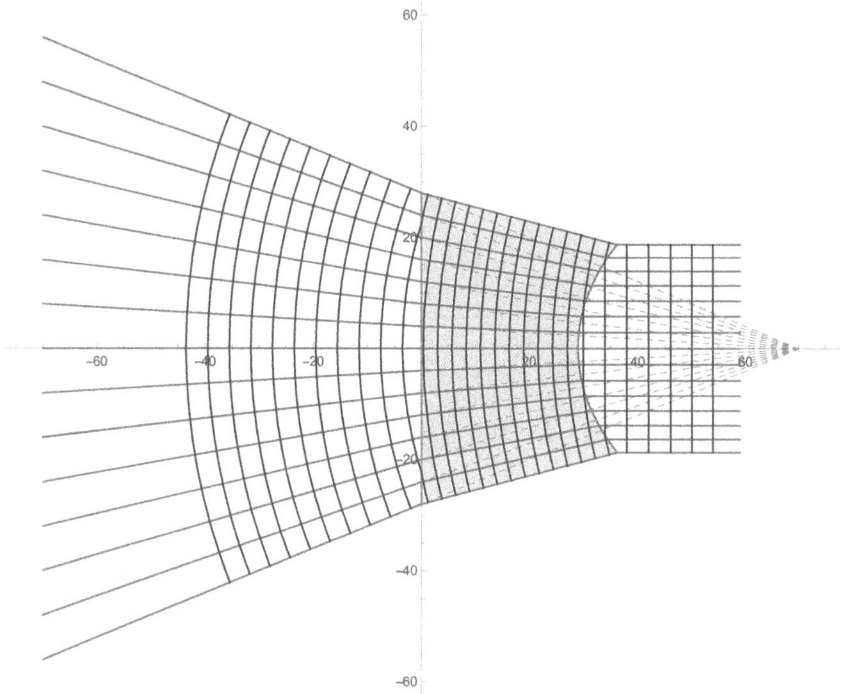

Figure 9.26. Design specifications: $n_o = n_i = 1$, $n = 1.5$, $z_o = 70$ mm, $\tau = 29$ mm, $z_i \to \infty$, $z_a = 0$, $z_b =$ equation (9.68) and $H(t, r_a) =$ equation (9.73).

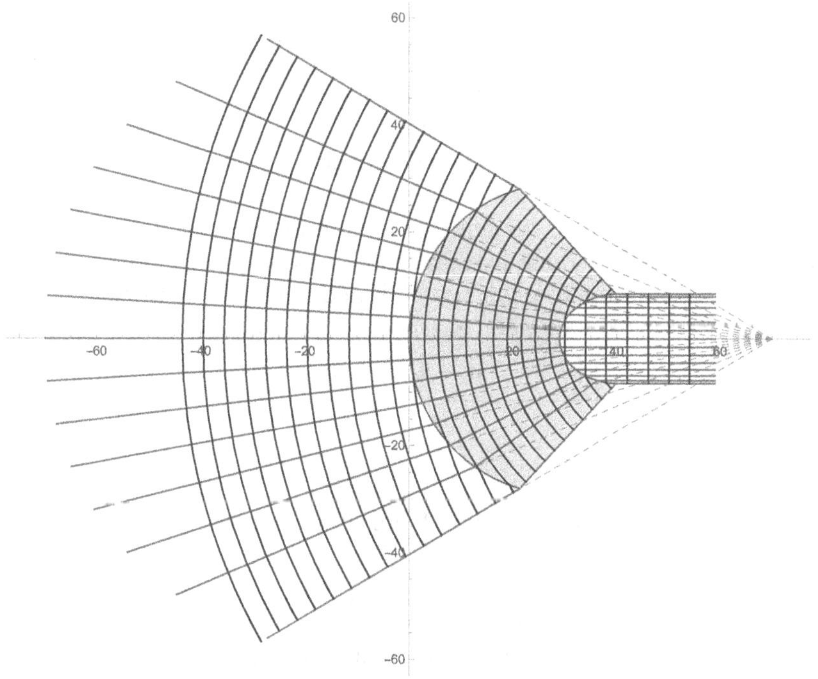

Figure 9.27. Design specifications: $n_o = n_i = 1$, $n = 1.5$, $z_o = 70$ mm, $\tau = 29$ mm, $z_i \to \infty$, $z_a = (29 - \sqrt{29^2 - r_a^2})$, $z_b =$ equation (9.68) and $H(t, r_a) =$ equation (9.73).

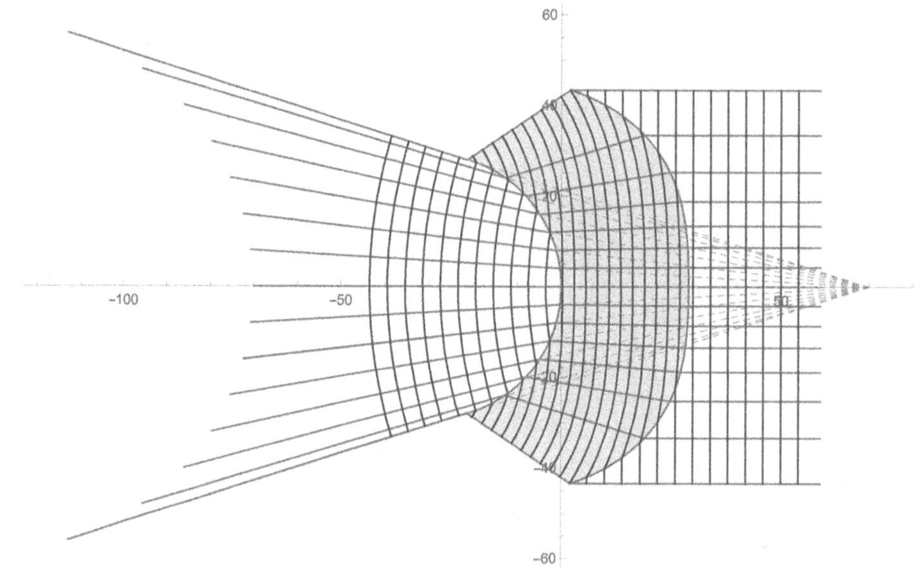

Figure 9.28. Design specifications: $n_o = n_i = 1$, $n = 1.5$, $z_o = 70$ mm, $\tau = 29$ mm, $z_i \to \infty$ $z_a = -(29 - \sqrt{29^2 - r_a^2})$, $z_b =$ equation (9.68) and $H(t, r_a) =$ equation (9.73).

9.5 The single-lens telescope

The last case is when $z_o \to -\infty$ and $z_i \to -\infty$. Thus, at the input/output the rays are collimated along the optical axis. We call this case the single-lens telescope.

To get the single-lens telescope general equation is simple. We need to compute the limit when $z_o \to -\infty$ in the cosine directors $\wp_r$, $\wp_z$ and in the parameters f equation (9.17), (9.18) and (9.24); or, respectively, use equations (9.50), (9.51) and (9.52).

Then, we need to compute the limit $z_i \to \infty$ in the Fermat principle of equation (9.6). But we have already done that procedure and only need to compute the limit $z_i \to \infty$ in ϑ of equation (9.67). Let's recall it and apply the mentioned limit over it,

$$\lim_{z_o \to -\infty}\left[\lim_{z_i \to \infty}(\vartheta)\right] = \lim_{z_o \to -\infty}\left[\frac{\left(-n_i + n\right)\tau - n_o z_o + n_i z_a + n_o \operatorname{sgn}(z_o)\sqrt{r_a^2 + (z_a - z_o)^2}}{n - n_i \wp_z}\right], \tag{9.76}$$

thus, when we apply the limit on $\lim_{z_o \to -\infty}\left[\lim_{z_i \to \infty}(\vartheta)\right]$, we have

$$\lim_{z_o \to -\infty}\left[\lim_{z_i \to \infty}(\vartheta)\right] = \frac{n_i(z_a - \tau) - z_a n_o + n\tau}{n - n_i\left[\lim_{z_o \to -\infty}(\wp_z)\right]}. \tag{9.77}$$

Thus, the second surface of the single-lens telescope is given by

$$\left\{\begin{aligned}\lim_{z_o\to-\infty}\left[\lim_{z_i\to\infty}(z_b)\right] &= z_a + \left\{\lim_{z_o\to-\infty}\left[\lim_{z_i\to\infty}(\vartheta)\right]\right\}\left[\lim_{z_o\to-\infty}(\wp_z)\right],\\ \lim_{z_o\to-\infty}\left[\lim_{z_i\to\infty}(r_b)\right] &= r_a + \left\{\lim_{z_o\to-\infty}\left[\lim_{z_i\to\infty}(\vartheta)\right]\right\}\left[\lim_{z_o\to-\infty}(\wp_r)\right].\end{aligned}\right. \tag{9.78}$$

Equation (9.78) corresponds to the second surface of the single-lens telescope if and only if the parameters $\left[\lim_{t_a\to-\infty}(\wp_r)\right]$, $\left[\lim_{t_a\to-\infty}(\wp_z)\right]$ and $\lim_{t_a\to-\infty}\left[\lim_{t_b\to\infty}(\vartheta)\right]$ are given by equations (9.50), (9.51) and (9.77), respectively.

Equation (9.78) is so simple; it is just a fraction as simple in comparison with the other cases when at least one or both object and image are finite. It is the general equation where every pair of surfaces that compose a single-lens telescope must have one surface at least with this shape.

9.5.1 The eikonal of the single-lens telescope

To get the eikonal of the single-lens telescope we need to apply a limit over the angles θ_1, θ_2, θ_3. The limits are when $z_i \to \infty$ and $z_o \to -\infty$.

Starting with θ_1, in equation (9.40) we apply the limits $z_i \to \infty$ and $z_o \to -\infty$ and we get

$$\lim_{z_i\to\infty}\left[\lim_{z_o\to-\infty}(\theta_1)\right] = \lim_{z_i\to\infty}\left\{\lim_{z_o\to-\infty}\arctan\left[\frac{r_a}{z_a(r_a) - z_o}\right]\right\} = 0. \tag{9.79}$$

Now we focus on θ_2, we apply the limits $z_i \to \infty$ and $z_o \to -\infty$ in equation (9.41),

$$\begin{aligned}\lim_{z_i\to\infty}\left[\lim_{z_o\to-\infty}(\theta_2)\right] &= \lim_{z_i\to\infty}\left\{\lim_{z_o\to-\infty}\arctan\left[\frac{r_b(r_a) - r_a}{z_b(r_a) - z_a(r_a)}\right]\right\}\\ &= \arctan\left\{\frac{\lim_{z_i\to\infty}\left[\lim_{z_o\to-\infty} r_b(r_a)\right] - r_a}{\lim_{z_i\to\infty}\left[\lim_{z_o\to-\infty} z_b(r_a)\right] - z_a(r_a)}\right\}.\end{aligned} \tag{9.80}$$

Then, in turn for θ_3, we apply the limits $z_i \to \infty$ and $z_o \to -\infty$ in equation (9.42),

$$\begin{aligned}\lim_{z_i\to\infty}\left[\lim_{z_o\to-\infty}(\theta_3)\right] &= \lim_{z_i\to\infty}\left\{\lim_{z_o\to-\infty}\arctan\left[\frac{r_b(r_a)}{z_b(r_a) - \tau - z_o}\right]\right\}\\ &= \arctan\left\{\frac{\lim_{z_i\to\infty}\left[\lim_{z_o\to-\infty} r_b(r_a)\right]}{\lim_{z_i\to\infty}\left[\lim_{z_o\to-\infty} z_b(r_a)\right] - \tau - z_o}\right\}.\end{aligned} \tag{9.81}$$

Therefore, the eikonal of the single-lens telescope $H(t, r_a)$ is given by

$$\lim_{z_i\to\infty}\left[\lim_{z_o\to-\infty} H(t, r_a)\right] = \begin{cases} [Z_o + t, r_a], & t < c_1, \\ \left[v(t - c_1)\cos\left(\lim_{z_i\to\infty}\left[\lim_{z_o\to-\infty}\theta_2\right]\right), \right. & \\ \left. v(t - c_1)\sin\left(\lim_{z_i\to\infty}\left[\lim_{z_o\to-\infty}\theta_2\right]\right)\right], & c_1 < t < c_2, \\ \left[\lim_{z_i\to\infty}\left[\lim_{z_o\to-\infty} z_b(r_a)\right] + (t - c_2)\cos\left(\lim_{z_i\to\infty}\left[\lim_{z_o\to-\infty}\theta_3\right]\right), \right. & \\ \left. \lim_{z_i\to\infty}\left[\lim_{z_o\to-\infty} r_b(r_a)\right] + (t - c_2)\cos\left(\lim_{z_i\to\infty}\left[\lim_{z_o\to-\infty}\theta_3\right]\right)\right], & t > c_2, \end{cases} \tag{9.82}$$

where Z_o is just a negative constant where the plot starts. The condition c_1 becomes

$$z_a(r_a) = t\cos\left(\lim_{z_i\to\infty}\left[\lim_{z_o\to-\infty}(\theta_1)\right]\right) + Z_o \Rightarrow c_1 = [z_a(r_a) - Z_o], \tag{9.83}$$

and c_2,

$$\lim_{z_i\to\infty}\left[\lim_{z_o\to-\infty} z_b(r_a)\right] = v(t - c_1)\cos\left(\lim_{z_i\to\infty}\left[\lim_{z_o\to-\infty}(\theta_2)\right]\right) \Rightarrow$$
$$c_2 = c_1 + \sec\left(\lim_{z_i\to\infty}\left[\lim_{z_o\to-\infty}(\theta_2)\right]\right)\frac{\lim_{z_i\to\infty}\left[\lim_{z_o\to-\infty} z_b(r_a)\right] - z_a(r_a)}{v}. \tag{9.84}$$

In the following section, we are going to show several eikonals and single-lens telescopes.

9.5.2 Gallery

Examples of single-lens telescopes are presented in the following gallery. Each figure shows the input values with the respective ray tracing. The figures of the gallery are 9.29–9.31.

9.6 End notes

In this chapter, we have demonstrated that all stigmatic singlet lenses that exist for real/virtual objects and real/virtual images are in a single equation, equation (9.39). We found the general equation that describes them; we also found that it is unique.

We tested several examples for several configurations: when the object is finite/infinite, and the image is finite/infinite. This is not only when the first surfaces are conics, but also when the first surfaces are other continuous functions. In all the cases, we obtained the expected results, when the rays do not cross each other.

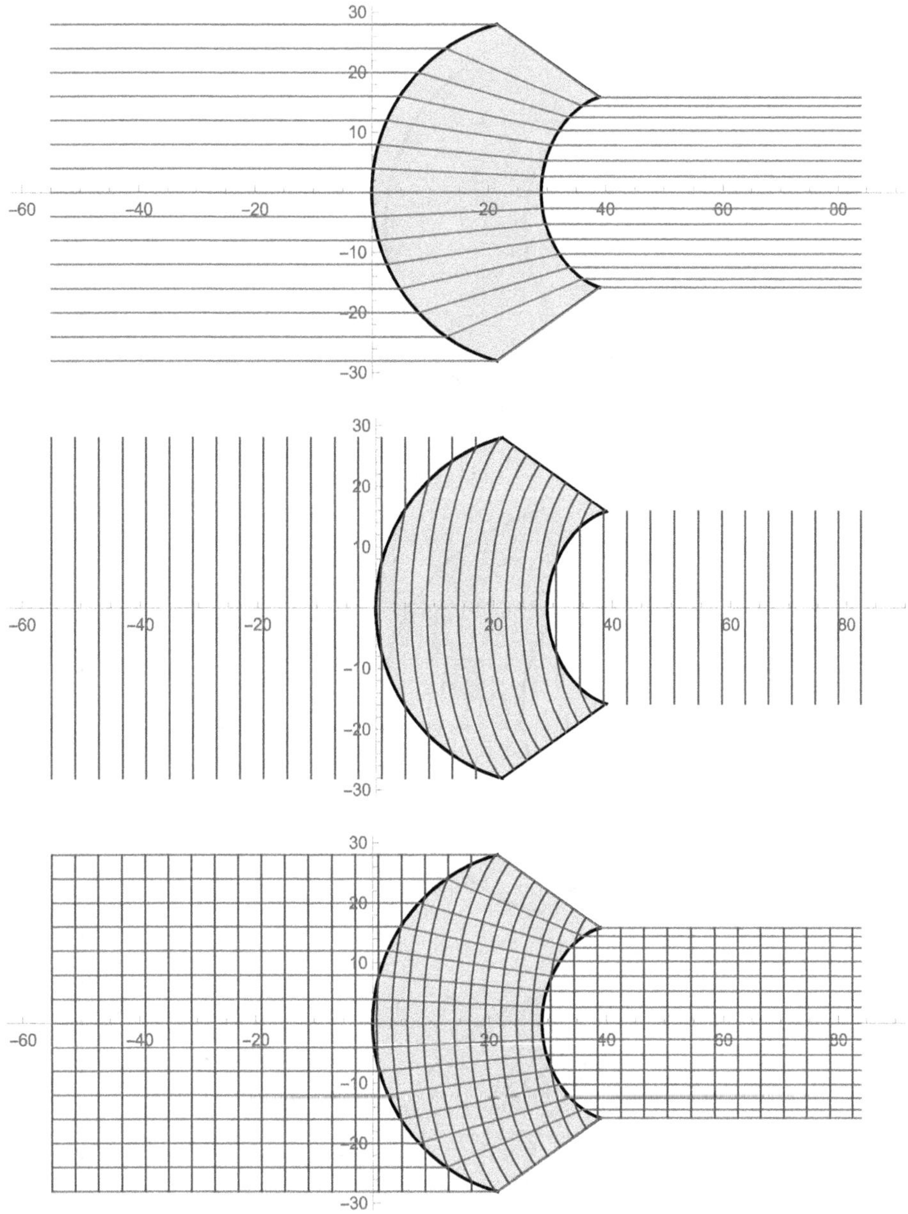

Figure 9.29. Design specifications: $n_o = n_i = 1$, $n = 1.5$, $z_o = -\infty$, $\tau = 29$ mm, $z_i \to \infty$, $z_a = 29 - \sqrt{29^2 - r_a^2}$, z_b = equation (9.78) and $H(t, r_a)$ = equation (9.82).

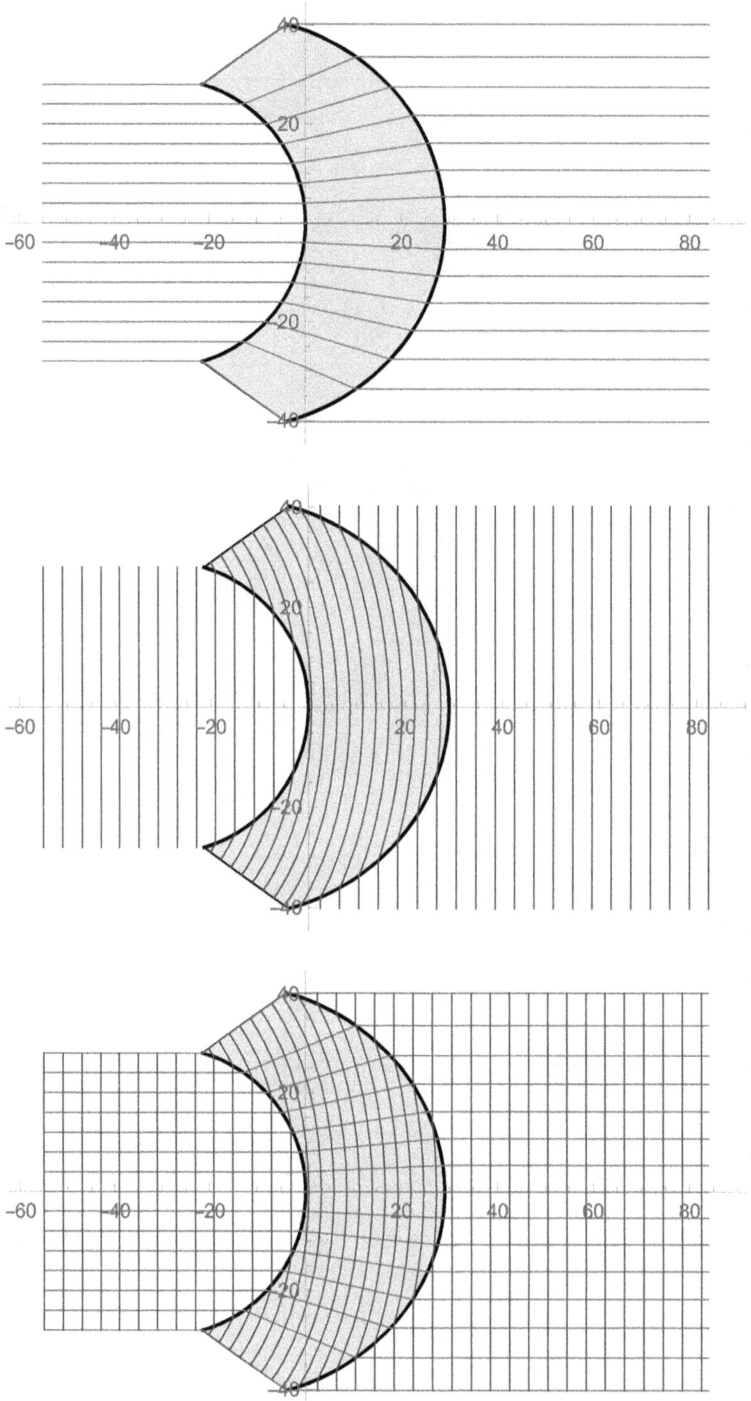

Figure 9.30. Design specifications: $n_o = n_i = 1$, $n = 1.5$, $z_o = -\infty$, $\tau = 29$ mm, $z_i \rightarrow \infty$, $z_a = -(29 - \sqrt{29^2 - r_a^2})$, $z_b =$ equation (9.78) and $H(t, r_a) =$ equation (9.82).

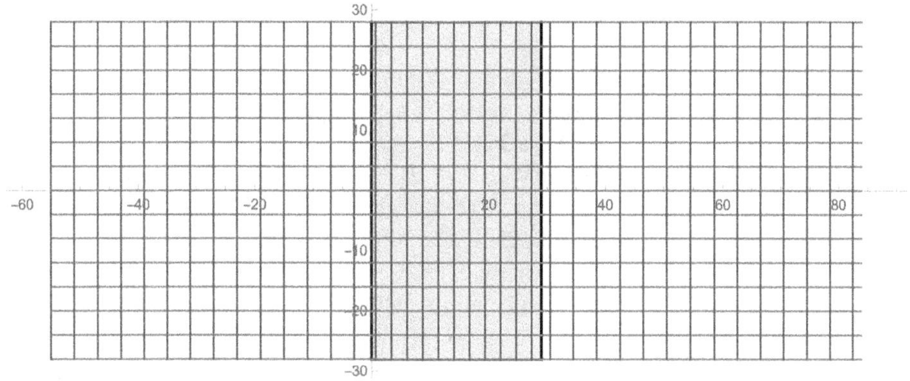

Figure 9.31. Design specifications: $n_o = n_i = 1$, $n = 1.5$, $z_o = -\infty$, $\tau = 29\,\text{mm}$, $z_i \to \infty$, $z_a = 0$, $z_b =$ equation (9.78) and $H(t, r_a) =$ equation (9.82).

References

Born M and Wolf E 2013 Principles of Optics: Electromagnetic Theory of Propagation *Interference and Diffraction of Light* (Amsterdam: Elsevier)

Chaves J 2016 *Introduction to Nonimaging Optics* 2nd edn (Boca Raton, FL: CRC Press)

Descartes R 1637a *De la nature des lignes courbes*

Descartes R 1637b *La Géométrie*

González-Acuña R G 2021 Surface solution to correct a freeform wavefront *Appl. Opt.* **60** 9887–91

González-Acuña R G and Chaparro-Romo H A 2018 General formula for bi-aspheric singlet lens design free of spherical aberration *App. Opt.* **57** 9341–5

González-Acuña R G and Guitiérrez-Vega J C 2018 Generalization of the axicon shape: the gaxicon *J. Opt. Soc. Am.* **A35** 1915–8

González-Acuña R G and Gutiérrez-Vega J C 2019 Analytic formulation of a refractive-reflective telescope free of spherical aberration *Opt. Eng.* **58** 085105

González-Acuña R G and Thibault S 2021 The general equation of the stigmatic lenses: its history and what we have learned from it *Int. Optical Design Conf.* (Washington, DC: Optical Society of America) 1207803 p

González-Acuña R G, Avendaño-Alejo M and Gutiérrez-Vega J C 2019a Singlet lens for generating aberration-free patterns on deformed surfaces *J. Opt. Soc. Am.* **A36** 925–9

González-Acuña R G, Chaparro-Romo H A and Gutiérrez-Vega J C 2019b General formula to design freeform singlet free of spherical aberration and astigmatism *Appl. Opt.* **58** 1010–5

González-Acuña R G, Chaparro-Romo H A and Gutiérrez-Vega J C 2020a Analytic aplanatic singlet lens: setting and design for three-point objects and images in the meridional plane *Opt. Eng.* **59** 055104

González-Acuña R G, Chaparro-Romo H A and Gutiérrez-Vega J C 2020b Analytic solution of the eikonal for a stigmatic singlet lens *Phys. Scr.*

González-Acuña R G, Chaparro-Romo H A and Gutiérrez-Vega J C 2020c *Analytical Lens Design* (Bristol: Institute of Physics Publishing)

González-Acuña R G, Chaparro-Romo H A and Gutiérrez-Vega J C 2021 General stigmatic surfaces *J. Opt. Soc. Am.* **A38** 298–302

González-Acuña R G, Chaparro-Romo H A and Gutíerrez-Vega J C 2019c *Single Lens Telescope* (arXiv:1903.11129)

González-Acuña R G and Gutiérrez-Vega J C 2019b General formula to eliminate spherical aberration produced by an arbitrary number of lenses *Opt. Eng.* **58** 1–6

González-Acuña R G and Gutiérrez-Vega J C 2019 General formula for aspheric collimator lens design free of spherical aberration *Current Developments in Lens Design and Optical Engineering XX* ed R B Johnson, V N Mahajan and S Thibault (Bellingham, WA: International Society for Optics and Photonics, SPIE) 181–4 pp

González Acuña R G and Gutiérrez-Vega J C 2019 General formula to design freeform collimator lens free of spherical aberration and astigmatism *Novel Optical Systems, Methods, and Applications XXII* **vol 11 105** (Bellingham, WA: International Society for Optics and Photonics) 111050A p

González Acuña R G and Gutiérrez-Vega J C 2019 General formula of the refractive telescope design free spherical aberration *Novel Optical Systems, Methods, and Applications XXII* **vol 11 105** ed C F Hahlweg and J R Mulley (Bellingham, WA: International Society for Optics and Photonics, SPIE) 162–6 p

Huygens C 1690 *Traité de la lumière*

Kingslake R and Johnson R B 2009 *Lens Design Fundamentals* (New York: Academic)

Luneburg R K 1964 *Mathematical Theory of Optics* (Berkeley, CA: University of California Press)

González-Acuña R G and Gutiérrez-Vega J C 2020 Analytic design of a spherochromatic singlet *J. Opt. Soc. Am.* **A37** 149–53

Newton I 1704 *Opticks: or, a Treatise of the Reflections, Refractions, Inflections & Colours of Light*

Silva-Lora A and Torres R 2023 Primary aberrations theory in optical systems composed of Cartesian refracting surfaces *Proc. R. Soc.* **A479** 20230186

Silva-Lora A and Torres R 2020 Explicit Cartesian oval as a superconic surface for stigmatic imaging optical systems with real or virtual source or image *Proc. R. Soc.* A476

Stavroudis O 2012 The Optics of Rays *Wavefronts, and Caustics* **vol 38** (Amsterdam: Elsevier)

Stavroudis O N 2006 *The Mathematics of Geometrical and Physical Optics: The K-function and Its Ramifications* (New York: Wiley)

Stavroudis O N and Feder D P 1954 Automatic computation of spot diagrams *J. Opt. Soc. Am.* **44** 163–70

Wassermann G D and Wolf E 1949 On the theory of aplanatic aspheric systems *Proc. Phys. Soc.* **B62** 2

Winston R, Miñano J C and Benitez P G *et al* 2005 *Nonimaging Optics* (Amsterdam: Elsevier)

Wolf E 1948 On the designing of aspheric surfaces *Proc. Phys. Soc.* **61** 494

Wolf E and Preddy W S 1947 On the determination of aspheric profiles *Proc. Phys. Soc.* **59** 704

Chapter 10

Aberrations in Cartesian ovals

We determine the aberration coefficients of Cartesian ovals. These coefficients describe the magnitude of spherical aberration, coma, astigmatism, field curvature and distortion that the Cartesian oval generates when the object does not match the stigmatic points.

10.1 Introduction

We have mentioned in the first chapter of this book that geometric aberrations refer to a characteristic of optical systems, including lenses, where light is dispersed across a certain area in space instead of being concentrated to a singular focal point. Aberrations are one of the primary shortcomings observed in optical systems.

The five fundamental aberration types, arising from the geometry of lenses or mirrors and relevant to systems working with monochromatic light, are termed Seidel aberrations. Coined in honor of Ludwig von Seidel's 1857 paper, these aberrations manifest in third-order optics, commonly referred to as Seidel optics.

The five Seidel aberrations are as follows: spherical aberration impacts rays originating from a point on the optical axis. Due to the varying paths, these rays pass through different sections of the lens, especially in the case of a spherical lens or one not precisely shaped to converge them, they fail to focus at a common point on the opposite side of the lens.

Coma affects rays from off-axis points. If spherical aberration is mitigated, different parts of the lens converge rays from the axis to a common focus. However, the location where the image of an off-axis point is formed may still change when different sections of the lens are considered.

Astigmatism is the aberration that influences rays from a point off the optical axis. As these rays traverse the lens towards their focal point in the image, they pass through a lens that, from their perspective, is tilted. Even if spherical aberration and coma are addressed, the rays in the plane of the tilt and those perpendicular to it may

doi:10.1088/978-0-7503-6423-2ch10

pass through distinct lens profiles, resulting in focal points at different distances from the lens.

Curvature of field is generated when light from every object point is accurately focused, the points of convergence may lie on a curved surface rather than a flat plane.

Despite correcting the aforementioned aberrations, the light from object points may converge on the image plane at an incorrect distance from the optical axis, deviating from a linear proportionality to the distance in the object. This aberration is called distortion. If the increase is faster than in the object, it leads to pincushion distortion; if slower, barrel distortion.

Classically, the five Seidel aberrations have been described in the context of spherical and aspherical surfaces. In this chapter, we are going to explore the Seidel aberrations generated by a Cartesian oval since we now have the general closed-form expression. For that, we are going to obtain closed-form mathematical expressions to describe each of them.

10.2 A change of notation for Cartesian ovals

In the realm of geometric optics, Descartes' ovoids or Cartesian surfaces are capable of flawlessly imaging a point source. We have studied these special kinds of surfaces in chapters 7 and 8. According to the notation given in chapter 8, we can express the Cartesian ovals as

$$z_a = \frac{\rho^2(c_{1_a}\rho^2 + c_{0_a})}{\sqrt{\rho^2\left(2b_{1_a} - c_{0_a}^2K_a\right) + 1} + b_{1_a}\rho^2 + 1}, \tag{8.1}$$

and

$$r_a = \operatorname{sgn}(\rho)\sqrt{\rho^2 - z_a(\rho)^2}, \tag{8.2}$$

where K_a is given by

$$K_a = \frac{\left(n^2 z_{in} - n_o^2 z_o\right)^2}{nn_o(nz_{in} - n_o z_o)(nz_o - n_o z_{in})}, \tag{8.3}$$

the expression for c_{0_a} is

$$c_{0_a} = \frac{nz_o - n_o z_{in}}{z_{in} z_o(n - n_o)}, \tag{8.4}$$

the expression for c_{1_a} is

$$c_{1_a} = \frac{(n - n_o)(n + n_o)^2}{4nn_o z_{in} z_o(nz_{in} - n_o z_o)} \tag{8.5}$$

and

$$b_{1_a} = \frac{(n + n_o)(n^2 z_{in} - n_o^2 z_o)}{2nn_o z_{in} z_o(nz_{in} - n_o z_o)}. \tag{8.6}$$

The above expressions are very useful for particular cases when the object is placed at the stigmatic points of the Cartesian oval and also when it is intended to design a lens with two surfaces that correspond to Cartesian ovals.

For our case, we want to see the aberration generated by a Cartesian oval when the object is at any point, thus the formulation above is not necessarily the most convenient for our purposes.

The following formulation for the Cartesian oval expresses the same surface, the Cartesian oval. But in a more convenient way for our goals in this chapter,

$$z_0 - \zeta_0 = \frac{1}{OG}\left(1 + S\rho^2 - \sqrt{1 + (2S - O^2G)\rho^2}\right), \tag{10.1}$$

where G, O and S are parameters. G can be seen as a generalization of the conic constant, or generalized Schwarzschild constant,

$$G = \frac{\left[n_1^2/(d_0 - \zeta_0) - n_0^2/(d_1 - \zeta_0)\right]^2}{n_0 n_1[n_1/(d_1 - \zeta_0) - n_0/(d_0 - \zeta_0)][n_1/(d_0 - \zeta_0) - n_0/(d_1 - \zeta_0)]}, \tag{10.2}$$

O is the paraxial curvature of the surface described by z_0.

$$O = \frac{n_1/(d_1 - \zeta_0) + n_0/(d_0 - \zeta_0)}{n_1 - n_0}, \tag{10.3}$$

and S is defined as

$$S = \frac{[(n_1 + n_0)/((d_1 - \zeta_0)(d_0 - \zeta_0))][n_1^2/(d_0 - \zeta_0) - n_0^2/(d_1 - \zeta_0)]}{2n_1 n_0[n_1/(d_0 - \zeta_0) - n_0/(d_1 - \zeta_0)]}. \tag{10.4}$$

S can be seen as the subfamily of Cartesian surfaces.

Please take the time to analyze figure 10.1, which shows a refracting surface with the parameters presented in equation (10.1). Let $\mathcal{O}$ be the origin of the coordinate system. Let ζ_0 be the axial distance from the vertex of the surface V_0 and the origin

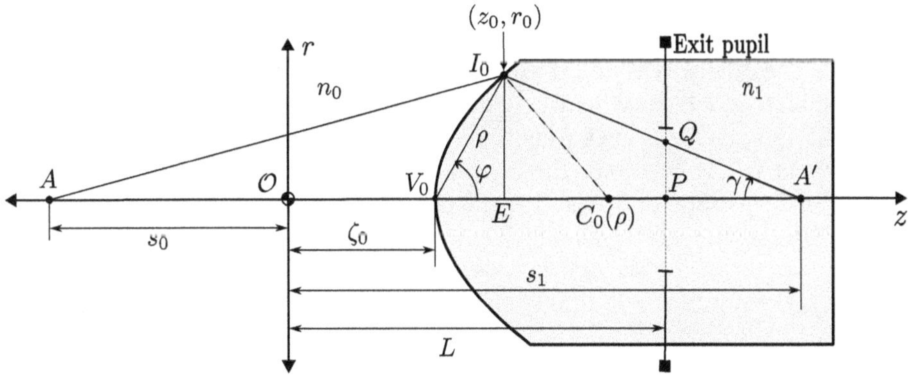

Figure 10.1. Refracting surface of a Cartesian oval. A ray emanates from a point object A, undergoes refraction at point I_0, and intersects with the optical axis at a new point, denoted A'.

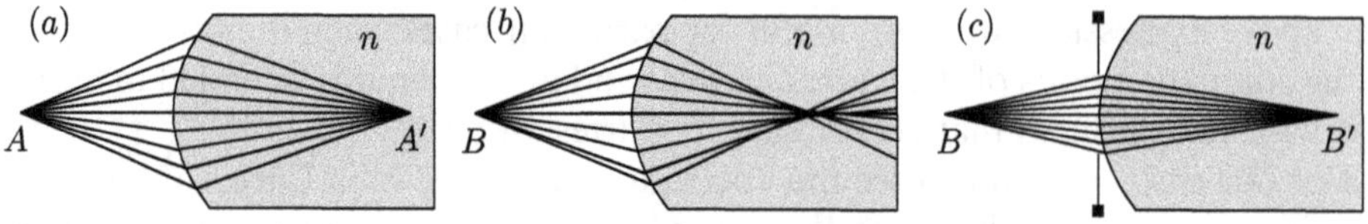

Figure 10.2. (a) A precisely stigmatic system that accurately converges all rays originating from point A. (b) In cases where the point object is not located at a stigmatic point, the surface is unable to produce a point image. (c) Stigmatism can be approximately restored by employing a physical aperture stop, restricting the luminous field accessible to the system.

$\mathcal{O}$. Thus, $\zeta_0 = \overline{\mathcal{O}V_0}$. In the new notation, the refraction indices are n_0 and n_1, and they separate the object and image space.

Observe that from figure 10.1, The distance between the vertex and the point I_0 is given by ρ. Now if this particular surface is capable of producing an ideal point image at A' from a point object in A, it is recognized as a Cartesian surface with distances to the object and image points, denoted s_0 and s_1, coinciding with the stigmatic points, d_0 and d_1. Notice that varying values of s_0 and s_1 give rise to geometric aberrations since the object will no be longer aligned with the stigmatic points d_0 and d_1.

Please notice that if the vertex of the Cartesian oval is placed in the origin V_0, then $\zeta_0 = \overline{\mathcal{O}V_0} = 0$ and s_0 and s_1 coincide with d_0 and d_1. Then, equation (10.1) is reduced to the expressions presented for the Cartesian oval in chapters 7 and 8. The alignment we are talking about is described in figure 10.2. In figure 10.2, (a) shows an optically precise stigmatic system that precisely converges all rays emanating from point A. (b) If the point object is not positioned at a stigmatic point, the surface fails to generate a point image. (c) Stigmatism can be approximately reinstated by utilizing a physical aperture stop to limit the luminous field accessible to the system. Notice that (c) is just an approximation, thus the spherical aberration is not zero.

10.3 On-axis aberrations

To obtain the expression of the on-axis aberration for a Cartesian oval we need to study figure 10.1 in detail. In this figure there are two rays. The first one is the axial ray $\overline{AV_0A'}$. The other ray is an arbitrary ray traveling with the path given by $\overline{AI_0A'}$. When points A and B align with the stigmatic points, all rays emanating from A converge at a shared focus in B, resulting in zero spherical aberration. In cases where this alignment is not achieved, the computation of spherical aberrations becomes necessary. Let's assume that A and B are not aligned with the stigmatic points and we need to compute the spherical aberration. In a broader sense, the spherical aberration and the rest of the aberrations are just expressions of the optical path; the difference is the optical path difference between $\overline{AV_0A'}$ and $\overline{AI_0A'}$ is as follows,

$$W_0 = [AI_0A'] - [Av_0A'] = n_0\overline{AI_0} + n_1\overline{I_0A'} - n_0\overline{AV_0} - n_1\overline{V_0A'}, \qquad (10.5)$$

where square brackets are employed to denote the association with the optical path length. Aberration functions are commonly formulated based on the transversal coordinate $r_0 = \overline{EI_0}$. However, in this study, we will express them in terms of the vertex-surface distance $\rho = \overline{V_0 I_0}$ (refer to figure 10.1). This approach simplifies the representation of aberrations concerning the exit pupil position L and the transversal coordinate of the pupil $r = PQ$.

Figure 10.1 illustrates a ray originating from a point object A, positioned at a distance s_0. This ray undergoes refraction at an optical surface at point I_0, proceeds through point Q on the exit pupil, and ultimately intersects the optical axis at point A', located at a distance s_1. The optical refracting surface separates two media with refractive indices n_0 and n_1, with its vertex positioned at a distance ζ_0. The parameter L represents the exit pupil position, and C_0 marks the intersection point of a line normal to point I_0 concerning the optical axis. In this context, considering the vertex-surface distance as ρ, we can express the ray paths in the object space (AI_0) and in the image space (I_0A') as follows. Utilizing the cosine theorem on triangle δAI_0V_0, the path AI_0 can be determined,

$$\overline{AI_0}^2 = (s_0 - \zeta_0)^2 + \rho^2 + 2(s_0 - \zeta_0)\rho\cos(\pi - \varphi). \tag{10.6}$$

This can be expressed as

$$\overline{AI_0}^2 = (s_0 - \zeta_0)^2\left[1 + \frac{\rho^2}{(s_0 - \zeta_0)^2} - \frac{2\rho\cos\varphi}{(s_0 - \zeta_0)}\right]. \tag{10.7}$$

Referring to figure 10.1, we find that $\rho\cos\varphi = z_0 - \zeta_0$, allowing us to express

$$\overline{AI_0}^2 = (s_0 - \zeta_0)^2\left[1 + \frac{\rho^2}{(s_0 - \zeta_0)^2} - \frac{2(z_0 - \zeta_0)}{(s_0 - \zeta_0)}\right]. \tag{10.8}$$

From this, the geometric path AI_0 can be expressed as

$$\overline{AI_0} = -(s_0 - \zeta_0)\left[1 + \frac{\rho^2}{(s_0 - \zeta_0)^2} - \frac{2(z_0 - \zeta_0)}{s_0 - \zeta_0}\right]^{1/2}, \tag{10.9}$$

where $|s_0 - \zeta_0| = -(s_0 - \zeta_0)$. The same procedure involving the cosine theorem can be obtained for $\overline{I_0A'}$,

$$\overline{I_0A'} = (s_1 - \zeta_0)\left[1 + \frac{\rho^2}{(s_1 - \zeta_0)^2} - \frac{2(z - \zeta_0)}{s_1 - \zeta_0}\right]^{1/2}. \tag{10.10}$$

Now the next step is to replace equation (10.9) and equation (10.10) in equation (10.5). But first, let's consider the Cartesian oval to be a power series expansion with respect to ρ,

$$z_0 - \zeta_0 = \frac{O}{2}\rho^2 + \frac{(2S - O^2G)^2}{8GO}\rho^4 + \mathcal{O}(\rho^5), \tag{10.11}$$

where $\mathcal{O}(\rho^5)$ is the symbol to represent all terms from the fifth power and beyond. The above expression is the fourth-order approximation of the Cartesian oval. Then, we replace the fourth-order approximation of the Cartesian oval in equation (10.9),

$$\overline{AI_0} \approx -(s_0 - \zeta_0)\left[1 + \frac{1}{s_0 - \zeta_0}\left(\frac{1}{s_0 - \zeta_0} - O\right)\rho^2 - \frac{(2S - O^2G)^2}{4GO(s_0 - \zeta_0)}\rho^4\right]^{1/2}, \tag{10.12}$$

and in equation (10.10)

$$\overline{I_0A'} \approx (s_1 - \zeta_0)\left[1 + \frac{1}{s_1 - \zeta_0}\left(\frac{1}{s_1 - \zeta_0} - O\right)\rho^2 - \frac{(2S - O^2G)^2}{4GO(s_1 - \zeta_0)}\rho^4\right]^{1/2}. \tag{10.13}$$

Now, we expand equation (10.12) considering only the terms up to a fourth-order,

$$\overline{AI_0} \approx -(s_0 - \zeta_0) - \frac{1}{2}\left(\frac{1}{s_0 - \zeta_0} - O\right)\rho^2 - \left[\frac{1}{8(s_0 - \zeta_0)}\left(\frac{1}{s_0 - \zeta_0} - O\right)^2 + \frac{(2S - O^2G)^2}{8GO}\right]\rho^4. \tag{10.14}$$

The same procedure applies for equation (10.13),

$$\overline{I_0A'} \approx -(s_1 - \zeta_0) - \frac{1}{2}\left(\frac{1}{s_1 - \zeta_0} - O\right)\rho^2 - \left[\frac{1}{8(s_1 - \zeta_0)}\left(\frac{1}{s_1 - \zeta_0} - O\right)^2 + \frac{(2S - O^2G)^2}{8GO}\right]\rho^4. \tag{10.15}$$

After this process, it is time to replace equation (10.12) and equation (10.13) in equation (10.5) and using $\overline{AV_0} = -(s_0 - \zeta_0)$ and $\overline{V_0A'} = -(s_1 - \zeta_0)$,

$$\begin{aligned} W_0 = {} & \frac{1}{2}\left[n_1\left(\frac{1}{s_1 - \zeta_0} - O\right) - n_0\left(\frac{1}{s_0 - \zeta_0} - O\right)\right]\rho^2 \\ & - \frac{1}{8}\left[\frac{n_1}{s_1 - \zeta_0}\left(\frac{1}{s_1 - \zeta_0} - O\right)^2 - \frac{n_0}{s_0 - \zeta_0}\left(\frac{1}{s_0 - \zeta_0} - O\right)^2\right]\rho^4 \\ & - \frac{(n_1 - n_0)(2S - O^2G)^2}{8GO}\rho^4. \end{aligned} \tag{10.16}$$

Observe the first term of the last expression. It is the Abbe invariant for Descartes' ovoids in a paraxial regime. Since that last expression corresponds to the optical path difference concerning the prediction of paraxial optics, then it must be zero for being the paraxial invariant,

$$n_1\left(\frac{1}{s_1 - \zeta_0} - O\right) - n_0\left(\frac{1}{s_0 - \zeta_0} - O\right) = 0. \tag{10.17}$$

Therefore, the optical path difference W_0 can be written as

$$W_0 = a_q\rho^4, \tag{10.18}$$

where

$$a_q = -\frac{1}{8}\left[\frac{n_1}{s_1-\zeta_0}\left(\frac{1}{s_1-\zeta_0}-O\right)^2 - \frac{n_0}{s_0-\zeta_0}\left(\frac{1}{s_0-\zeta_0}-O\right)^2\right] - \frac{(n_1-n_0)(2S-O^2G)^2}{8GO}, \tag{10.19}$$

a_q is the coefficient that measures the contribution to spherical aberration by a Descartes' ovoid. Notice that we can express a_q in two terms, as follows,

$$a_q = a_s + \alpha_0. \tag{10.20}$$

Let a_s be the coefficient of spherical aberration for spherical refracting surfaces,

$$a_s = -\frac{n_1^2}{8}\left(\frac{1}{s_1-\zeta_0}-O\right)^2\left[\frac{1}{n_1(s_1-\zeta_0)}-\frac{1}{n_0(s_0-\zeta_0)}\right], \tag{10.21}$$

and let α_0 be the coefficient of the contribution to the spherical aberration given by the phase difference between the spherical and Cartesian surfaces,

$$\alpha_0 = -\frac{(n_1-n_0)(2S-O^2G)^2}{8GO}. \tag{10.22}$$

Keep in mind that the distances s_0 and s_1 refer to the object and image distances, respectively. These distances may not align with the distances of the stigmatic points on the Cartesian surface, denoted d_0 and d_1, where spherical aberration is zero. So, replacing equations (10.2), (10.3) and (10.4) in a_s

$$a_s = -\frac{n_1[n_0/(s_1-\zeta_0)-n_1/(s_0-\zeta_0)][n_0(d_0-s_1)/(d_0-\zeta_0)-n_1(d_1-s_1)/(d_1-\zeta_0)]^2}{8n_0(n_0-n_1)^2(s_1-\zeta_0)^2}, \tag{10.23}$$

and in α_0,

$$\alpha_0 = \frac{n_0 n_1(d_0-d_1)^2[n_0/(d_1-\zeta_0)-n_1/(d_0-\zeta_0)]}{8(n_0-n_1)^2(d_0-\zeta_0)^2(d_1-\zeta_0)^2}. \tag{10.24}$$

Now, if the object and the image are just placed in the stigmatic points, this means that we assume that for this particular case $s_0 = d_0$ and $s_1 = d_1$, then a_s is reduced to

$$a_s = -\frac{n_0 n_1(d_0-d_1)^2[n_0/(d_1-\zeta_0)-n_1/(d_0-\zeta_0)]}{8(n_0-n_1)^2(d_0-\zeta_0)^2(d_1-\zeta_0)^2}, \tag{10.25}$$

which gives a predicted interesting result,

$$a_s = -\alpha_0, \tag{10.26}$$

thus,

$$a_q = a_s + \alpha_0. \tag{10.27}$$

The system is free of spherical aberration if and only if the object and the image are just placed in the stigmatic points, $s_0 = d_0$ and $s_1 = d_1$. It is a very interesting result since we are here just considering the fourth-order approximation of the Cartesian oval. For sure, the Cartesian oval with an object and image in its stigmatic points is free of spherical aberration.

10.4 Off-axis aberrations

While it is crucial to assess on-axis aberrations for a comprehensive understanding, evaluating a system's imaging performance with extended objects requires the characterization of off-axis aberrations. When dealing with off-axis point objects, the initial step involves computing the disparity in sag between a spherical surface and an aspheric surface. Subsequently, the additional phase introduced in the aberration function needs to be determined.

In this section, we focus on the off-axis aberrations. Figure 10.3 shows the trajectory of three rays emanating from an off-axis point object, refracted by a Cartesian refracting surface. The first ray is given by $\overline{BV'VB'}$. The second ray is the chief ray and its path is $\overline{BI_1'I_1B'}$. Finally, the third ray is the marginal ray whose path is $\overline{BI_0'I_0B'}$. The initiation of the aberration function formulation for the Cartesian refracting surface begins with addressing the supplementary impact arising from phase differences at $\overline{I_0I_0'}$ and $\overline{I_1I_1'}$, compared with the travels of the same rays when the surface is a spherical surface, from where it is obtained

$$\Delta W_q = -n_0\overline{I_0I_0'} + n_1\overline{I_0I_0'} + n_0\overline{I_1I_1'} - n_1\overline{I_1I_1'}, \tag{10.28}$$

collecting terms

$$\Delta W_q = \left(n_1 - n_0\right)\left(\overline{I_0I_0'} - \overline{I_1I_1'}\right). \tag{10.29}$$

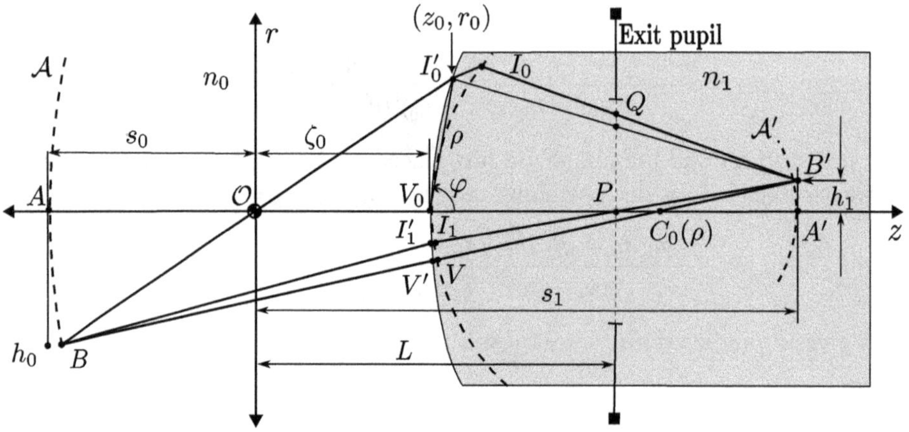

Figure 10.3. For off-axis point objects in the scenario of a Cartesian refracting surface, the surface delineated by the dotted line is part of a sphere centered at C_0. The surface portrayed by the solid line is a Cartesian surface, aligning its axial curvature with that of the spherical surface.

The terms $\overline{I_0 I_0'}$ and $\overline{I_1 I_1'}$ are derived by subtracting the equation of a sphere from the fourth-order expansion of Cartesian surfaces, thus,

$$\overline{I_0 I_0'} = \frac{O}{2}\overline{V_0 I_0'}^2 - \left[\frac{O}{2}\overline{V_0 I_0'}^2 + \frac{(2S - O^2 G)^2}{8GO}\overline{V_0 I_0'}^4\right], \tag{10.30}$$

and

$$\overline{I_1 I_1'} = \frac{O}{2}\overline{V_0 I_1'}^2 - \left[\frac{O}{2}\overline{V_0 I_1'}^2 + \frac{(2S - O^2 G)^2}{8GO}\overline{V_0 I_1'}^4\right]. \tag{10.31}$$

Therefore, ΔW_q becomes

$$\Delta W_q = -\frac{(n_1 - n_0)(2S - O^2 G)^2}{8GO}\left(\overline{V_0 I_0'}^4 - \overline{V_0 I_1'}^4\right), \tag{10.32}$$

where $\overline{V_0 I_0'}$. The projection of the exit pupil onto the Cartesian refracting the surface can be considered as shown in figure 10.4, using the law of cosines,

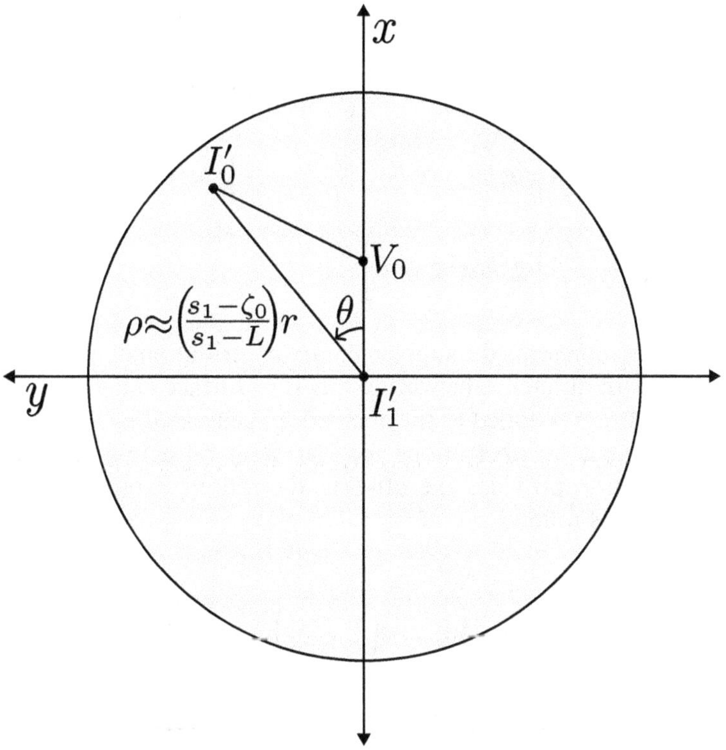

Figure 10.4. Projection of the exit pupil onto the Cartesian refracting surface is achieved with the chief ray represented by $\overline{BI'B'}$. Consequently, point I serves as the central point at which the aberration function attains a zero value.

$$\overline{V_0 I_0'}^2 = \left(\frac{s_1 - \zeta_0}{s_1 - L}\right)^2 r^2 + \overline{V_0 I_1'}^2 - 2\left(\frac{s_1 - \zeta_0}{s_1 - L}\right) r \overline{V_0 I_1'} \cos\theta. \tag{10.33}$$

From figure 10.4, we can use the relation between triangles $\Delta V_0 I_1' P$ and $\Delta B' A' P$ to get $\overline{V_0 I_1'}$,

$$\frac{\overline{V_0 I_1'}}{L - \zeta_0} \approx \frac{h_1}{s_1 - L}, \tag{10.34}$$

from which we obtain

$$\overline{V_0 I_1'} \approx g_0 h_1, \tag{10.35}$$

where

$$g_0 = \frac{L - \zeta_0}{s_1 - L}. \tag{10.36}$$

By substituting equation (10.35) into (10.33) and subsequently incorporating the result into (10.32), while taking into account that α_0 is determined by equation (10.22), we arrive at

$$\begin{aligned} \Delta W_q = \alpha_0 \Bigg(& \left(\frac{s_1 - \zeta_0}{s_1 - L}\right)^4 r^4 + 2\left(\frac{s_1 - \zeta_0}{s_1 - L}\right)^2 g_0^2 h_1^2 r^2 \\ & - 4\left(\frac{s_1 - \zeta_0}{s_1 - L}\right) r^3 g_0 h_1 r^3 \cos\theta - 4\left(\frac{s_1 - \zeta_0}{s_1 - L}\right) g_0^3 h_1^3 r \cos\theta \\ & + 4\left(\frac{s_1 - \zeta_0}{s_1 - L}\right)^2 g_0^2 h_1^2 r^2 \cos^2\theta \Bigg). \end{aligned} \tag{10.37}$$

The last expression represents the supplementary contribution of aberrations, arising from the optical path difference between a spherical surface and a Cartesian surface. When this contribution is added to the aberrations generated by a spherical surface, the result yields the total aberrations for Cartesian refracting surfaces. In other words, the summation provides the off-axis aberrations expression for Cartesian refracting surfaces, as follows,

$$\begin{aligned} W &= W_s + \Delta W_q, \\ &= \left(a_{ss} + \left(\frac{s_1 - \zeta_0}{s_1 - L}\right)^4 \alpha_0\right) r^4 + \left(a_{cs} - 4\left(\frac{s_1 - \zeta_0}{s_1 - L}\right)^3 \alpha_0 g_0\right) h_1 r^3 \cos\theta \\ &+ \left(a_{as} + 4\left(\frac{s_1 - \zeta_0}{s_1 - L}\right)^2 \alpha_0 g_0^2\right) h_1^2 r^2 \cos^2\theta + \left(a_{ds} + 2\left(\frac{s_1 - \zeta_0}{s_1 - L}\right)^2 \alpha_0 g_0^2\right) h_1^2 r^2 \\ &+ \left(a_{ts} - 4\left(\frac{s_1 - \zeta_0}{s_1 - L}\right) \alpha_0 g_0^3\right) h_1^3 r \cos\theta. \end{aligned} \tag{10.38}$$

In the above expression, the initial term corresponds to spherical aberration, the second term signifies coma, the third term represents astigmatism, the fourth term denotes field curvature and the fifth term accounts for distortion,

$$W = a_{sq}r^4 + a_{cq}h_1r^3\cos\theta + a_{aq}h_1^2r^2\cos^2\theta + a_{dq}h_1^2r^2 + a_{tq}h_1^3r\cos\theta. \quad (10.39)$$

So for spherical aberration

$$a_{sq} = a_{ss} + \left(\frac{s_1 - \zeta_0}{s_1 - L}\right)^4 \alpha_0, \quad (10.40)$$

for coma

$$a_{cq} = a_{cs} - 4\left(\frac{s_1 - \zeta_0}{s_1 - L}\right)^3 \alpha_0 g_0, \quad (10.41)$$

for astigmatism

$$a_{aq} = a_{as} + 4\left(\frac{s_1 - \zeta_0}{s_1 - L}\right)^2 \alpha_0 g_0^2, \quad (10.42)$$

field curvature

$$a_{dq} = \frac{1}{2}a_{aq}, \quad (10.43)$$

and finally, for distortion

$$a_{tq} = a_{ts} - 4\left(\frac{s_1 - \zeta_0}{s_1 - L}\right)\alpha_0 g_0^3, \quad (10.44)$$

where the coefficients a_{ss}, a_{cs}, a_{as}, a_{ds} and a_{ts} are the well-known coefficients generated by a spherical surface

$$a_{cs} = 4a_{ss}b_0 \quad (10.45)$$

$$a_{as} = 4a_{ss}b_0^2 \quad (10.46)$$

$$a_{ds} = \frac{1}{2}a_{ss} \quad (10.47)$$

$$a_{ts} = 4a_{ss}b_0^3 \quad (10.48)$$

where

$$b_0 = \frac{\zeta_0 + R - L}{s_1 - \zeta_0 - R}, \quad (10.49)$$

Table 10.1. Primary aberrations of the Cartesian refracting surface.

Aberration	**Expression**
spherical aberration	$a_{sq}\, r^4$
coma	$a_{cq}\, h_1\, r^3 \cos\theta$
astigmatism	$a_{aq}\, h_1^2\, r^2 \cos^2\theta$
field curvature	$a_{dq}\, h_1^2\, r^2$
distortion	$a_{tq}\, h_1^3\, r \cos\theta$

and

$$a_{ss} = a\left(\frac{s_1 - \zeta_0}{s_1 - L}\right)^4. \tag{10.50}$$

The derived expressions represent the fundamental aberrations for a refractive Cartesian surface and are consolidated in table 10.1. Among these, Expression (10.39) stands out as the most significant outcome in this study, offering two crucial contributions. Firstly, it addresses the challenge of formulating aberrations that facilitate the measurement of their impact on image degradation in optical systems comprising Cartesian surfaces. Secondly, it highlights the uniqueness of using form parameters identical to those employed in precisely expressing the geometries of rigorous stigmatic surfaces (G, O and S).

10.5 End notes

In this chapter, we have derived explicit expressions for the primary aberrations of Cartesian ovals. For that, we change the notation of the Cartesian ovals to a more generalized expression. In this new precise formulation for Cartesian surfaces, we were able to find the aberration function for on-axis point objects and off-axis objects. The expression of the on-axis object leads us to express the spherical aberration coefficient as a parameter of the Cartesian oval.

Furthermore, the primary aberrations for an off-axis point object, including spherical aberration, coma, astigmatism, curvature of the field and distortion, were determined and detailed. These aberrations are expressed in terms of their respective coefficients, which are defined by the parameters related to the Cartesian oval, like G, O and S, to mention a few. All these coefficients play a crucial role in formulating an aberration function for a Cartesian oval surface.

References

Estrada J C V, Calle Á H B and Hernández D M 2013 Explicit representations of all refractive optical interfaces without spherical aberration *J. Opt. Soc. Am.* **A30** 1814–24

González-Acuña R G 2021 Surface solution to correct a freeform wavefront *Appl. Opt.* **60** 9887–91

González-Acuña R G and Chaparro-Romo H A 2018 General formula for bi-aspheric singlet lens design free of spherical aberration *App. Opt.* **57** 9341–5

González-Acuña R G and Guitiérrez-Vega J C 2018 Generalization of the axicon shape: the gaxicon *J. Opt. Soc. Am.* **A35** 1915–8

González-Acuña R G and Gutiérrez-Vega J C 2019 Analytic formulation of a refractive-reflective telescope free of spherical aberration *Opt. Eng.* **58** 085105

González-Acuña R G and Thibault S 2021 The general equation of the stigmatic lenses: its history and what we have learned from it *Int. Optical Design Conf.* (Washington, DC: Optical Society of America) 1207803 p

González-Acuña R G, Avendaño-Alejo M and Gutiérrez-Vega J C 2019a Singlet lens for generating aberration-free patterns on deformed surfaces *J. Opt. Soc. Am.* **A36** 925–9

González-Acuña R G, Chaparro-Romo H A and Gutiérrez-Vega J C 2019b General formula to design freeform singlet free of spherical aberration and astigmatism *Appl. Opt.* **58** 1010–5

González-Acuña R G, Chaparro-Romo H A and Gutiérrez-Vega J C 2020a Analytic aplanatic singlet lens: setting and design for three-point objects and images in the meridional plane *Opt. Eng.* **59** 055104

González-Acuña R G, Chaparro-Romo H A and Gutiérrez-Vega J C 2020b Analytic solution of the eikonal for a stigmatic singlet lens *Phys. Scr.*

González-Acuña R G, Chaparro-Romo H A and Gutiérrez-Vega J C 2020c *Analytical Lens Design* (Bristol: Institute of Physics Publishing)

González-Acuña R G, Chaparro-Romo H A and Gutiérrez-Vega J C 2021 General stigmatic surfaces *J. Opt. Soc. Am.* **A38** 298–302

González-Acuña R G, Chaparro-Romo H A and Gutíerrez-Vega J C 2019c *Single Lens Telescope* (arXiv:1903.11129)

González-Acuña R G and Gutiérrez-Vega J C 2019a Analytic formulation of a refractive-reflective telescope free of spherical aberration *Opt. Eng.* **58** 1–5

González-Acuña R G and Gutiérrez-Vega J C 2019b General formula to eliminate spherical aberration produced by an arbitrary number of lenses *Opt. Eng.* **58** 1–6

González-Acuña R G and Gutiérrez-Vega J C 2019 General formula for aspheric collimator lens design free of spherical aberration *Current Developments in Lens Design and Optical Engineering XX* ed R B Johnson, V N Mahajan and S Thibault (Bellingham, WA: International Society for Optics and Photonics, SPIE) 181–4 pp

González-Acuña R G and Gutiérrez-Vega J C 2020 Analytic design of a spherochromatic singlet *J. Opt. Soc. Am.* **A37** 149–53

González Acuña R G and Gutiérrez-Vega J C 2019 General formula to design freeform collimator lens free of spherical aberration and astigmatism *Novel Optical Systems, Methods, and Applications XXII* **vol 11 105** (Bellingham, WA: International Society for Optics and Photonics) 111050A p

González Acuña R G and Gutiérrez-Vega J C 2019 General formula of the refractive telescope design free spherical aberration *Novel Optical Systems, Methods, and Applications XXII* **vol 11 105** ed C F Hahlweg and J R Mulley (Bellingham, WA: International Society for Optics and Photonics, SPIE) 162–6 p

Silva-Lora A and Torres R 2023 Primary aberrations theory in optical systems composed of Cartesian refracting surfaces *Proc. R. Soc.* **A479** 20230186

Silva-Lora A and Torres R 2020 Explicit Cartesian oval as a superconic surface for stigmatic imaging optical systems with real or virtual source or image *Proc. R. Soc.* A476

Valencia-Estrada J C and Malacara-Doblado D 2014 Parastigmatic corneal surfaces *Appl. Opt.* **53** 3438–47

Valencia-Estrada J C, Flores-Hernández R B and Malacara-Hernández D 2015 Singlet lenses free of all orders of spherical aberration *Proc. R. Soc.* **A471** 20140608

IOP Publishing

Stigmatic Optics (Second Edition)

Rafael G González-Acuña and Héctor A Chaparro-Romo

Chapter 11

The stigmatic lens and the Cartesian ovals

In previous chapters we have studied stigmatic lenses. In chapter 8 we studied stigmatic lenses generated by Cartesian ovals with the Silva–Torres equation. In chapter 9 we deduced the general equation for stigmatic lenses and further deduced that it is unique. In this chapter we will compare the lenses of chapters 8 and 9, and with this study we will verify the uniqueness of the designs, the general equation of stigmatic lenses and stigmatism in general.

11.1 Introduction

Up to this point in this treatise, we have studied the origin of the foundations of geometric optics. Starting from the eikonal, we defined the ray and observed that when all the rays of an object point converge at an image point, we have stigmatism.

Then we began to study the stigmatic systems: conical mirrors and stigmatic refractive surfaces, Cartesian ovals. We explored different ways of describing Cartesian ovals. With Cartesian ovals, we designed stigmatic lenses. Later, in chapter 9, we found the general equation of the stigmatic lens.

The procedure to find the stigmatic lens tells us that the solution is unique. We have to be precise; the equation is general because it covers all cases of lenses that are stigmatic. Now given an initial configuration, there is only a second surface that makes the lens stigmatic. For example, given a function z_a, z_o, τ, z_i, n_o, n, n_i there is only a single z_b, r_b capable of making the system stigmatic.

Although it is evident with the procedure of the previous chapter, readers might not find it easy to see because the equations are enormous and can cause doubt, which is natural. In this chapter, we will generate three stigmatic lenses and compare their output surfaces. If the surfaces are equal, it means that, as we expected, the general equation for stigmatic lenses given a configuration gives a single solution.

The three types of astigmatic lenses start from very different procedures, and in the end, we compare for finite objects and finite images if the second surfaces are equal.

doi:10.1088/978-0-7503-6423-2ch11

Additionally, we will make the same comparison for when the object is very far from the first surface of the stigmatic lens.

11.2 Comparison of different stigmatic lenses made by Cartesian ovals

The three systems that we are going to compare are the Silva–Torres stigmatic lenses from chapter 8. The stigmatic lenses with the general equation of chapter 9 with a first entrance surface of a Cartesian oval proposed by the Valencia–Calle model in chapter 6. Finally, we will take a hybrid version of the Silva–Torres Cartesian oval from chapter 6 and the general equation of stigmatic lenses from chapter 9.

Since all the equations presented from chapters 6 to 9 are unusually long, we will show stigmatic lenses in sections. After the lenses are presented, the next section will compare the final surface of the lens.

The idea is that since each model will have a Cartesian oval, it is the same, but described in different ways; the models must give as a second surface a single Cartesian oval.

In the section where the models are compared, no equation is given since the notation in many of the terms is repeated and can be confusing, only the number of the equation used and the section to which it belongs is quoted.

In this way, we would confirm that all the models are consistent and all arrive at the same curve and that nature, for this specific problem, only accepts one solution. All roads lead to Rome.

Lastly, before presenting a comparison of the three models, a hybrid between the Silva–Torres Cartesian oval and the general equation of stigmatic lenses will be deduced.

11.3 Cartesian ovals in a parametric form

We start with the Silva–Torres stigmatic lenses from chapter 8. So, we first recall the parameter of the first surface, equations (8.1)–(8.13). We start with the parameters inside equations (8.1) and (8.8).

The first term is equation (8.3),

$$K_a = \frac{(n^2 z_{in} - n_o^2 z_o)^2}{n n_o (n z_{in} - n_o z_o)(n z_o - n_o z_{in})}, \tag{8.3}$$

where n is the refraction index inside the lens, z_o is the distance from the object to the vertex of the first surface and z_{in} is the image formed by the first surface (see figure 11.1).

We continue with the other parameters, which are

$$c_{0_a} = \frac{n z_o - n_o z_{in}}{z_{in} z_o (n - n_o)}. \tag{8.4}$$

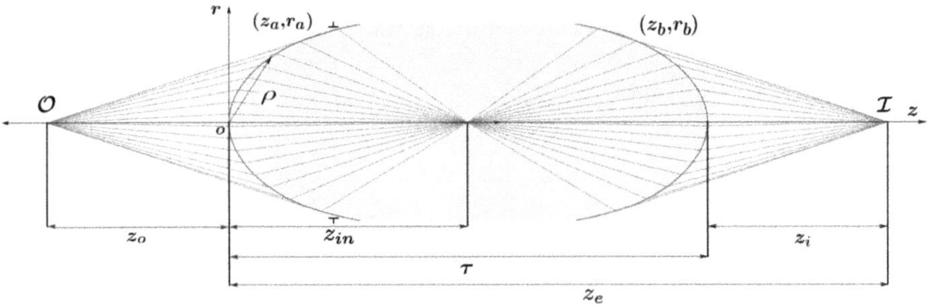

Figure 11.1. Diagram of a lens generated by two Cartesian ovals. The first surface is (z_a, r_a) and the second Cartesian oval is (z_b, r_b). The refraction index in the object space is n_o, inside the glass the refraction index is n and in the image space is n_i. The origin is placed in the vertex of the first surface. ρ is the distance from the origin to a point in the first surface. The distance from the object to the first surface is z_o; the distance from the first surface to the image generated by the first surface is z_{in}. The central thickness of the lens is τ. From the origin to the image, the gap is given by z_e and for the distance from the second surface to the image, the length is z_i.

We take c_{1_a} from equation (8.5),

$$c_{1_a} = \frac{(n - n_o)(n + n_o)^2}{4nn_oz_{in}z_o(nz_{in} - n_oz_o)}, \tag{8.5}$$

and b_{1_a} from equation (8.6),

$$b_{1_a} = \frac{(n + n_o)(n^2z_{in} - n_o^2z_o)}{2nn_oz_{in}z_o(nz_{in} - n_oz_o)}. \tag{8.6}$$

With these parameters we can now properly recall the first surface, equation (8.1),

$$z_a = \frac{\rho^2(c_{1_a}\rho^2 + c_{0_a})}{\sqrt{\rho^2\left(2b_{1_a} - c_{0_a}^2K_a\right) + 1} + b_{1_a}\rho^2 + 1}, \tag{8.1}$$

and its radius, equation (8.2),

$$r_a = \text{sgn}(\rho)\sqrt{\rho^2 - z_a(\rho)^2}. \tag{8.2}$$

Remember that n_o is the object area, n_i is the refractive index at the image point and we take $n_o = n_1$.

We continue with the parameters of the second surface; we take K_b from equation (8.10),

$$K_b = \frac{[n_o^2(z_e - \tau) - n^2(z_{in} - \tau)]^2}{nn_o[n(z_e - \tau) - n_o(z_{in} - \tau)][n(z_{in} - \tau) - n_o(z_e - \tau)]}. \tag{8.10}$$

Remember that z_e is the distance from the origin of the coordinate system to the image formed by the second surface and τ is the central thickness of the lens (see figure 11.1).

c_{0_b} is from equation (8.11),

$$c_{0_b} = \frac{n(z_e - \tau) - n_o(z_{in} - \tau)}{(n - n_o)(z_e - \tau)(z_{in} - \tau)}. \tag{8.11}$$

Then, from equation (8.12), we have

$$c_{1_b} = \frac{(n - n_o)(n + n_o)^2}{4nn_o(z_e - \tau)(z_{in} - \tau)[n(z_{in} - \tau) - n_o(z_e - \tau)]}, \tag{8.12}$$

finally, we recall equation (8.13),

$$b_{1_b} = \frac{(n + n_o)[n^2(z_{in} - \tau) - n_o^2(z_e - \tau)]}{2nn_o(z_e - \tau)(z_{in} - \tau)[n(z_{in} - \tau) - n_o(z_e - \tau)]}. \tag{8.13}$$

The second surface is given by equation (8.7), so we recall it,

$$z_{b0} = \frac{\rho^2(c_{1_b}\rho^2 + c_{0_b})}{\sqrt{\rho^2\left(2b_{1_b} - c_{0_b}^2 K_b\right) + 1} + b_{1_b}\rho^2 + 1}, \tag{8.7}$$

where,

$$z_b = \tau + z_{b0}(\rho), \tag{8.8}$$

and the radius of the second surface is given by equation (8.9),

$$r_b = \text{sgn}(\rho)\sqrt{\rho^2 - z_{b0}(\rho)^2}\,. \tag{8.9}$$

We are going to use the equations of this section to compare them with the stigmatic lenses of the following sections.

11.4 Cartesian ovals in an explicit form as a first surface and general equation of stigmatic lenses

Now it is time to look at the hybrid model when the first surface is given by an explicit expression of the Cartesian oval described by the paradigm of chapter 6. Therefore, we take the first surface as of Case E 6.3.5, equation (6.63),

$$z_a = c_2\frac{r^2}{2f} + c_4\frac{r^4}{8f^3} + c_6\frac{2r^6}{32f^5} + c_8\frac{5r^8}{128f^7} + c_{11}\frac{2r^{10}}{215f^9} + c_{12}\frac{14f^{12}}{2048f^{11}} \cdots = \sum_{k=1}^{\infty} c_{2k}\frac{I_k r^{2k}}{(2f)^{2k-1}}, \tag{6.63}$$

where the coefficients are given by equation (6.65)

$$\begin{cases} c_2 = -\dfrac{\alpha n + 1}{\alpha(n-1)}, \\ c_4 = -\dfrac{\alpha^3 n^2 + \left(\alpha^3 + 2\alpha^2 - 2\alpha - 1\right)n - 1}{\alpha^3(n-1)^2}, \\ c_6 = -\dfrac{\alpha^5 n^3 + \left(2\alpha^5 + 3\alpha^4 - 3\alpha^3 + \alpha^2 + 3\alpha + 1\right)n^2 + \left(\alpha^5 + 3\alpha^4 + \alpha^3 - 3\alpha^2 + 3\alpha + 2\right)n + 1}{\alpha^5(n-1)^3}, \\ c_8 = -\dfrac{1}{\alpha^7(n-1)^4}\left(\alpha^7 n^4 + \left(3\alpha^7 + 4\alpha^6 - 4\alpha^5 + 2\alpha^4 + 2\alpha^3 - 4\alpha^2 - 4\alpha + 1\right)n^3\right) \\ -\dfrac{1}{\alpha^7(n-1)^4}\left(3\alpha^7 + 8\alpha^6 - 8\alpha^4 + 8\alpha^3 - 8\alpha - 3\right)n^2 \\ -\dfrac{1}{\alpha^7(n-1)^4}\left(\left(\alpha^7 + 4\alpha^6 + 4\alpha^5 - 2\alpha^4 - 2\alpha^3 + 4\alpha^2 - 4\alpha - 3\right)n - 1\right). \end{cases} \tag{6.65}$$

The expression of (z_b, r_b) is given by equation (9.39),

$$\begin{cases} z_b = z_a + \vartheta\wp_z, \\ r_b = r_a + \vartheta\wp_r \end{cases}, \tag{9.39}$$

where

$$\vartheta = \frac{-\left[z_f n + n_i^2\left(r_a\wp_r + z_T\wp_z\right)\right] \pm \sqrt{\left(n_i^2 - n^2\right)\left[n_i^2\left(r_a^2 + z_T^2\right) - z_f^2\right] + \left[z_f n + n_i^2\left(r_a\wp_r + z_T\wp_z\right)\right]^2}}{n_i^2 - n^2}, \tag{9.38}$$

n is the refraction index inside the lens, $n_0 = n_i$ are the refraction index outside the lens, $z_o = \alpha f$ is the distance from the object to the vertex of the first surface, τ is the central thickness of the lens and z_i is the distance from the second surface to the image (see figure 11.2).

We continue with z_f by recalling equation (9.24),

$$z_f = -n_o z_o + n\tau + n_i z_i + \operatorname{sgn}(z_o) n_0 \sqrt{r_a^2 + (z_o - z_a)^2}, \tag{9.24}$$

and z_τ by recalling equation (9.27),

$$z_T = -\tau + z_a - z_i. \tag{9.27}$$

Finally, $\wp_r$, $\wp_z$ are the director cosine of the ray inside the lens,

$$\wp_r = \frac{n_o[(z_a - z_o)z_a' + r_a]}{-\operatorname{sgn}(z_o) n \sqrt{r_a^2 + (z_o - z_a)^2}\left(1 + z_a'^2\right)} - z_a' \frac{\sqrt{1 - \dfrac{n_o^2[r_a + (z_a - z_o)z_a']^2}{n^2[r_a^2 + (z_o - z_a)^2]\left(1 + z_a'^2\right)}}}{\sqrt{1 + z_a'^2}}, \tag{9.17}$$

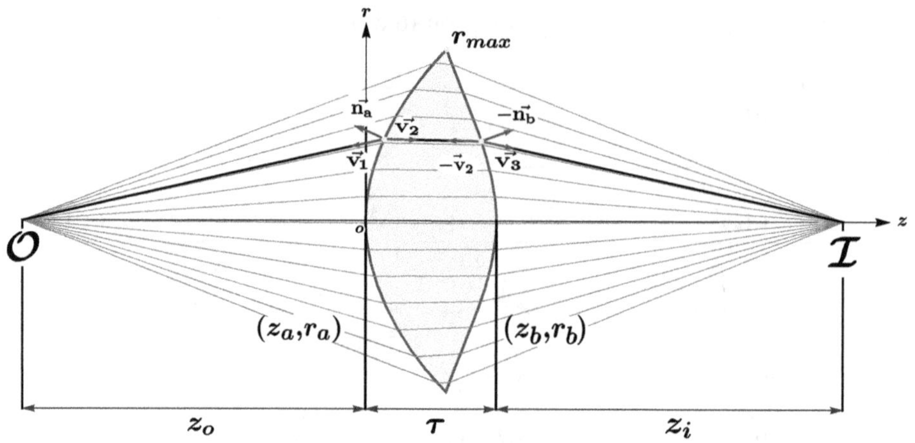

Figure 11.2. Diagram of an on-axis stigmatic singlet. The first is (z_a, r_a) and second surfaces are (z_a, r_a) and (z_b, r_b). Here, in this chapter, we take the first surface as a polynomial expansion of a Cartesian oval. The gap from the object to the first is z_o, the axial thickness of the lens is τ, and the gap from the second surface to the image is z_i. The refraction index in the object space is n_o. The refraction index of the lens is n and in the image space, the refraction index is n_i. Notice that the origin is placed in the vertex of the first surface, the polynomial expansion of a Cartesian oval.

and

$$\wp_z = \frac{n_o[r_a + (z_a - z_o)z_a']z_a'}{-\text{sgn}(z_o)n\sqrt{r_a^2 + (z_o - z_a)^2}\left(1 + z_a'^2\right)} + \frac{\sqrt{1 - \dfrac{n_o^2[r_a + (z_a - z_o)z_a']^2}{n^2[r_a^2 + (z_o - z_a)^2]\left(1 + z_a'^2\right)}}}{\sqrt{1 + z_a'^2}}. \tag{9.18}$$

Notice that the following condition holds, $\wp_r^2 + \wp_z^2 = 1$.
These are the representative equations for the stigmatic lens of this section.

11.5 Cartesian ovals in a parametric form as a first surface and general equation of stigmatic lenses

Since we do not deduce the expression for a Cartesian oval in a parametric form as a first surface and general equation of stigmatic lenses in this section, we are going to demonstrate it.

We start with the notation: n is the refraction index inside the lens; $n_0 = n_i$ are the refraction index outside; z_o is the gap between the object and the first surface; the central thickness of the lens is given by τ; z_o is the gap between the image and the second surface (see figure 11.3).

11.5.1 First surface

Now, let's focus on the first surface. The first surface is a Cartesian oval under the paradigm of Silva–Torres. Therefore, it is in a parametric form where the independent variable is ρ.

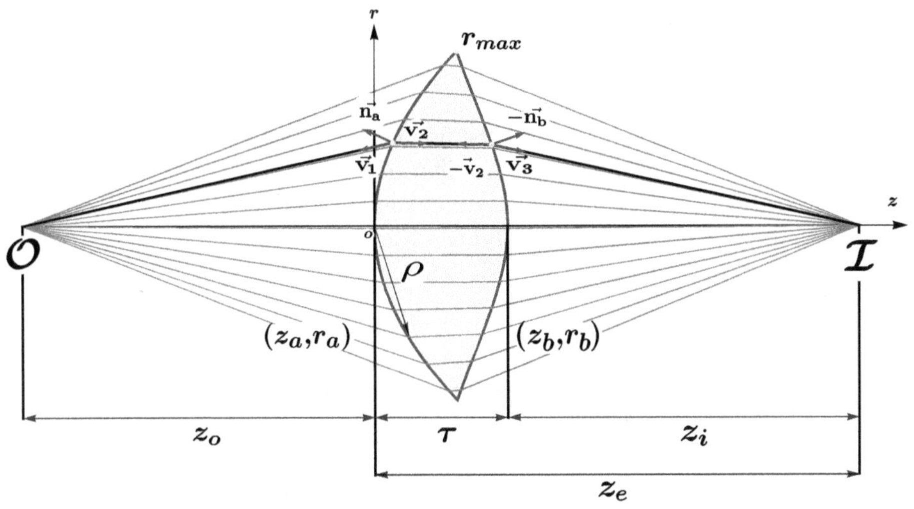

Figure 11.3. An on-axis stigmatic singlet. The first surface is (z_a, r_a) and second surface is (z_a, r_a) and (z_b, r_b). The gap from the object to the first is z_o, the axial thickness of the lens is τ, and the length from the second surface to the image is z_i. The refraction index around the object is n_o. The refraction index of the lens is n and around the image, the refraction index is n_i.

So, for K_a we recall equation (8.3),

$$K_a = \frac{(n^2 z_{in} - n_o^2 z_o)^2}{nn_o(nz_{in} - n_o z_o)(nz_o - n_o z_{in})}. \tag{8.3}$$

For c_{0_a}, we use equation (8.4),

$$c_{0_a} = \frac{nz_o - n_o z_{in}}{z_{in} z_o (n - n_o)}, \tag{8.4}$$

then, we recall equation (8.5),

$$c_{1_a} = \frac{(n - n_o)(n + n_o)^2}{4nn_o z_{in} z_o (nz_{in} - n_o z_o)}. \tag{8.5}$$

Finally, we use equation (8.6) for b_{1_a}

$$b_{1_a} = \frac{(n + n_o)(n^2 z_{in} - n_o^2 z_o)}{2nn_o z_{in} z_o (nz_{in} - n_o z_o)}. \tag{8.6}$$

Given the parameters inside the first surface, we can recall it as a Cartesian oval under the paradigm of Silva–Torres with equation (8.1),

$$z_a = \frac{\rho^2(c_{1_a}\rho^2 + c_{0_a})}{\sqrt{\rho^2\left(2b_{1_a} - c_{0_a}^2 K_a\right) + 1} + b_{1_a}\rho^2 + 1}, \tag{8.1}$$

where the radius is equation (8.2),

$$r_a = \text{sgn}(\rho)\sqrt{\rho^2 - z_a(\rho)^2}\,. \tag{8.2}$$

Notice that z_{in} is the image formed by the first surface. Since the second surface is in terms of a Cartesian oval under the paradigm of Silva–Torres, the general equation of stigmatic lenses z_{in} will not be as relevant as in the model of Silva–Torres. Hence, it is not presented in the diagram of figure 11.3.

11.5.2 Second surface

Notice the first surface in a parametric form, where the independent variable is ρ. The general equation of stigmatic lenses presented in chapter 9 receives first surfaces explicit expressions, not parametric ones. Therefore, we need to adapt it such that the first surface is an expression in a parametric form, where the independent variable is ρ. In this section, we are going to deduce the adaptation above of the general equation of stigmatic lenses.

We begin by recalling Snell's law at the first surface,

$$\boxed{\vec{\mathbf{v}}_2 = \frac{n_o}{-\text{sgn}(z_o)n}\left[\vec{\mathbf{v}}_1 - (\vec{\mathbf{n}}_a\cdot\vec{\mathbf{v}}_1)\vec{\mathbf{n}}_a\right] - \vec{\mathbf{n}}_a\sqrt{1 + \frac{n_o^2}{n^2}(\vec{\mathbf{n}}_a\times\vec{\mathbf{v}}_1)^2} \quad \text{for} \quad \vec{\mathbf{v}}_2, \vec{\mathbf{v}}_1, \vec{\mathbf{n}}_a \in \mathbb{R}^2}, \tag{11.1}$$

where $\vec{\mathbf{v}}_1$ is the unit vector of the incident ray, $\vec{\mathbf{v}}_2$ is unit vector of the refracted ray and finally, $\vec{\mathbf{n}}_a$ is the normal vector of the first surface. Notice the term $-\text{sgn}(z_o)$; it comes from the fact that the object can be real or virtual (see figure 11.3).

The unit vectors at the first surface are

$$\vec{\mathbf{v}}_1 = \frac{r_a\vec{\mathbf{e}}_1 + (z_a - z_o)\vec{\mathbf{e}}_2}{\sqrt{r_a^2 + (z_a - z_o)^2}}, \quad \vec{\mathbf{v}}_2 = \frac{(r_b - r_a)\vec{\mathbf{e}}_1 + (z_b - z_a)\vec{\mathbf{e}}_2}{\sqrt{(r_b - r_a)^2 + (z_b - z_a)^2}}, \quad \vec{\mathbf{n}}_a = \frac{z_a'\vec{\mathbf{e}}_1 - r_a'\vec{\mathbf{e}}_2}{\sqrt{r_a'^2 + z_a'^2}}, \tag{11.2}$$

where $\vec{\mathbf{e}}_1$ is for the r direction and $\vec{\mathbf{e}}_2$ is for the z direction.

Also notice that the normal vector has been changed because now the derivatives are in respect to ρ and not with respect to r_a.

Thus, $r_a' = \frac{\partial r_a}{\partial \rho}$ and $z_a' = \frac{\partial z_a}{\partial \rho}$.

We need to replace the unit vectors of equation (11.2) in equation (11.1). We start with $(\vec{\mathbf{n}}_a \times \vec{\mathbf{v}}_1)$,

$$(\vec{\mathbf{n}}_a \times \vec{\mathbf{v}}_1) = \frac{[r_a r_a' + (z_a - z_o)z_a']}{\sqrt{r_a^2 + (z_a - z_o)^2}\sqrt{r_a'^2 + z_a'^2}}(\vec{\mathbf{e}}_1 \times \vec{\mathbf{e}}_2), \tag{11.3}$$

squaring

$$(\vec{\mathbf{n}}_a \times \vec{\mathbf{v}}_1)^2 = -\frac{[r_a r_a' + (z_a - z_o)z_a']^2}{[r_a^2 + (z_a - z_o)^2](r_a'^2 + z_a'^2)}. \tag{11.4}$$

Then, the square root of equation (11.1) is

$$-\vec{\mathbf{n}}_a\sqrt{1+\frac{n_o^2}{n^2}(\vec{\mathbf{n}}_a\times\vec{\mathbf{v}}_1)^2}=-\frac{(z_a'\vec{\mathbf{e}}_1-r_a'\vec{\mathbf{e}}_2)}{\sqrt{r_a'^2+z_a'^2}}\sqrt{1-\frac{[r_ar_a'+(z_a-z_o)z_a']^2}{n^2\left[r_a^2+(z_a-z_o)^2\right]\left(r_a'^2+z_a'^2\right)}}. \tag{11.5}$$

Let's focus on the other terms of equation (11.1), $\vec{\mathbf{v}}_1-(\vec{\mathbf{n}}_a\cdot\vec{\mathbf{v}}_1)\vec{\mathbf{n}}_a$,

$$\vec{\mathbf{v}}_1-(\vec{\mathbf{n}}_a\cdot\vec{\mathbf{v}}_1)\vec{\mathbf{n}}_a=\frac{r_a\vec{\mathbf{e}}_1+(z_a-z_o)\vec{\mathbf{e}}_2}{\sqrt{r_a^2+(z_a-z_o)^2}}-\left[\frac{r_az_a'-(z_a-z_o)r_a'}{\sqrt{r_a^2+(z_a-z_o)^2}\sqrt{r_a'^2+z_a'^2}}\right]\frac{(z_a'\vec{\mathbf{e}}_1-r_a'\vec{\mathbf{e}}_2)}{\sqrt{r_a'^2+z_a'^2}}. \tag{11.6}$$

Manipulating and multiplying by $n_o/(-\text{sgn}(z_o)n$

$$\begin{aligned}\frac{n_o}{-\text{sgn}(z_o)n}\left[\vec{\mathbf{v}}_1-(\vec{\mathbf{n}}_a\cdot\vec{\mathbf{v}}_1)\vec{\mathbf{n}}_a\right]&=\frac{n_o[(z_a-z_o)z_a'+r_ar_a']}{-\text{sgn}(z_o)n\sqrt{r_a^2+(z_o-z_a)^2}\,(1+z_a'^2)}\vec{\mathbf{e}}_1\\&+\frac{n_o(r_a+(z_a-z_o)z_a')z_a'}{-\text{sgn}(z_o)n\sqrt{r_a^2+(z_o-z_a)^2}\,(1+z_a'^2)}\vec{\mathbf{e}}_2.\end{aligned} \tag{11.7}$$

Summing equations (11.5) and (11.7), we have $\vec{\mathbf{v}}_2$ of equation tal as we wanted,

$$\begin{aligned}\vec{\mathbf{v}}_2=&\frac{n_o[(z_a-z_o)z_a'+r_ar_a']r_a'}{-\text{sgn}(z_o)n\sqrt{r_a^2+(z_o-z_a)^2}\left(r_a'^2+z_a'^2\right)}\vec{\mathbf{e}}_1+\frac{n_o[r_ar_a'+(z_a-z_o)z_a']z_a'}{-\text{sgn}(z_o)n\sqrt{r_a^2+(z_o-z_a)^2}\left(r_a'^2+z_a'^2\right)}\vec{\mathbf{e}}_2\\&-\frac{(z_a'\vec{\mathbf{e}}_1-r_a'\vec{\mathbf{e}}_2)}{\sqrt{r_a'^2+z_a'^2}}\sqrt{1-\frac{n_o^2[r_ar_a'+(z_a-z_o)z_a']^2}{n^2\left[r_a^2+(z_o-z_a)^2\right]\left(r_a'^2+z_a'^2\right)}}.\end{aligned} \tag{11.8}$$

Separating the coordinates of the last equation, $\vec{\mathbf{e}}_1$ and $\vec{\mathbf{e}}_2$,

$$\begin{aligned}\frac{r_b-r_a}{\sqrt{(z_b-z_a)^2+(r_b-r_a)^2}}=&\frac{n_o[(z_a-z_o)z_a'+r_ar_a']r_a'}{-\text{sgn}(z_o)n\sqrt{r_a^2+(z_o-z_a)^2}\,(r_a'^2+z_a'^2)}\\&-z_a'\frac{\sqrt{1-\frac{n_o^2[r_ar_a'+(z_a-z_o)z_a']^2}{n^2[r_a^2+(z_o-z_a)^2](r_a'^2+z_a'^2)}}}{\sqrt{r_a'^2+z_a'^2}},\end{aligned} \tag{11.9}$$

and

$$\begin{aligned}\frac{z_b-z_a}{\sqrt{(z_b-z_a)^2+(r_b-r_a)^2}}=&\frac{n_o[r_ar_a'+(z_a-z_o)z_a']z_a'}{-\text{sgn}(z_o)n\sqrt{r_a^2+(z_o-z_a)^2}\,(r_a'^2+z_a'^2)}\\&+r_a'\frac{\sqrt{1-\frac{n_o^2[r_ar_a'+(z_a-z_o)z_a']^2}{n^2[r_a^2+(z_o-z_a)^2](r_a'^2+z_a'^2)}}}{\sqrt{r_a'^2+z_a'^2}}.\end{aligned} \tag{11.10}$$

Notice that in the left side of the equations, again, there are only parameters that we know. Thus, we assign them to the new parameters $\wp_r$ and $\wp_z$,

$$\wp_r = \frac{n_o[(z_a - z_o)z_a' + r_a r_a']r_a'}{-\text{sgn}(z_o)n\sqrt{r_a^2 + (z_o - z_a)^2}\left(r_a'^2 + z_a'^2\right)} - z_a'\frac{\sqrt{1 - \frac{n_o^2[r_a r_a' + (z_a - z_o)z_a']^2}{n^2[r_a^2 + (z_o - z_a)^2]\left(r_a'^2 + z_a'^2\right)}}}{\sqrt{r_a'^2 + z_a'^2}}, \tag{11.11}$$

and

$$\wp_z = \frac{n_o[r_a r_a' + (z_a - z_o)z_a']z_a'}{-\text{sgn}(z_o)n\sqrt{r_a^2 + (z_o - z_a)^2}\left(r_a'^2 + z_a'^2\right)} + r_a'\frac{\sqrt{1 - \frac{n_o^2[r_a r_a' + (z_a - z_o)z_a']^2}{n^2[r_a^2 + (z_o - z_a)^2]\left(r_a'^2 + z_a'^2\right)}}}{\sqrt{r_a'^2 + z_a'^2}}. \tag{11.12}$$

It is important to remark that $\wp_r$ and $\wp_z$ are the cosine directors traveling inside the lens. $\wp_r$ is the cosine director with direction r and $\wp_z$ is the cosine director with direction z. Thus, $\wp_r^2 + \wp_z^2 = 1$.

For derivation purposes, we assign a name to the distance traveled by each ray inside the lens as

$$\vartheta \equiv \sqrt{(z_b - z_a)^2 + (r_b - r_a)^2}. \tag{11.13}$$

Replacing equation (11.13) in equations (11.11) and (11.12),

$$\frac{z_b - z_a}{\vartheta} = \wp_z, \tag{11.14}$$

and

$$\frac{r_b - r_a}{\vartheta} = \wp_r. \tag{11.15}$$

Solving for z_b and r_b,

$$z_b = z_a + \vartheta\wp_z, \tag{11.16}$$

and

$$r_b = r_a + \vartheta\wp_r. \tag{11.17}$$

We get the same structure of chapter 8; we just need to know what ϑ is. For that, we will follow a similar procedure. Thus, we recall the Fermat principle. The Fermat principle optical path length of the axial ray is equal to the optical path length of the non-axial ray,

$$\begin{aligned} -n_o z_o + n\tau + n_i z_i &= n\sqrt{(r_b - r_a)^2 + (z_b - z_a)^2} \\ &+ n_i\sqrt{r_b^2 + (-\tau + z_b - z_i)^2} + \text{sgn}(z_o)n_o\sqrt{r_a^2 + (z_o - z_a)^2}. \end{aligned} \tag{11.18}$$

Let's assign the following parameter to clean the last equation,

$$z_f = -n_o z_o + n\tau + n_i z_i + \mathrm{sgn}(z_o) n_0 \sqrt{r_a^2 + (z_o - z_a)^2}, \tag{11.19}$$

replacing equation (11.19) in equation (11.18),

$$z_f = n\sqrt{(r_b - r_a)^2 + (z_b - z_a)^2} + n_i \sqrt{r_b^2 + (-\tau + z_b - z_i)^2}, \tag{11.20}$$

replacing equations (11.17) and (11.16) in equation (11.20),

$$z_f = n\vartheta + n_i \sqrt{(r_a + \vartheta \wp_r^2 + (-\tau + z_a - z_i + \vartheta \wp_z)^2}, \tag{11.21}$$

we assign a new parameter to clean equation (11.21),

$$z_\tau = -\tau + z_a - z_i, \tag{11.22}$$

then, we replace equation (11.22) in equation (11.21),

$$z_f = n\vartheta + n_i \sqrt{(r_a + \vartheta \wp_r)^2 + (z_\tau + \vartheta \wp_z)^2}, \tag{11.23}$$

squaring both sides of equation (11.23) and manipulating

$$\left(z_f - n\vartheta\right)^2 = n_i^2 [(r_a + \vartheta \wp_r)^2 + (z_\tau + \vartheta \wp_z)^2], \tag{11.24}$$

expanding,

$$z_f^2 - 2z_f n\vartheta + n^2\vartheta^2 = n_i^2 [r_a^2 + 2r_a \vartheta \wp_r + \vartheta^2 \wp_r^2 + z_\tau^2 + 2z_\tau \vartheta \wp_z + \vartheta^2 \wp_z^2], \tag{11.25}$$

solving for ϑ,

$$\vartheta = \frac{-[z_f n + n_i^2 (r_a \wp_r + z_\tau \wp_z)] \pm \sqrt{\left(n_i^2 - n^2\right)\left[n_i^2\left(r_a^2 + z_\tau^2\right) - z_f^2\right] + \left[z_f n + n_i^2\left(r_a \wp_r + z_\tau \wp_z\right)\right]^2}}{n_i^2 - n^2}, \tag{11.26}$$

then, we get almost the same equation. The sagitta is

$$z_b = z_a + \vartheta \wp_z \tag{11.27}$$

and the radius,

$$r_b = r_a + \vartheta \wp_r. \tag{11.28}$$

The difference between equations (11.27), (11.28) and equation (9.39) is that the first ones are for the first surface in a parametric form. Equation (9.39) is for the first surface in an explicit form.

The procedure is almost the same as the one performed in chapter 9, but now the independent variable is ρ instead of r_a.

The equations of this section that we use in the following comparison are (11.27), (11.28), (11.19), (11.22), (11.17) and (11.16).

The following expressions are the extended versions of (11.27) and (11.28). We show them to raise awareness of the complexity and length of these expressions, but we always refer to them as equations (11.27) and (11.28).

$$
\begin{aligned}
z_b(\rho) = {} & z_a(\rho) \\
& + \frac{1}{n^2-n_i^2}\left(\frac{n_o[z'_a(\rho)(r_a(\rho)r'_a(\rho)+(z_a(\rho)-z_o)z'_a(\rho))]}{n\sqrt{r_a(\rho)^2+(z_o-z_a(\rho))^2}\left(r'_a(\rho)^2+z'_a(\rho)^2\right)} + \frac{r'_a(\rho)\sqrt{1-\frac{n_o^2[r_a(\rho)r'_a(\rho)+(z_a(\rho)-z_o)z'_a(\rho)]^2}{n^2\left[r_a(\rho)^2+(z_o-z_a(\rho))^2\right]\left(r'_a(\rho)^2+z'_a(\rho)^2\right)}}}{\sqrt{r'_a(\rho)^2+z'_a(\rho)^2}}\right) \\
& \left(n\left(n_i z_i - n_o z_o + n\tau + n_o\,\mathrm{Sign}\left[z_o\right]\sqrt{r_a(\rho)^2+(z_o-z_a(\rho))^2}\right) \right. \\
& + n_i^2 \left\{ \left[r_a(\rho) \times \left(\frac{n_o[z'_a(\rho)(r_a(\rho)r'_a(\rho)+(z_a(\rho)-z_o)z'_a(\rho))]}{n\sqrt{r_a(\rho)^2+(z_o-z_a(\rho))^2}\left(r'_a(\rho)^2+z'_a(\rho)^2\right)} - \frac{z'_a(\rho)\sqrt{1-\frac{n_o^2[r_a(\rho)r'_a(\rho)+(z_a(\rho)-z_o)z'_a(\rho)]^2}{n^2\left[r_a(\rho)^2+(z_o-z_a(\rho))^2\right]\left(r'_a(\rho)^2+z'_a(\rho)^2\right)}}}{\sqrt{r'_a(\rho)^2+z'_a(\rho)^2}} \right) \right] \right. \\
& \left. + \left[\left(z_a(\rho)-z_i-\tau\right) \times \left(\frac{n_o[z'_a(\rho)(r_a(\rho)r'_a(\rho)+(z_a(\rho)-z_o)z'_a(\rho))]}{n\sqrt{r_a(\rho)^2+(z_o-z_a(\rho))^2}\left(r'_a(\rho)^2+z'_a(\rho)^2\right)} + \frac{r'_a(\rho)\sqrt{1-\frac{n_o^2[r_a(\rho)r'_a(\rho)+(z_a(\rho)-z_o)z'_a(\rho)]^2}{n^2\left[r_a(\rho)^2+(z_o-z_a(\rho))^2\right]\left(r'_a(\rho)^2+z'_a(\rho)^2\right)}}}{\sqrt{r'_a(\rho)^2+z'_a(\rho)^2}} \right) \right] \right\} \\
& - n_i\,\mathrm{Sign}\left[n_1\right] \times \\
& \left. \sqrt{ \begin{aligned}
& \left(n_i z_i - n_o z_o + n\tau + n_o\,\mathrm{Sign}\left[z_o\right]\sqrt{r_a(\rho)^2+(z_o-z_a(\rho))^2}\right)^2 + n^2\left(r_a(\rho)^2+(z_a(\rho)-z_i-\tau)^2\right) \\
& \times \Bigg(-n_i^2 \left\{ \left[\left(z_a(\rho)-z_i-\tau\right) \times \left(\frac{n_o[z'_a(\rho)(r_a(\rho)r'_a(\rho)+(z_a(\rho)-z_o)z'_a(\rho))]}{n\sqrt{r_a(\rho)^2+(z_o-z_a(\rho))^2}\left(r'_a(\rho)^2+z'_a(\rho)^2\right)} - \frac{z'_a(\rho)\sqrt{1-\frac{n_o^2[r_a(\rho)r'_a(\rho)+(z_a(\rho)-z_o)z'_a(\rho)]^2}{n^2\left[r_a(\rho)^2+(z_o-z_a(\rho))^2\right]\left(r'_a(\rho)^2+z'_a(\rho)^2\right)}}}{\sqrt{r'_a(\rho)^2+z'_a(\rho)^2}} \right) \right] \right. \\
& \qquad \left. + \left[-r_a(\rho) \times \left(\frac{n_o[z'_a(\rho)(r_a(\rho)r'_a(\rho)+(z_a(\rho)-z_o)z'_a(\rho))]}{n\sqrt{r_a(\rho)^2+(z_o-z_a(\rho))^2}\left(r'_a(\rho)^2+z'_a(\rho)^2\right)} + \frac{r'_a(\rho)\sqrt{1-\frac{n_o^2[r_a(\rho)r'_a(\rho)+(z_a(\rho)-z_o)z'_a(\rho)]^2}{n^2\left[r_a(\rho)^2+(z_o-z_a(\rho))^2\right]\left(r'_a(\rho)^2+z'_a(\rho)^2\right)}}}{\sqrt{r'_a(\rho)^2+z'_a(\rho)^2}} \right) \right] \right\}^2 \\
& \quad + 2n\left(n_i z_i - n_o z_o + n\tau + n_o\,\mathrm{Sign}\left[z_o\right]\sqrt{r_a(\rho)^2+(z_o-z_a(\rho))^2}\right) \\
& \quad \times \left\{ \left[r_a(\rho) \times \left(\frac{n_o[z'_a(\rho)(r_a(\rho)r'_a(\rho)+(z_a(\rho)-z_o)z'_a(\rho))]}{n\sqrt{r_a(\rho)^2+(z_o-z_a(\rho))^2}\left(r'_a(\rho)^2+z'_a(\rho)^2\right)} - \frac{z'_a(\rho)\sqrt{1-\frac{n_o^2[r_a(\rho)r'_a(\rho)+(z_a(\rho)-z_o)z'_a(\rho)]^2}{n^2\left[r_a(\rho)^2+(z_o-z_a(\rho))^2\right]\left(r'_a(\rho)^2+z'_a(\rho)^2\right)}}}{\sqrt{r'_a(\rho)^2+z'_a(\rho)^2}} \right) \right] \right. \\
& \qquad \left. + \left[\left(z_a(\rho)-z_i-\tau\right) \times \left(\frac{n_o[z'_a(\rho)(r_a(\rho)r'_a(\rho)+(z_a(\rho)-z_o)z'_a(\rho))]}{n\sqrt{r_a(\rho)^2+(z_o-z_a(\rho))^2}\left(r'_a(\rho)^2+z'_a(\rho)^2\right)} + \frac{r'_a(\rho)\sqrt{1-\frac{n_o^2[r_a(\rho)r'_a(\rho)+(z_a(\rho)-z_o)z'_a(\rho)]^2}{n^2\left[r_a(\rho)^2+(z_o-z_a(\rho))^2\right]\left(r'_a(\rho)^2+z'_a(\rho)^2\right)}}}{\sqrt{r'_a(\rho)^2+z'_a(\rho)^2}} \right) \right] \right\} \Bigg)
\end{aligned} } \right)
\end{aligned}
$$

$$
\begin{aligned}
&r_b(\rho)=r_a(\rho)\\
&+\frac{1}{n^2-n_i^2}\left(\frac{n_o[z_a'(\rho)(r_a(\rho)r_a'(\rho)+(z_a(\rho)-z_o)z_a'(\rho))]}{n\sqrt{r_a(\rho)^2+(z_o-z_a(\rho))^2}\left(r_a'(\rho)^2+z_a'(\rho)^2\right)}-\frac{z_a'(\rho)\sqrt{1-\frac{n_o^2[r_a(\rho)r_a'(\rho)+(z_a(\rho)-z_o)z_a'(\rho)]^2}{n^2\left[r_a(\rho)^2+(z_o-z_a(\rho))^2\right]\left(r_a'(\rho)^2+z_a'(\rho)^2\right)}}}{\sqrt{r_a'(\rho)^2+z_a'(\rho)^2}}\right)\\
&\times\Bigg(n\left(n_i z_i-n_o z_o+n\tau+n_o\,\mathrm{Sign}\left[z_o\right]\sqrt{r_a(\rho)^2+(z_o-z_a(\rho))^2}\right)\\
&\quad+n_i^2\left\{\begin{array}{l}\left[r_a(\rho)\times\left(\frac{n_o[z_a'(\rho)(r_a(\rho)r_a'(\rho)+(z_a(\rho)-z_o)z_a'(\rho))]}{n\sqrt{r_a(\rho)^2+(z_o-z_a(\rho))^2}\left(r_a'(\rho)^2+z_a'(\rho)^2\right)}-\frac{z_a'(\rho)\sqrt{1-\frac{n_o^2[r_a(\rho)r_a'(\rho)+(z_a(\rho)-z_o)z_a'(\rho)]^2}{n^2\left[r_a(\rho)^2+(z_o-z_a(\rho))^2\right]\left(r_a'(\rho)^2+z_a'(\rho)^2\right)}}}{\sqrt{r_a'(\rho)^2+z_a'(\rho)^2}}\right)\right]\\
+\\
\left[\left(z_a(\rho)-z_i-\tau\right)\times\left(\frac{n_o[z_a'(\rho)(r_a(\rho)r_a'(\rho)+(z_a(\rho)-z_o)z_a'(\rho))]}{n\sqrt{r_a(\rho)^2+(z_o-z_a(\rho))^2}\left(r_a'(\rho)^2+z_a'(\rho)^2\right)}+\frac{r_a'(\rho)\sqrt{1-\frac{n_o^2[r_a(\rho)r_a'(\rho)+(z_a(\rho)-z_o)z_a'(\rho)]^2}{n^2\left[r_a(\rho)^2+(z_o-z_a(\rho))^2\right]\left(r_a'(\rho)^2+z_a'(\rho)^2\right)}}}{\sqrt{r_a'(\rho)^2+z_a'(\rho)^2}}\right)\right]\end{array}\right\}\\
&\quad-n_i\,\mathrm{Sign}\left[n_1\right]\times\\
&\quad\sqrt{\begin{array}{l}\left(n_i z_i-n_o z_o+n\tau+n_o\,\mathrm{Sign}\left[z_o\right]\sqrt{r_a(\rho)^2+(z_o-z_a(\rho))^2}\right)^2+n^2\left(r_a(\rho)^2+(z_a(\rho)-z_i-\tau)^2\right)\\
-n_i^2\left\{\begin{array}{l}\left[\left(z_a(\rho)-z_i-\tau\right)\times\left(\frac{n_o[z_a'(\rho)(r_a(\rho)r_a'(\rho)+(z_a(\rho)-z_o)z_a'(\rho))]}{n\sqrt{r_a(\rho)^2+(z_o-z_a(\rho))^2}\left(r_a'(\rho)^2+z_a'(\rho)^2\right)}-\frac{z_a'(\rho)\sqrt{1-\frac{n_o^2[r_a(\rho)r_a'(\rho)+(z_a(\rho)-z_o)z_a'(\rho)]^2}{n^2\left[r_a(\rho)^2+(z_o-z_a(\rho))^2\right]\left(r_a'(\rho)^2+z_a'(\rho)^2\right)}}}{\sqrt{r_a'(\rho)^2+z_a'(\rho)^2}}\right)\right]\\
+\\
\left[-r_a(\rho)\times\left(\frac{n_o[z_a'(\rho)(r_a(\rho)r_a'(\rho)+(z_a(\rho)-z_o)z_a'(\rho))]}{n\sqrt{r_a(\rho)^2+(z_o-z_a(\rho))^2}\left(r_a'(\rho)^2+z_a'(\rho)^2\right)}+\frac{r_a'(\rho)\sqrt{1-\frac{n_o^2[r_a(\rho)r_a'(\rho)+(z_a(\rho)-z_o)z_a'(\rho)]^2}{n^2\left[r_a(\rho)^2+(z_o-z_a(\rho))^2\right]\left(r_a'(\rho)^2+z_a'(\rho)^2\right)}}}{\sqrt{r_a'(\rho)^2+z_a'(\rho)^2}}\right)\right]\end{array}\right\}^2\\
+2n\left(n_i z_i-n_o z_o+n\tau+n_o\,\mathrm{Sign}\left[z_o\right]\sqrt{r_a(\rho)^2+(z_o-z_a(\rho))^2}\right)\\
\times\left\{\begin{array}{l}\left[r_a(\rho)\times\left(\frac{n_o[z_a'(\rho)(r_a(\rho)r_a'(\rho)+(z_a(\rho)-z_o)z_a'(\rho))]}{n\sqrt{r_a(\rho)^2+(z_o-z_a(\rho))^2}\left(r_a'(\rho)^2+z_a'(\rho)^2\right)}-\frac{z_a'(\rho)\sqrt{1-\frac{n_o^2[r_a(\rho)r_a'(\rho)+(z_a(\rho)-z_o)z_a'(\rho)]^2}{n^2\left[r_a(\rho)^2+(z_o-z_a(\rho))^2\right]\left(r_a'(\rho)^2+z_a'(\rho)^2\right)}}}{\sqrt{r_a'(\rho)^2+z_a'(\rho)^2}}\right)\right]\\
+\\
\left[(z_a(\rho)-z_i-\tau)\times\left(\frac{n_o[z_a'(\rho)(r_a(\rho)r_a'(\rho)+(z_a(\rho)-z_o)z_a'(\rho))]}{n\sqrt{r_a(\rho)^2+(z_o-z_a(\rho))^2}\left(r_a'(\rho)^2+z_a'(\rho)^2\right)}+\frac{r_a'(\rho)\sqrt{1-\frac{n_o^2[r_a(\rho)r_a'(\rho)+(z_a(\rho)-z_o)z_a'(\rho)]^2}{n^2\left[r_a(\rho)^2+(z_o-z_a(\rho))^2\right]\left(r_a'(\rho)^2+z_a'(\rho)^2\right)}}}{\sqrt{r_a'(\rho)^2+z_a'(\rho)^2}}\right)\right]\end{array}\right\}\end{array}}\Bigg)
\end{aligned}
$$

11.6 Illustrative comparison

We have three different models of stigmatic lenses with very different procedures to obtain the solution.

The method in section 11.3 starts from a very extensive procedure presented in chapters 8 and 7.

In section 11.4, we took the first surface from chapter 6 and the second surface from chapter 9. How the first surface of chapter 6 is obtained compared to chapters 8 and 7 is very different. The first comes from solving a differential equation and the second from an algebraic manipulation of a consequence of the Fermat principle.

It is essential to mention that all the presented equations of the first and, above all, of the second surface are massive equations that at first glance are not seen to give the same or similar result.

In the following figures, figures 11.4–11.7, a series of lenses described by the methods of the three sections is shown. The continuous gray line is for the refractive surfaces of section 11.3. The dashed purple line is for the refraction surfaces computed using the equations of 11.4. Finally, the dotted blue line represents the lines obtained by the equations of section 11.5. Note that for all examples, the gray, purple and blue lines are overwritten.

This result is no accident and confirms what we predicted in chapter 9. Given the input values, the curves of the stigmatic lenses are unique.

We have three very different paradigms, the functions that describe them are very complex, lengthy and varied, but when they are plotted, they give the same curve.

Although in this section we limit ourselves to the examples already mentioned, we have compared around a hundred curves with different design specifications. In all cases, the same curve is always obtained.

In the following four sections, we will demonstrate that this equality is not unique in the case that objects and images are finite. We will take the object from infinity minus and make the same comparison.

Following the same order, the next section will be for the collector proposed by Silva–Torres, section 11.7. In section 11.8, it will be the turn for the hybrid collector lens model between the Cartesian oval described by Valencia–Calle and the general equation of stigmatic lenses. Section 11.9 presents the collector lens described by the hybrid method that makes up the first surface of a Silva–Torres Cartesian oval and the general equation for stigmatic lenses. Finally, in section 11.10, there is another illustrative comparison.

11.7 Cartesian ovals in a parametric form for an object at minus infinity

We begin the new comparison with the model of Silva–Torres of a stigmatic lens for $z_o \to -\infty$.

Therefore, we take $\lim_{z_o \to -\infty} K_a$ from equation (8.14),

$$\lim_{z_o \to -\infty} K_a = -\frac{n_o^2}{n^2}, \tag{8.14}$$

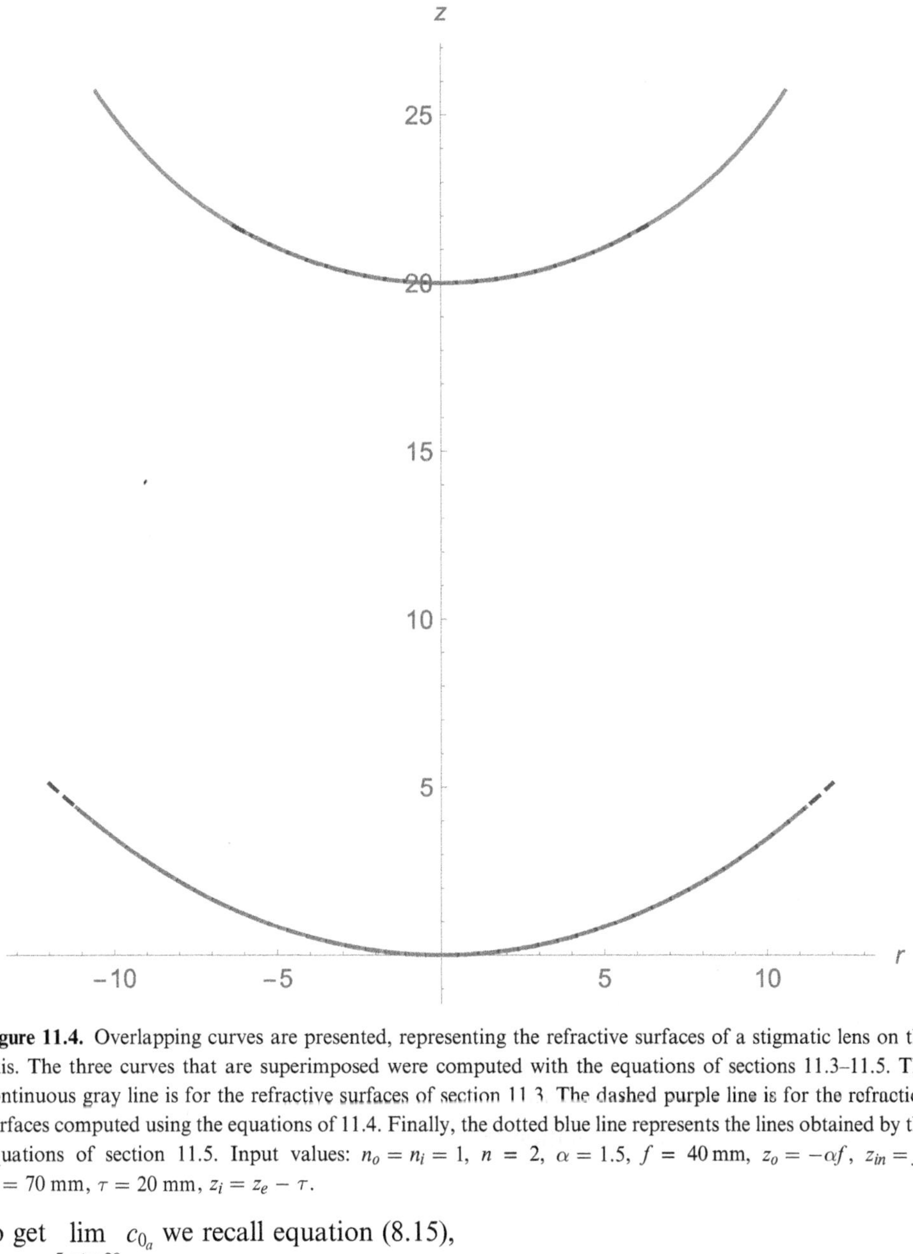

Figure 11.4. Overlapping curves are presented, representing the refractive surfaces of a stigmatic lens on the axis. The three curves that are superimposed were computed with the equations of sections 11.3–11.5. The continuous gray line is for the refractive surfaces of section 11.3. The dashed purple line is for the refraction surfaces computed using the equations of 11.4. Finally, the dotted blue line represents the lines obtained by the equations of section 11.5. Input values: $n_o = n_i = 1$, $n = 2$, $\alpha = 1.5$, $f = 40\,\text{mm}$, $z_o = -\alpha f$, $z_{in} = f$, $z_e = 70\,\text{mm}$, $\tau = 20\,\text{mm}$, $z_i = z_e - \tau$.

to get $\lim_{z_o \to -\infty} c_{0_a}$ we recall equation (8.15),

$$\lim_{z_o \to -\infty} c_{0_a} = \frac{n}{z_{in}(n - n_o)}, \tag{8.15}$$

$\lim_{z_o \to -\infty} c_{1_a}$ is given by equation (8.16),

$$\lim_{z_o \to -\infty} c_{1_a} = 0, \tag{8.16}$$

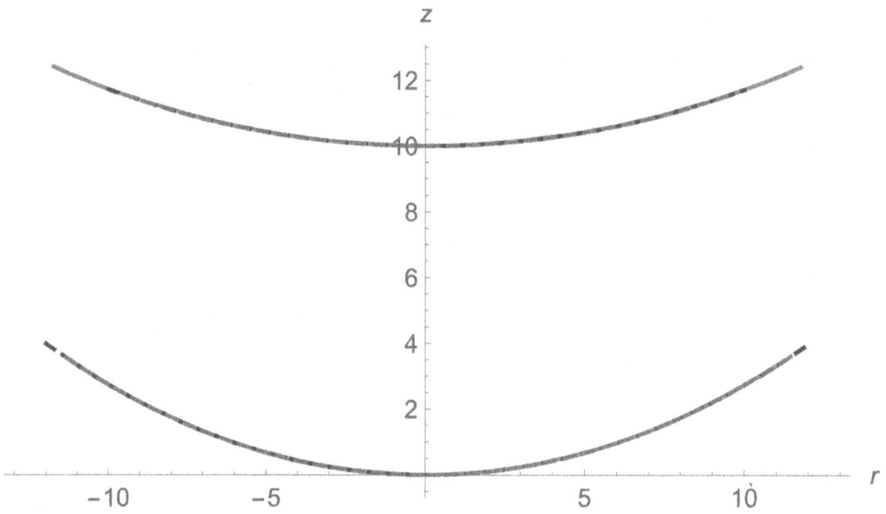

Figure 11.5. Input values: $n_o = n_i = 1$, $n = 2$, $\alpha = 1.5$, $f = 50\,\text{mm}$, $z_o = -\alpha f$, $z_{in} = f$, $z_e = 70\,\text{mm}$, $\tau = 10\,\text{mm}$, $z_i = z_e - \tau$.

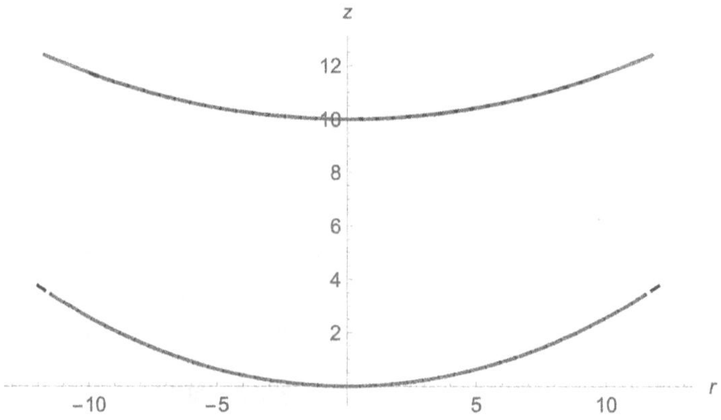

Figure 11.6. Input values: $n_o = n_i = 1$, $n = 2$, $\alpha = 2$, $f = 50\,\text{mm}$, $z_o = -\alpha f$, $z_{in} = f$, $z_e = 70\,\text{mm}$, $\tau = 10\,\text{mm}$, $z_i = z_e - \tau$.

and $\lim\limits_{z_o \to -\infty} b_{1_a}$ is taken from equation (8.17),

$$\lim_{z_o \to -\infty} b_{1_a} = 0, \tag{8.17}$$

the sagitta when $\lim\limits_{z_o \to -\infty}$ is described with equation (8.17),

$$\lim_{z_o \to -\infty} z_a = \frac{\left(\lim\limits_{z_o \to -\infty} c_{0_a}\right)\rho^2}{\sqrt{1 - \left(\lim\limits_{z_o \to -\infty} c_{0_a}\right)^2 \left(\lim\limits_{z_o \to -\infty} K_a\right)\rho^2} + 1}, \tag{8.18}$$

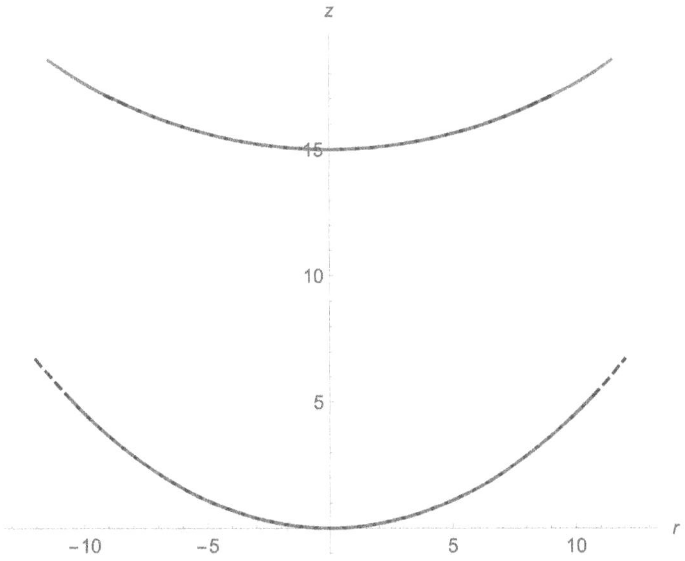

Figure 11.7. Coinciding curves are shown describing the refractive surfaces of a stigmatic lens. The three curves that are overlapped were computed with the equations of sections 11.3–11.5. The constant gray line is for the refractive surfaces of section 11.3. The dashed purple line is for the refraction surfaces of 11.4. The dotted blue line represents the lines of the equations of section 11.5. Input values: $n_o = n_i = 1$, $n = 1.5$, $\alpha = 1.45$, $f = 50$ mm, $z_o = -\alpha f$, $z_{in} = f$, $z_e = 70$ mm, $\tau = 15$ mm, $z_i = z_e - \tau$.

and the radius, equation (8.19),

$$\lim_{z_o \to -\infty} r_a = \mathrm{sgn}(\rho)\sqrt{\rho^2 - \lim_{z_o \to -\infty}(z_a^2)}. \tag{8.19}$$

The parameters of the second surface are the following

$$K_b = \frac{[n_o^2(z_i - \tau) - n^2(z_{in} - \tau)]^2}{nn_o[n(z_i - \tau) - n_o(z_{in} - \tau)][n(z_{in} - \tau) - n_o(z_i - \tau)]} \tag{8.10}$$

$$c_{0_b} = \frac{n(z_i - \tau) - n_o(z_{in} - \tau)}{(n - n_o)(z_i - \tau)(z_{in} - \tau)}, \tag{8.11}$$

we recall equation (8.12),

$$c_{1_b} = \frac{(n - n_o)(n + n_o)^2}{4nn_o(z_i - \tau)(z_{in} - \tau)[n(z_{in} - \tau) - n_o(z_i - \tau)]}, \tag{8.12}$$

finally, $\lim_{z_o \to -\infty} b_{1_b}$ is recalled with equation (8.13),

$$b_{1_b} = \frac{(n + n_o)(n^2(z_{in} - \tau) - n_o^2(z_i - \tau))}{2nn_o(z_i - \tau)(z_{in} - \tau)[n(z_{in} - \tau) - n_o(z_i - \tau)]}. \tag{8.13}$$

The second surface of the stigmatic lens proposed by Silva–Torres is

$$z_{b_0} = \frac{\rho^2(c_{1_b}\rho^2 + c_{0_b})}{\sqrt{\rho^2\left(2b_{1_b} - c_{0_b}^2 K_b\right) + 1} + b_{1_b}\rho^2 + 1}, \tag{8.7}$$

where

$$z_b = \tau + z_{b_0}(\rho), \tag{8.8}$$

and

$$r_b = \operatorname{sgn}(\rho)\sqrt{\rho^2 - z_{b_0}(\rho)^2}. \tag{8.9}$$

The equations of this section, equations (8.7)–(8.19), are the ones used in section 10.10.

11.8 Cartesian ovals in an explicit form for an object at minus infinity

Now we introduce the Cartesian oval when $z_o \to -\infty$ is in its explicit form, with equation (6.44),

$$z_a = \frac{n\left[\left(n - 1\right)f - \operatorname{sign}(f)\sqrt{(n - 1)n^2[(n - 1)f^2 - (n + 1)r^2]}\right]}{n^2 - 1}. \tag{6.44}$$

Given the Cartesian oval when the limit $z_o \to -\infty$ is in its explicit form, the second surface is described by equation (9.55),

$$\begin{cases} \lim\limits_{z_o \to -\infty}(z_b) = z_a + \left[\lim\limits_{z_o \to -\infty}(\vartheta)\right]\left[\lim\limits_{z_o \to -\infty}(\wp_z)\right], \\ \lim\limits_{z_o \to -\infty}(r_b) = r_a + \left[\lim\limits_{z_o \to -\infty}(\vartheta)\right]\left[\lim\limits_{z_o \to -\infty}(\wp_r)\right] \end{cases}, \tag{9.55}$$

where $\lim\limits_{z_o \to -\infty}(\vartheta)$ is described by equation (9.53),

$$\lim_{z_o \to -\infty}(\vartheta) = \frac{-\beta \pm \sqrt{\beta^2 + \left(n^2 - n_i^2\right)\left\{n_i^2 r_a^2 + n_i^2 z_\tau^2 - \left[\lim\limits_{z_o \to -\infty}(z_f)\right]^2\right\}}}{(n_i^2 - n^2)}, \tag{9.53}$$

where

$$\beta \equiv \left\{\left[\lim_{z_o \to -\infty}(z_f)\right]n + n_i^2 r_a\left[\lim_{z_o \to -\infty}(\wp_r)\right] + n_i^2 z_\tau\left[\lim_{z_o \to -\infty}(\wp_z)\right]\right\}, \tag{9.54}$$

and

$$\lim_{z_o \to -\infty} z_f = -z_a n_o + n_i z_i + n\tau. \tag{9.52}$$

Finally, to get the cosine directors when $\lim_{z_o \to -\infty}$, we recall equations (9.50) and (9.51),

$$\lim_{z_o \to -\infty} \wp_z = \frac{z_a' \left(n_o - n\sqrt{(z_a')^2 + 1}\sqrt{\frac{(n^2 - n_o^2)(z_a')^2 + n^2}{n^2[(z_a')^2 + 1]}} \right)}{n[(z_a')^2 + 1]}, \tag{9.50}$$

and

$$\lim_{z_o \to -\infty} \wp_z = \frac{\sqrt{\frac{(n^2 - n_o^2)(z_a')^2 + n^2}{n^2((z_a')^2 + 1)}}}{\sqrt{(z_a')^2 + 1}} + \frac{n_o(z_a')^2}{n(z_a')^2 + n}. \tag{9.51}$$

The above equations are those implemented in the comparison of section 10.10.

11.9 Cartesian ovals in a parametric form as a first surface and general equation of stigmatic lenses for an object at minus infinity

Finally, we arrive at our last model, when the first surface is a Cartesian oval of Silva–Torres and the second surface is the general equation of stigmatic lenses.

We start by recalling $\lim_{z_o \to -\infty} K_a$ with equation (8.14),

$$\lim_{z_o \to -\infty} K_a = -\frac{n_o^2}{n^2}, \tag{8.14}$$

$\lim_{z_o \to -\infty} c_{0_a}$ is with equation (8.15),

$$\lim_{z_o \to -\infty} c_{0_a} = \frac{n}{z_{in}(n - n_o)}. \tag{8.15}$$

If we recall equation (8.16), we recall $\lim_{z_o \to -\infty} c_{1_a}$,

$$\lim_{z_o \to -\infty} c_{1_a} = 0 \tag{8.16}$$

and

$$\lim_{z_o \to -\infty} b_{1_a} = 0. \tag{8.17}$$

The sagitta of the first surface when $z_o \to -\infty$ is

$$\lim_{z_o \to -\infty} z_a = \frac{\left(\lim_{z_o \to -\infty} c_{0_a}\right)\rho^2}{\sqrt{1 - \left(\lim_{z_o \to -\infty} c_{0_a}\right)^2 \left(\lim_{z_o \to -\infty} K_a\right)\rho^2} + 1}, \tag{8.18}$$

and its radius is

$$\lim_{z_o \to -\infty} r_a = \operatorname{sgn}(\rho)\sqrt{\rho^2 - \lim_{z_o \to -\infty} (z_a^2)}. \tag{8.19}$$

Now, the second surface is when limit $z_o \to -\infty$ is applied over equations (11.27) and (11.28),

$$\begin{cases} \lim_{z_o \to -\infty} (z_b) = \lim_{z_o \to -\infty} (z_a) + \left[\lim_{z_o \to -\infty} (\vartheta)\right]\left[\lim_{z_o \to -\infty} (\wp_z)\right], \\ \lim_{z_o \to -\infty} (r_b) = \lim_{z_o \to -\infty} (r_a) + \left[\lim_{z_o \to -\infty} (\vartheta)\right]\left[\lim_{z_o \to -\infty} (\wp_r)\right] \end{cases}, \tag{11.29}$$

where $\lim_{z_o \to -\infty} (\vartheta)$ is the limit when applied over equation (11.26),

$$\lim_{z_o \to -\infty} (\vartheta) = \frac{\beta - \sqrt{\alpha + \beta^2}}{n^2 - n_i^2}. \tag{11.30}$$

α is

$$\alpha \equiv (n^2 - n_i^2)\left\{n_i^2\left[\lim_{z_o \to -\infty} (r_a^2)\right] + z_\tau^2 - \lim_{z_o \to -\infty} \left(z_f^2\right)\right\}. \tag{11.31}$$

β is

$$\beta \equiv \lim_{z_o \to -\infty} (z_f)n + n_i^2\left[\lim_{z_o \to -\infty} (r_a) \lim_{z_o \to -\infty} (\wp_r) + z_\tau \lim_{z_o \to -\infty} (\wp_z)\right]. \tag{11.32}$$

z_f when $z_o \to -\infty$ is given by

$$\lim_{z_o \to -\infty} z_f = \lim_{z_o \to -\infty} (n\tau + n_i z_i - n_o z_a). \tag{11.33}$$

z_τ when $z_o \to -\infty$ is given by

$$\lim_{z_o \to -\infty} z_\tau = \lim_{z_o \to -\infty} (-\tau + z_a - z_i). \tag{11.34}$$

Finally, when $z_o \to -\infty$, the cosine directors of equations (11.11) and (11.12) are given by

$$\lim_{z_o \to -\infty} \wp_r = \lim_{z_o \to -\infty} \left\{ \frac{n_o\left[\left(z_a - z_o\right)z_a' + r_a r_a'\right]r_a'}{-\operatorname{sgn}(z_o)n\sqrt{r_a^2 + (z_o - z_a)^2}\left(r_a'^2 + z_a'^2\right)} - z_a' \frac{\sqrt{1 - \dfrac{n_o^2\left[r_a r_a' + \left(z_a - z_o\right)z_a'\right]^2}{n^2\left[r_a^2 + (z_o - z_a)^2\right]\left(r_a'^2 + z_a'^2\right)}}}{\sqrt{r_a'^2 + z_a'^2}} \right\} \tag{11.35}$$

and

$$\lim_{z_o \to -\infty} \wp_z = \lim_{z_o \to -\infty} \left\{ \frac{n_o\left[r_a r_a' + (z_a - z_o) z_a'\right] z_a'}{-\operatorname{sgn}(z_o) n \sqrt{r_a^2 + (z_o - z_a)^2}\left(r_a'^2 + z_a'^2\right)} + r_a' \frac{\sqrt{1 - \frac{n_o^2\left[r_a r_a' + (z_a - z_o) z_a'\right]^2}{n^2\left[r_a^2 + (z_o - z_a)^2\right]\left(r_a'^2 + z_a'^2\right)}}}{\sqrt{r_a'^2 + z_a'^2}} \right\} \tag{11.36}$$

computing the aforementioned limit over equations (11.35) and (11.36),

$$\lim_{z_o \to -\infty} \wp_r = \frac{z_a'\left(n_o r_a' - n\sqrt{r_a'^2 + z_a'^2}\sqrt{\frac{(n^2 - n_o^2) z_a'^2 + n^2 r_a'^2}{n^2(r_a'^2 + z_a'^2)}}\right)}{n(r_a'^2 + z_a'^2)} \tag{11.37}$$

and

$$\lim_{z_o \to -\infty} \wp_z = \frac{r_a'\sqrt{\frac{(n^2 - n_o^2) z_a'^2 + n^2 r_a'^2}{n^2(r_a'^2 + z_a'^2)}}}{\sqrt{r_a'^2 + z_a'^2}} + \frac{n_o z_a'^2}{n(r_a'^2 + z_a'^2)}. \tag{11.38}$$

Equations (8.14)–(8.19), (11.29)–(11.34), (11.37) and (11.38) of this section are the ones implemented to design a singlet stigmatic collector lens with the paradigm proposed in this section.

11.10 Illustrative comparison

Once again, we compare the three paradigms presented in the previous three sections. This time, the comparison is for collector lenses, that is $z_o \to -\infty$.

In figures 11.8–11.11 are lenses described by the methods of the previous three sections. The constant gray line is for the refractive surfaces of section 11.7. The dashed purple line is for the refraction surfaces of 11.8. The dotted blue line represents the lines of the equations of section 11.9. Again, all examples of the gray, purple and blue lines are overwritten.

We repeat that this is not a coincidence; we have tested hundreds of examples. This result only confirms that nature has made stigmatic systems unique for each

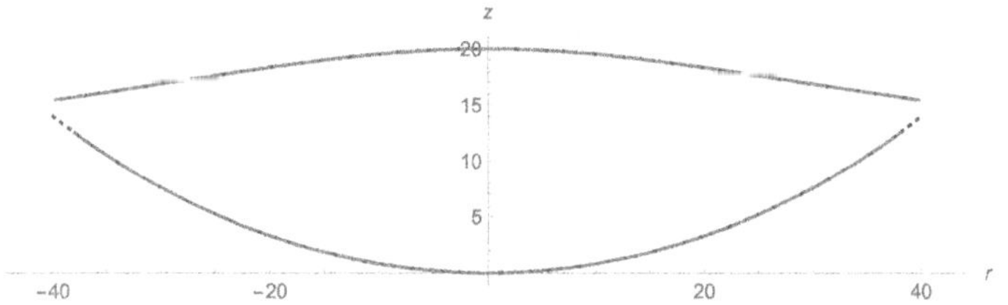

Figure 11.8. Input values: $n_o = n_i = 1$, $n = 1.7$, $z_o = -\infty$, $z_{in} = 150$ mm, $z_e = 70$ mm, $\tau = 20$ mm, $z_i = z_e - \tau$.

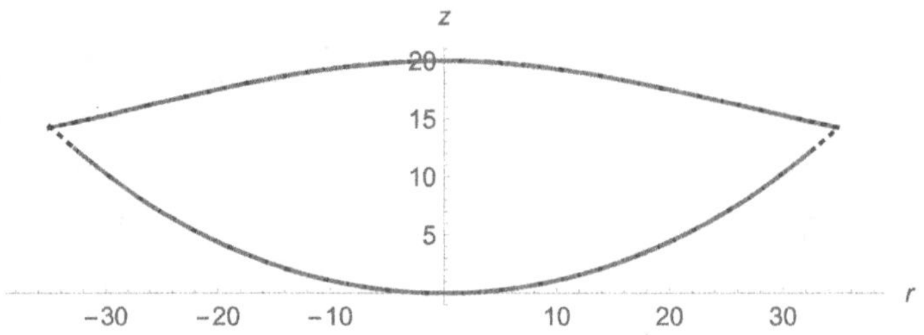

Figure 11.9. Input values: $n_o = n_i = 1$, $n = 1.5$, $z_o = -\infty$, $z_{in} = 140$ mm, $z_e = 70$ mm, $\tau = 20$ mm, $z_i = z_e - \tau$.

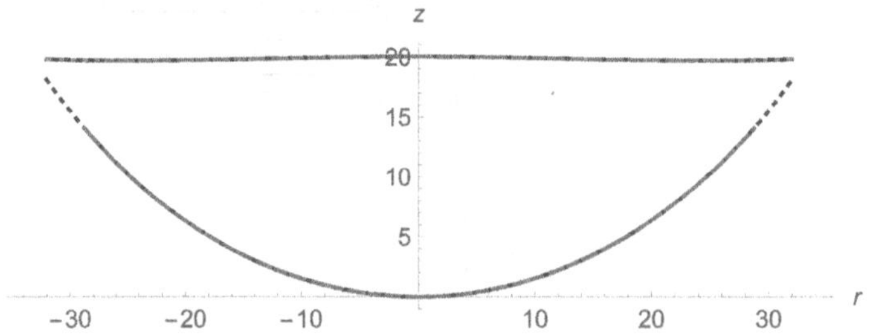

Figure 11.10. Input values: $n_o = n_i = 1$, $n = 1.5$, $z_o = -\infty$, $z_{in} = 100$ mm, $z_e = 70$ mm, $\tau = 20$ mm, $z_i = z_e - \tau$.

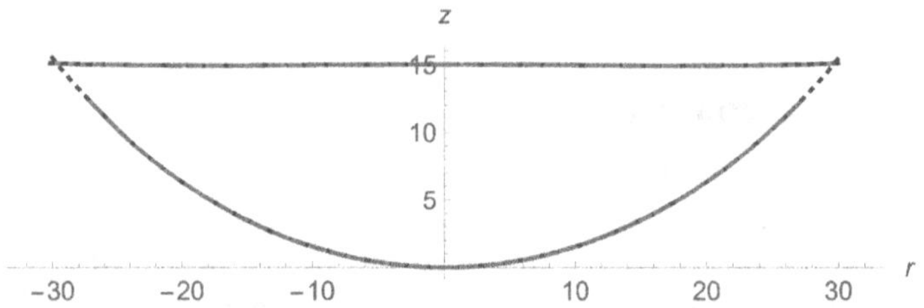

Figure 11.11. Input values: $n_o = n_i = 1$, $n = 1.5$, $z_o = -\infty$, $z_{in} = 100$ mm, $z_e = 70$ mm, $\tau = 15$ mm, $z_i = z_e - \tau$.

particular configuration. There is an infinite number of stigmatic lenses, but for a given first surface, provided indexes of refraction, the distance between a given object and image, and given a central thickness, there is only one solution.

Given a distance z_i and a $z_o \to -\infty$ there is only a single reflective surface that causes the rays emerging from the object at $z_o \to -\infty$ to converge on the image at z_i. That curve is a parabola given by

$$z_a = \frac{r_a^2}{4z_i}. \tag{11.39}$$

There are no others, nature does not allow that! This is what we discovered in chapter 5, and in chapter 7 we reaffirmed it.

The implications of this comparison are not trivial and show a profound characteristic nature of light under SVEA and stigmatism as an implication of it. In the following section, we discuss these implications in more detail.

11.11 Implications

The implications of the uniqueness of stigmatic surfaces are not left in the mirror or single-lens stigmatic systems.

In our previous treatise, *Analytical lens design*, in chapters 11 and 12 we showed the generalization of the general equation of stigmatic lenses (see chapter 9).

The generalization consists in giving an arbitrary series of the refractive surfaces as it should be one last refractive surface of the system for the system to become stigmatic on the axis. Not only is the shape of the refractive surfaces known, but also their central thicknesses, and their indexes of refraction.

The procedure for finding said last refractive surface is very similar to that presented in chapter 9. Which, as in this chapter, predicts that the last refractive surface is unique for a given configuration. In other words, given the shapes of the refractive surfaces, their central thickness and the index of refraction, there is only one additional refractive surface such that the system is stigmatic.

Resolving the system for the last refractive surface does not limit the model in general. Since, for example, with the comparisons prescribed, we can say that given the input parameters and the first surface of the lens, there is the only one second surface that makes up for the first with the ability to make the lens stigmatic.

So now we talk about a pair of surfaces that a stigmatic lens makes. For example, if our first surface is a sphere, with the general equation of stigmatic lenses, we can design an aspheric surface such that, given the already mentioned first surface, the system is stigmatic.

But the geometric optics is reversible, that is to say, given the aforementioned stigmatic surface, there is a spherical surface that makes the system stigmatic on the axis.

The same thing happens if we have more surfaces. Given a series of refractive surfaces, there is only one that can group with the previous refractive surfaces such that the complete system is stigmatic on the axis. So once you see that the system is stigmatic on the axis, there is no way to distinguish which surface does the work; all of them work together to make the system stigmatic on the axis.

The standard for optical design is the use of commercial optical design software to design all kinds of lenses and many systems with a lot of lenses. Commercial optical design software uses a series of algorithms that optimize an optical system to help the user to reduce aberrations of the optical system in question. In general, what is expected is that the system is more or less stigmatic on the optical axis. The filter to accept a design, among other considerations, is that at least the spot diagram is inside the air disk for the object that is on-axis that produced an on-axis image.

To achieve this, commercial optical design software uses purely numerical algorithms that approximate the system to the fulfillment of the requirement above.

On the other hand, it is essential to note that in the optical design industry, the use of mostly spherical surfaces is encouraged, due to the simplicity of their production.

The exciting thing here is that, if we start from a system made by pure spheres and such that it has already been optimized to be more or less stigmatic on the axis, we can describe the outermost surface as we described in chapters 11 and 12 of *Analytical lens design* and that the refractive surface will look similar to a sphere.

The equations presented in chapters 11 and 12 of *Analytical lens design* will approximate, more or less, the last refractive spherical surface of the system above, depending on how stigmatic the system is.

The same is true for a lens made in commercial software or designed with the methodology in chapter 9. Given the input values, the second surface of the lens designed in commercial software will be so similar to the second surface of the lens in chapter 9, like how stigmatic it is. All of this is because the nature of stigmatism is confirmed with the comparisons made in this chapter.

It may be that mathematical expressions of the commercial software look very different from the expression of chapter 9. But the curves will look very similar if the expression of the commercial software is close to being stigmatic. All this is taking the section of the curves that are exposed by the axis object and that converge in the axis image.

Therefore, we can understand the existence and uniqueness of stigmatism.

When a problem of initial value mathematically models a physical situation, the existence and uniqueness of the solution is of the utmost importance, since it is surely expected to have a solution, because physically something must happen. On the other hand, the solution is supposed to be unique, because if we repeat the experiment under identical conditions, we can expect the same results, as long as the model is deterministic. Therefore, when considering an initial value problem of stigmatism, it is natural to ask:

Existence: Will there be a solution to the problem? From the results of chapters 6, 7, 8, 9 and 10 we know that stigmatic lenses exist. Uniqueness: If there is a solution, will it be unique? From equation (9.39), the general equation of stigmatic lenses, *we know that stigmatism is unique*. The results of this chapter just confirm it. Determination: If there is a solution, how do we determine it? Geometrical optics is deterministic.

11.12 End notes

In this chapter, we compared a series of stigmatic lenses generated by Cartesian ovals that are described by different approaches. But it does not imply that the model is the one chosen, given the same input values for all the standards compared, the results are precisely the same.

This implies that stigmatism, given the input values, supports a single solution.

References

Estrada J C V, Calle Á H B and Hernández D M 2013 Explicit representations of all refractive optical interfaces without spherical aberration *J. Opt. Soc. Am.* **A30** 1814–24

González-Acuña R G 2021 Surface solution to correct a freeform wavefront *Appl. Opt.* **60** 9887–91

González-Acuña R G and Chaparro-Romo H A 2018 General formula for bi-aspheric singlet lens design free of spherical aberration *App. Opt.* **57** 9341–5

González-Acuña R G and Guitiérrez-Vega J C 2018 Generalization of the axicon shape: the gaxicon *J. Opt. Soc. Am.* **A35** 1915–8

González-Acuña R G and Gutiérrez-Vega J C 2019 Analytic formulation of a refractive-reflective telescope free of spherical aberration *Opt. Eng.* **58** 085105

González-Acuña R G and Thibault S 2021 The general equation of the stigmatic lenses: its history and what we have learned from it *Int. Optical Design Conf.* (Washington, DC: Optical Society of America) 1207803 p

González-Acuña R G, Avendaño-Alejo M and Gutiérrez-Vega J C 2019a Singlet lens for generating aberration-free patterns on deformed surfaces *J. Opt. Soc. Am.* **A36** 925–9

González-Acuña R G, Chaparro-Romo H A and Gutiérrez-Vega J C 2019b General formula to design freeform singlet free of spherical aberration and astigmatism *Appl. Opt.* **58** 1010–5

González-Acuña R G, Chaparro-Romo H A and Gutiérrez-Vega J C 2020a Analytic aplanatic singlet lens: setting and design for three-point objects and images in the meridional plane *Opt. Eng.* **59** 055104

González-Acuña R G, Chaparro-Romo H A and Gutiérrez-Vega J C 2020b Analytic solution of the eikonal for a stigmatic singlet lens *Phys. Scr.*

González-Acuña R G, Chaparro-Romo H A and Gutiérrez-Vega J C 2020c *Analytical Lens Design* (Bristol: Institute of Physics Publishing)

González-Acuña R G, Chaparro-Romo H A and Gutiérrez-Vega J C 2021 General stigmatic surfaces *J. Opt. Soc. Am.* **A38** 298–302

González-Acuña R G, Chaparro-Romo H A and Gutíerrez-Vega J C 2019c *Single Lens Telescope* (arXiv:1903.11129)

González-Acuña R G and Gutiérrez-Vega J C 2019a Analytic formulation of a refractive-reflective telescope free of spherical aberration *Opt. Eng.* **58** 1–5

González-Acuña R G and Gutiérrez-Vega J C 2019b General formula to eliminate spherical aberration produced by an arbitrary number of lenses *Opt. Eng.* **58** 1–6

González-Acuña R G and Gutiérrez-Vega J C 2019c General formula for aspheric collimator lens design free of spherical aberration *Current Developments in Lens Design and Optical Engineering XX* ed R B Johnson, V N Mahajan and S Thibault (Bellingham, WA: International Society for Optics and Photonics, SPIE) 181–4 pp

González Acuña R G and Gutiérrez-Vega J C 2019 General formula to design freeform collimator lens free of spherical aberration and astigmatism *Novel Optical Systems, Methods, and Applications XXII* **vol 11 105** (Bellingham, WA: International Society for Optics and Photonics) 111050A p

González Acuña R G and Gutiérrez-Vega J C 2019 General formula of the refractive telescope design free spherical aberration *Novel Optical Systems, Methods, and Applications XXII* **vol 11 105** ed C F Hahlweg and J R Mulley (Bellingham, WA: International Society for Optics and Photonics, SPIE) 162–6 p

González-Acuña R G and Gutiérrez-Vega J C 2020 Analytic design of a spherochromatic singlet *J. Opt. Soc. Am.* **A37** 149–53

Silva-Lora A and Torres R 2023 Primary aberrations theory in optical systems composed of Cartesian refracting surfaces *Proc. R. Soc.* **A479** 20230186

Silva-Lora A and Torres R 2020 Explicit Cartesian oval as a superconic surface for stigmatic imaging optical systems with real or virtual source or image *Proc. R. Soc.* A476

Valencia-Estrada J C and Malacara-Doblado D 2014 Parastigmatic corneal surfaces *Appl. Opt.* **53** 3438–47

Valencia-Estrada J C, Flores-Hernández R B and Malacara-Hernández D 2015 Singlet lenses free of all orders of spherical aberration *Proc. R. Soc.* **A471** 20140608

IOP Publishing

Stigmatic Optics (Second Edition)

Rafael G González-Acuña and Héctor A Chaparro-Romo

Chapter 12

Algorithms for stigmatic design

12.1 Programs for chapter 6

12.1.1 Case: real finite object—real finite image

```
" ClearAll["Global`*"]
(** Definition of set variables **)
```

$$C2 = \frac{n+1}{n-1};$$

$$C4 = \frac{n+1}{n-1};$$

$$C6 = \frac{(n+1)(n^2+6n+1)}{(n-1)^3};$$

$$C8 = \frac{n+1}{n-1};$$

$$C10 = \frac{(n+1)(7n^4+124n^3+122n^2+124n+7)}{(n-1)^5};$$

$$C12 = \frac{(n+1)(3n^4-44n^3-46n^2-44n+3)}{(n-1)^5};$$

$$za[ra\_] := C2\,\frac{ra^2}{2f} + C4\,\frac{ra^4}{8\,f^3} + C6\,\frac{2ra^6}{32f^5} + C8\,\frac{5ra^8}{128f^7} + C10\,\frac{2ra^{10}}{512f^9} + C12\,\frac{14ra^{12}}{2048f^{11}};$$

$$\varrho z[ra\_] := \frac{no(za'[ra]\,(ra+(-zo+za[ra])\,za'[ra]))}{n\sqrt{ra^2+(zo-za[ra])^2}\,(1+(za'[ra])^2)} + \frac{1}{\sqrt{1+(za'[ra])^2}}\sqrt{\left(1-\frac{no^2(ra+(-zo+za[ra])\,za'[ra])^2}{n^2\,(ra^2+(zo-za[ra])^2)\,(1+(za'[ra])^2)}\right)}$$

$$\varrho r[ra\_] := \frac{no(ra+(-zo+za[ra])\,za'[ra])}{n\sqrt{ra^2+(zo-za[ra])^2}\,(1+(za'[ra])^2)} - \frac{za'[ra]}{\sqrt{1+(za'[ra])^2}}\sqrt{\left(1-\frac{no^2(ra+(-zo+za[ra])\,za'[ra])^2)}{n^2\,(ra^2+(zo-za[ra])^2)\,(1+(za'[ra])^2)}\right)}$$

```
(** Homotopy for Eikonal **)

θ1[ra_] := ArcTan[(ra)/(−zo+za[ra])];

θ2[ra_] := ArcTan[(ra)/(za[ra] −zi)];

c1[ra_] := Sec[θ1[ra]](−zo+za[ra]);


(** Setting variable's **)

c = 1;
v = c/n;

no = 1;
ni = no;
n = 1.5;

rmax = 16;
density = 4;
zo = −40;
zi = 40;
f = −zo


(** Plot Section **)

SurfaceV := ParametricPlot[{{za[ra],ra},{zb[ra],rb[ra]}},{ra,−rmax−1,rmax+1},
        PlotStyle →{Black,Red}, AspectRatio→Automatic, AxesLabel →{r,z}]


LensV := Graphics[{EdgeForm[Directive[{Black,Thin}]],
```

```
        Directive[Gray,Opacity[0.3]],
        Polygon[Join[#[[1,1]],#[[2,1]],Reverse[#[[4,1]]],Reverse[#[[3,1]]]]&
        [Cases[SurfaceV,_Line,∞]]]}];


WavesV := Table[{
          ParametricPlot[{
             Piecewise[{
                   {Cos[θ1[ra]]tt+zo,tt<=c1[ra]},
                   {v Cos[θ2[ra]](tt−c1[ra])+za[ra],c1[ra]<tt<c2[ra]}
                  }],

             Piecewise[{
                   {Sin[θ1[ra]]tt,tt<=c1[ra]},
                   {v Sin[θ2[ra]](tt−c1[ra])+ra,c1[ra]<tt<c2[ra]}
                   }]},
          {ra,−rmax,rmax},
          PlotRange→All,
          PlotStyle →{Blue, Thickness[0.002]}]},
          {tt,0,−zo+ n zi,density}];

 RaysV := Table[{
        ParametricPlot[{
           Piecewise[{
                {Cos[θ1[ra]]tt+zo,tt<=c1[ra]},
                {v Cos[θ2[ra]](tt−c1[ra])+za[ra],c1[ra]<tt<c2[ra]},
             }],

           Piecewise[{
                {Sin[θ1[ra]]tt,tt<=c1[ra]},
                {v Sin[θ2[ra]](tt−c1[ra])+ra,c1[ra]<tt<c2[ra]},
              }]},

        {tt,0,−zo+ n  zi},
        PlotRange→All,PlotStyle →{Lighter[Purple], Thickness[0.002]}]},
        {ra,−rmax,rmax,density}];


Show[SurfaceV,LensV,WavesV,RaysV,
   PlotRange→All,ImageSize→{640,Automatic}]
```

12.1.2 Case: Real infinity object—real finite image

```
(** Definition of set variables **)
```

$$\text{za[ra\_]:=}\frac{\text{n((n-1)f-Sign[f]}\sqrt{\text{(n-1)((n-1)f\^{}2-(n+1)ra}^2\text{))}}}{\text{(n}^2\text{-1)}};$$

$$\varrho\text{z[ra\_]:=}\frac{\text{za}'\text{[ra]}^2}{\text{n+n za}'\text{[ra]}^2}+\frac{1}{\sqrt{1+\text{za}'\text{[ra]}^2}}\sqrt{1+\frac{1}{n^2}\left(-1+\frac{1}{1+\text{za}'\text{[ra]}^2}\right)};$$

$$\varrho\text{r[ra\_]:=}\frac{\text{za}'\text{[ra]}}{\text{n }(1+\text{za}'\text{[ra]}^2)}\left(1-\text{n}\sqrt{1+\text{za}'\text{[ra]}^2}\sqrt{1+\frac{1}{n^2}\left(-1+\frac{1}{1+\text{za}'\text{[ra]}^2}\right)}\right);$$

```
(** Homotopy for Eikonal **)

θ1[ra_] := ArcTan[Limit[(ra)/(−zo+za[ra]),zo→−Infinity]]

θ2[ra_] := ArcTan[(ra)/(za[ra]−zi )];

c1[ra_] := (−Zo+za[ra]);


(** Setting variable's **)

c = 1;
v = c/n;

no = 1;
ni = no;
n = 1.5;

rmax = 16;
density = 4;
(* zo −> −Infinity *)
Zo = −55;
zi = 40;
f=zi;


(** Plot Section **)

SurfaceV := ParametricPlot[{{za[ra],ra},{zb[ra],rb[ra]}},{ra,−rmax−1,rmax+1},
          PlotStyle →{Black,Red}, AspectRatio→Automatic, AxesLabel →{r,z}]

LensV := Graphics[{EdgeForm[Directive[{Black,Thin}]],
          Directive[Gray,Opacity[0.3]],
          Polygon[Join[#[[1,1]],#[[2,1]],Reverse[#[[4,1]]],Reverse[#[[3,1]]]]&
          [Cases[SurfaceV,_Line,∞]]]}];

WavesV := Table[{
            ParametricPlot[{
               Piecewise[{
                     {Cos[θ1[ra]]tt+Zo,tt<=c1[ra]},
```

```
                {v Cos[θ2[ra]](tt−c1[ra])+za[ra],c1[ra]<tt<c2[ra]},
               }],

          Piecewise[{
                {ra,tt<=c1[ra]},
                {v Sin[θ2[ra]](tt−c1[ra])+ra,c1[ra]<tt<c2[ra]},
             }]},

        {ra,−rmax,rmax},
        PlotRange→All,
        PlotStyle →{Blue, Thickness[0.002]}]},
        {tt,0,−Zo+n zi,density}];

RaysV := Table[{
      ParametricPlot[{
        Piecewise[{
             {Cos[θ1[ra]]tt+Zo,tt<=c1[ra]},
             {v Cos[θ2[ra]](tt−c1[ra])+za[ra],c1[ra]<tt<c2[ra]},
         }],

        Piecewise[{
             {ra,tt<=c1[ra]},
             {v Sin[θ2[ra]](tt−c1[ra])+ra,c1[ra]<tt<c2[ra]},
             }]},

      {tt,0,−Zo+ n zi},
      PlotRange→All,PlotStyle →{Lighter[Purple], Thickness[0.002]}]},
      {ra,−rmax,rmax,density}];


Show[SurfaceV,LensV,WavesV,RaysV,
   PlotRange→All,ImageSize→{640,Automatic}]
```

12.1.3 Case: Real infinity object—virtual finite image

"

```
(** Definition of set variables **)
```

$$za[ra\_]:=\frac{n((n-1)f-Sign[f]\sqrt{(n-1)((n-1)f\hat{}2-(n+1)ra^2)})}{(n^2-1)};$$

$$\varrho z[ra\_]:=\frac{za'[ra]^2}{n+n\ za'[ra]^2}+\frac{1}{\sqrt{1+za'[ra]^2}}\left(\sqrt{1+\frac{1}{n^2}\left(-1+\frac{1}{1+za'[ra]^2}\right)}\right);$$

$$\varrho r[ra\_]:=\frac{za'[ra]}{n\ (1+za'[ra]^2)}\left(1-n\ \sqrt{1+za'[ra]^2}\ \sqrt{1+\frac{1}{n^2}\left(-1+\frac{1}{1+za'[ra]^2}\right)}\right);$$

```
(** Homotopy for Eikonal **)

θ1[ra_] := ArcTan[Limit[(ra)/(−zo+za[ra]),zo→−Infinity]]

θ2[ra_] := ArcTan[(ra)/(za[ra]−zi )];

c1[ra_] := (−Zo+za[ra]);


(** Setting variable's **)

c = 1;
v = c/n;

no = 1;
ni = no;
n = 1.5;

rmax = 16;
density = 4;
(* zo −> −Infinity *)
Zo = −55;
zi = −40;
f = zi;



(** Plot Section **)

SurfaceV := ParametricPlot[{{za[ra],ra},{zb[ra],rb[ra]}},{ra,−rmax−1,rmax+1},
         PlotStyle →{Black,Red}, AspectRatio→Automatic, AxesLabel →{r,z}]

LensV := Graphics[{EdgeForm[Directive[{Black,Thin}]],
          Directive[Gray,Opacity[0.3]],
          Polygon[Join[#[[1,1]],#[[2,1]],Reverse[#[[4,1]]],Reverse[#[[3,1]]]]&
          [Cases[SurfaceV,_Line,∞]]]}];

 WavesV := Table[{
            ParametricPlot[{
               Piecewise[{
                     {Cos[θ1[ra]]tt+Zo,tt<=c1[ra]},
                     {v Cos[θ2[ra]](tt−c1[ra])+za[ra],c1[ra]<tt<c2[ra]},
                   }],
               Piecewise[{
                     {ra,tt<=c1[ra]},
                     {v Sin[θ2[ra]](tt−c1[ra])+ra,c1[ra]<tt<c2[ra]},
                     }]},
            {ra,−rmax,rmax},
```

```
            PlotRange→All,
            PlotStyle →{Blue, Thickness[0.002]}]},
            {tt,0,−Zo −n zi,density}];

RaysV := Table[{
        ParametricPlot[{
            Piecewise[{
                {Cos[θ1[ra]]tt+Zo,tt<=c1[ra]},
                {v Cos[θ2[ra]](tt−c1[ra])+za[ra],c1[ra]<tt<c2[ra]},
                }],

            Piecewise[{
                {ra,tt<=c1[ra]},
                {v Sin[θ2[ra]](tt−c1[ra])+ra,c1[ra]<tt<c2[ra]},
            }]},

        {tt,0,−Zo − n zi},
        PlotRange→All,PlotStyle →{Lighter[Purple], Thickness[0.002]}]},
        {ra,−rmax,rmax,density}];


Show[SurfaceV,LensV,WavesV,RaysV,
    PlotRange→All,ImageSize→{640,Automatic}]
```

12.1.4 Case: Real finite object—virtual finite image

```
" ClearAll["Global`*"]
 (** Definition of set variables **)
```

$$\text{za}[\text{ra\_}]:=\frac{\text{n}((\text{n}-1)\text{f}-\text{Sign}[\text{f}]\sqrt{(\text{n}-1)((\text{n}-1)\text{f}\hat{\ }2-(\text{n}+1)\text{ra}^2)})}{(\text{n}^2-1)};$$

$$\varrho\text{z}[\text{ra\_}]:=\frac{\text{za}'[\text{ra}]^2}{\text{n}+\text{n za}'[\text{ra}]^2}+\frac{1}{\sqrt{1+\text{za}'[\text{ra}]^2}}\sqrt{1+\frac{1}{\text{n}^2}(-1+\frac{1}{1+\text{za}'[\text{ra}]^2}});$$

$$\varrho\text{r}[\text{ra\_}]:=\frac{\text{za}'[\text{ra}]}{\text{n }(1+\text{za}'[\text{ra}]^2)}(1-\text{n }\sqrt{1+\text{za}'[\text{ra}]^2}\ \sqrt{1+\frac{1}{\text{n}^2}(-1+\frac{1}{1+\text{za}'[\text{ra}]^2}}));$$

```
(** Homotopy for Eikonal **)

θ1[ra_] := ArcTan[(ra)/(−zo+za[ra])];

θ2[ra_] := ArcTan[(rb[ra]−ra)/(zb[ra]−za[ra] )];

c1[ra_] := Sec[θ1[ra]](−zo+za[ra]);

(** Setting variable's **)

c = 1;
v = c/n;

no = 1;
ni = no;
n = 1.5;

rmax = 16;
density = 4;
zo = −55;
zi = −40;
f =zi;

(** Plot Section **)

SurfaceV := ParametricPlot[{{za[ra],ra},{zb[ra],rb[ra]}},{ra,−rmax−1,rmax+1},
          PlotStyle →{Black,Red}, AspectRatio→Automatic, AxesLabel →{r,z}]

LensV := Graphics[{EdgeForm[Directive[{Black,Thin}]],
          Directive[Gray,Opacity[0.3]],
          Polygon[Join[#[[1,1]],#[[2,1]],Reverse[#[[4,1]]],Reverse[#[[3,1]]]]&
          [Cases[SurfaceV,_Line,∞]]]}];

 WavesV := Table[{
          ParametricPlot[{
             Piecewise[{
                   {Cos[θ1[ra]]tt+zo,tt<=c1[ra]},
                   {v Cos[θ2[ra]](tt−c1[ra])+za[ra],c1[ra]<tt<c2[ra]},
                 }],

             Piecewise[{
                   {ra,tt<=c1[ra]},
                   {v Sin[θ2[ra]](tt−c1[ra])+ra,c1[ra]<tt<c2[ra]},
                }]},
          { ra,−rmax,rmax},
          PlotRange→All,
          PlotStyle →{Blue, Thickness[0.002]}]},
          {tt,0,−zo − n zi+0.25,density}];
```

```
RaysV := Table[{
        ParametricPlot[{
           Piecewise[{
                {Cos[θ1[ra]]tt+zo,tt<=c1[ra]},
                {v Cos[θ2[ra]](tt−c1[ra])+za[ra],c1[ra]<tt<c2[ra]},
               }],

           Piecewise[{
                {ra,tt<=c1[ra]},
                {v Sin[θ2[ra]](tt−c1[ra])+ra,c1[ra]<tt<c2[ra]},
              }]},

        {tt,0,−zo+ n zi+0.25},
        PlotRange→All,PlotStyle →{Lighter[Purple], Thickness[0.002]}]},
        { ra,−rmax,rmax,density}];


Show[SurfaceV,LensV,WavesV,RaysV,
    PlotRange→All,ImageSize→{640,Automatic}]
```

12.1.5 Case: Real finite object—real infinite image

```
" ClearAll["Global`*"]
(** Definition of set variables **)
```

$$za[ra\_]:=\frac{(n-1)f-\mathrm{Sign}[f]\sqrt{(n-1)((n-1)f^2+(n+1)ra^2)}}{(n^2-1)}$$

$$\varrho z[ra\_]:=\frac{za'[ra]\ (ra+(-zo+za[ra])\ za'[ra])}{-\mathrm{Sign}[zo]n\ \sqrt{ra^2+(zo-za[ra])^2}\ (1+(za'[ra])^2)}$$

$$+\frac{1}{\sqrt{1+(za'[ra])^2}}\sqrt{1-\frac{(ra+(-zo+za[ra])\ za'[ra])^2}{n^2\ (ra^2+(zo-za[ra])^2)\ (1+(za'[ra])^2)}}$$

$$\varrho r[ra\_]:=\frac{ra+(-zo+za[ra])\ za'[ra]}{-\mathrm{Sign}[zo]n\ \sqrt{ra^2+(zo-za[ra])^2}\ (1+(za'[ra])^2)}$$

$$-\frac{za'[ra]}{\sqrt{1+(za'[ra])^2}}\sqrt{1-\frac{(ra+(-zo+za[ra])\ za'[ra])^2}{n^2\ (ra^2+(zo-za[ra])^2)\ (1+(za'[ra])^2)}}$$

```
(** Homotopy for Eikonal **)

 θ1[ra_] := ArcTan[(ra)/(−zo+za[ra])];

 c1[ra_] := Sec[θ1[ra]](−zo+za[ra]);

(** Setting variable's **)

c = 1;
v = c/n;

no = 1;
ni = no;
n = 1.5;

rmax = 16;
density = 4;
zo = −40;
(* zi --> Infinity*)
Zi = 20;
 f = zo;


(** Plot Section **)

SurfaceV := ParametricPlot[{{za[ra],ra},{zb[ra],rb[ra]}},{ra,−rmax−1,rmax+1},
          PlotStyle →{Black,Red}, AspectRatio→Automatic, AxesLabel →{r,z}]

LensV := Graphics[{EdgeForm[Directive[{Black,Thin}]],
          Directive[Gray,Opacity[0.3]],
          Polygon[Join[#[[1,1]],#[[2,1]],Reverse[#[[4,1]]],Reverse[#[[3,1]]]]&
          [Cases[SurfaceV,_Line,∞]]]}];

 WavesV := Table[{
              ParametricPlot[{
                 Piecewise[{
                        {Cos[θ1[ra]]tt+zo,tt<=c1[ra]},
                        {v Cos[θ2[ra]](tt−c1[ra])+za[ra],c1[ra]<tt<c2[ra]},
                        {zb[ra]+(tt−c2[ra])Cos[θ3[ra]],tt>=c2[ra]}}],

                 Piecewise[{
                        {Sin[θ1[ra]]tt,tt<=c1[ra]},
                        {v Sin[θ2[ra]](tt−c1[ra])+ra,c1[ra]<tt<c2[ra]},
                        {rb[ra]+(tt−c2[ra])Sin[θ3[ra]],tt>=c2[ra]}}]},
              {ra,−rmax,rmax},
              PlotRange→All,
              PlotStyle →{Blue, Thickness[0.002]}]},
              {tt,0,−zo + n Zi + 0.25,density}];

 RaysV := Table[{
          ParametricPlot[{
              Piecewise[{
```

```
            {Cos[θ1[ra]]tt+zo,tt<=c1[ra]},
            {v Cos[θ2[ra]](tt−c1[ra])+za[ra],c1[ra]<tt<c2[ra]},
            {zb[ra]+(tt−c2[ra])Cos[θ3[ra]],tt>=c2[ra]}}],

        Piecewise[{
            {Sin[θ1[ra]]tt,tt<=c1[ra]},
            {v Sin[θ2[ra]](tt−c1[ra])+ra,c1[ra]<tt<c2[ra]},
            {rb[ra]+(tt−c2[ra])Sin[θ3[ra]],tt>=c2[ra]}}]},

    {tt,0,−zo + n Zi + 0.25},
    PlotRange→All,PlotStyle →{Lighter[Purple], Thickness[0.002]}]},
    {ra,−rmax,rmax,density}];


Show[SurfaceV,LensV,WavesV,RaysV,
    PlotRange→All,ImageSize→{640,Automatic}]
```

12.1.6 Case: Virtual finite object—real infinite image

```
" ClearAll["Global`*"]
(** Definition of set variables **)
```

$$za[ra\_]:=\frac{(n-1)f-Sign[f]\sqrt{(n-1)((n-1)f^2+(n+1)ra^2)}}{(n^2-1)}$$

$$\varrho z[ra\_]:=\frac{za'[ra]\ (ra+(-zo+za[ra])\ za'[ra])}{-Sign[zo]n\ \sqrt{ra^2+(zo-za[ra])^2}\ (1+(za'[ra])^2)}+\frac{1}{\sqrt{1+(za'[ra])^2}}\sqrt{1-\frac{(ra+(-zo+za[ra])\ za'[ra])^2}{n^2\ (ra^2+(zo-za[ra])^2)\ (1+(za'[ra])^2)}}$$

$$\varrho r[ra\_]:=\frac{ra+(-zo+za[ra])\ za'[ra]}{-Sign[zo]n\ \sqrt{ra^2+(zo-za[ra])^2}\ (1+(za'[ra])^2)}-\frac{za'[ra]}{\sqrt{1+(za'[ra])^2}}\sqrt{1-\frac{(ra+(-zo+za[ra])\ za'[ra])^2}{n^2\ (ra^2+(zo-za[ra])^2)\ (1+(za'[ra])^2)}}$$

```
(** Homotopy for Eikonal **)
```

```
θ1[ra_] := ArcTan[(ra)/(−zo+za[ra])];

 c1[ra_] := (zo + n  Zi) + Sec[θ1[ra]](−zo + za[ra]);

(** Setting variable's **)

c = 1;
v = c/n;

no = 1;
ni = no;
n = 1.5;

rmax = 16;
density = 4;
zo = 40;
(* zi −> Infinity*)
Zi = 20;
f = zo;


(** Plot Section **)

SurfaceV := ParametricPlot[{{za[ra],ra},{zb[ra],rb[ra]}},{ra,−rmax−1,rmax+1},
          PlotStyle →{Black,Red}, AspectRatio→Automatic, AxesLabel →{r,z}]


LensV := Graphics[{EdgeForm[Directive[{Black,Thin}]],
           Directive[Gray,Opacity[0.3]],
           Polygon[Join[#[[1,1]],#[[2,1]],Reverse[#[[4,1]]],Reverse[#[[3,1]]]]&
           [Cases[SurfaceV,_Line,∞]]]}];

 WavesV := Table[{
             ParametricPlot[{
                Piecewise[{
                      {Cos[θ1[ra]](tt−(n zo+n τ))+zo,tt<=c1[ra]},
                      {v Cos[θ2[ra]](tt−c1[ra])+za[ra],c1[ra]<tt<c2[ra]},
                      }],

                Piecewise[{
                      {Sin[θ1[ra]](tt−(n zo+n τ)),tt<=c1[ra]},
                      {v Sin[θ2[ra]](tt−c1[ra])+ra,c1[ra]<tt<c2[ra]},
                      }]},
             {ra,−rmax,rmax},
             PlotRange→All,
             PlotStyle →{Blue, Thickness[0.002]}]},
             {tt,0, zo + n  Zi ,density}];

  RaysV := Table[{
          ParametricPlot[{
             Piecewise[{
                  {Cos[θ1[ra]](tt−(n zo+n τ))+zo,tt<=c1[ra]},
```

```
            {v Cos[θ2[ra]](tt−c1[ra])+za[ra],c1[ra]<tt<c2[ra]},
        }],

        Piecewise[{
            {Sin[θ1[ra]](tt−(n zo+n τ)),tt<=c1[ra]},
            {v Sin[θ2[ra]](tt−c1[ra])+ra,c1[ra]<tt<c2[ra]},
        }]},

    {tt,0, zo + n  Zi + 0.25},
    PlotRange→All,PlotStyle →{Lighter[Purple], Thickness[0.002]}]},
    {ra,−rmax,rmax,density}];


Show[SurfaceV,LensV,WavesV,RaysV,
    PlotRange→All,ImageSize→{640,Automatic}]
```

12.1.7 Case: Virtual finite object—virtual finite image

```
" ClearAll["Global`*"]
 (** Definition of set variables **)
```

$$C2 = -\frac{\alpha n+1}{\alpha(n-1)};$$

$$C4 = -\frac{\alpha^3 n^2+(\alpha^3+2\alpha^2-2\alpha-1)n+1}{\alpha^3(n-1)^2};$$

$$C6 = -\frac{\alpha^5 n^3+(2\alpha^5+3\alpha^4-3\alpha^3+\alpha^2+3\alpha+1)n^2+(\alpha^5+3\alpha^4+\alpha^3-3\alpha^2+3\alpha+2)n-1}{\alpha^5(n-1)^3};$$

$$C8 = -\frac{1}{\alpha^7(n-1)^4}(\alpha^7 n^4+(3\alpha^7+4\alpha^6-4\alpha^5+2\alpha^4+2\alpha^3-4\alpha^2-4\alpha+1)n^3$$

$$+(3\alpha^7+8\alpha^6-8\alpha^4+8\alpha^3-8\alpha-3)n^2+(\alpha^7+4\alpha^6+4\alpha^5-2\alpha^4-2\alpha^3+4\alpha^2-4\alpha-3)n-1);$$

$$za[ra\_] := C2\,\frac{ra^2}{2f} + C4\,\frac{ra^4}{8\,f^3} + C6\,\frac{2ra^6}{32f^5} + C8\,\frac{5ra^8}{128f^7};$$

$$\varrho z[ra\_]:=\frac{za'[ra]\,(ra+(-zo+za[ra])\,za'[ra])}{-\mathrm{Sign}[zo]n\,\sqrt{ra^2+(zo-za[ra])^2}\,(1+(ra'[ra])^2)}$$

$$+\frac{1}{\sqrt{1+(\text{za}'[\text{ra}])^2}}\sqrt{1-\frac{(\text{ra}+(-\text{zo}+\text{za}[\text{ra}])\ \text{za}'[\text{ra}])^2}{\text{n}^2\ (\text{ra}^2+(\text{zo}-\text{za}[\text{ra}])^2)\ (1+(\text{za}'[\text{ra}])^2)}};$$

$$\varrho\text{r}[\text{ra\_}]:=\frac{\text{ra}+(-\text{zo}+\text{za}[\text{ra}])\ \text{za}'[\text{ra}]}{-\text{Sign}[\text{zo}]\text{n}\ \sqrt{\text{ra}^2+(\text{zo}-\text{za}[\text{ra}])^2}\ (1+(\text{za}'[\text{ra}])^2)}$$

$$-\frac{\text{za}'[\text{ra}]}{\sqrt{1+(\text{za}'[\text{ra}])^2}}\sqrt{1-\frac{(\text{ra}+(-\text{zo}+\text{za}[\text{ra}])\ \text{za}'[\text{ra}])^2}{\text{n}^2\ (\text{ra}^2+(\text{zo}-\text{za}[\text{ra}])^2)\ (1+(\text{za}'[\text{ra}])^2)}};$$

```
(** Homotopy for Eikonal **)

θ1[ra_] := ArcTan[(ra)/(−zo+za[ra])];

θ2[ra_] := ArcTan[(ra)/(za[ra] −zi)];

c1[ra_] := (zo + n zi) + Sec[θ1[ra]](−zo+za[ra]);


(** Setting variable's **)

c = 1;
v = c/n;

no = 1;
ni = no;
n = 1.5;


α = 1.5;
rmax = 16;
density = 4;
zo = α  40;
zi = −40;
f = zo


(** Plot Section **)

SurfaceV := ParametricPlot[{{za[ra],ra},{zb[ra],rb[ra]}},{ra,−rmax−1,rmax+1},
          PlotStyle →{Black,Red}, AspectRatio→Automatic, AxesLabel →{r,z}]


LensV := Graphics[{EdgeForm[Directive[{Black,Thin}]],
           Directive[Gray,Opacity[0.3]],
```

```
        Polygon[Join[#[[1,1]],#[[2,1]],Reverse[#[[4,1]]],Reverse[#[[3,1]]]]&
        [Cases[SurfaceV,_Line,∞]]]}];

WavesV := Table[{
          ParametricPlot[{
             Piecewise[{
                   {Cos[θ1[ra]](tt−(zo + n zi))+zo,tt<=c1[ra]},
                   {v Cos[θ2[ra]](tt−c1[ra])+za[ra],c1[ra]<tt<c2[ra]},
                  }],

             Piecewise[{
                   {Sin[θ1[ra]](tt−(zo + n zi)),tt<=c1[ra]},
                   {v Sin[θ2[ra]](tt−c1[ra])+ra,c1[ra]<tt<c2[ra]},
                   }]},
          {ra,−rmax,rmax},
          PlotRange→All,
          PlotStyle →{Blue, Thickness[0.002]}]},
          {tt,0,zo+n τ − n zi+0.25,density}];

 RaysV := Table[{
        ParametricPlot[{
           Piecewise[{
               {Cos[θ1[ra]](tt−(zo + n zi))+zo,tt<=c1[ra]},
               {v Cos[θ2[ra]](tt−c1[ra])+za[ra],c1[ra]<tt<c2[ra]},
             }],

           Piecewise[{
               {Sin[θ1[ra]](tt−(zo + n zi)),tt<=c1[ra]},
               {v Sin[θ2[ra]](tt−c1[ra])+ra,c1[ra]<tt<c2[ra]},
             }]},

        {tt,0,zo+ n − n zi+0.25},
        PlotRange→All,PlotStyle →{Lighter[Purple], Thickness[0.002]}]},
        {ra,−rmax,rmax,density}];


Show[SurfaceV,LensV,WavesV,RaysV,
    PlotRange→All,ImageSize→{640,Automatic}]
```

12.2 Programs for chapter 7

12.2.1 Case 1: Real finite object—real finite image

"ClearAll["Global'*"]

(** Definition of set variables **)

$$\text{Ka} = \frac{(\text{n}^2\ \text{zin}-\text{no}^2\ \text{zo})^2}{\text{n no (n zin}-\text{no zo) (n zo}-\text{no zin)}};$$

$$\text{coa} = \frac{\text{n zo}-\text{no zin}}{\text{zin zo(n}-\text{no)}};$$

$$\text{c1a} = \frac{(\text{n}-\text{no})(\text{n}+\text{no})^2}{4\ \text{n no zin zo(n zin}-\text{no zo)}};$$

$$\text{b1a} = \frac{(\text{n}+\text{no})(\text{n}^2\text{zin}-\text{no}^2\text{zo})}{2\ \text{n no zin zo(n zin}-\text{no zo)}};$$

$$\text{za}[\rho\_] := \frac{\rho^2\ (\text{coa}+\text{c1a}\ \rho^2)}{1+\text{b1a}\ \rho^2+\sqrt{1+(2\ \text{b1a}-\text{coa}^2\ \text{Ka})\ \rho^2}}$$

$$\text{ra}[\rho\_] := \text{Piecewise}[\{\{\sqrt{\rho^2-\text{za}[\rho]^2},\rho\geq 0\},\{-\sqrt{\rho^2-\text{za}[\rho]^2},\rho<0\}\}]$$

$$\varrho\text{z}[\rho\_] := \frac{\text{no(za}'[\rho]\ (\text{ra}[\rho]\text{ra}'[\rho]+(-\text{zo}+\text{za}[\rho])\ \text{za}'[\rho]))}{\text{n}\ \sqrt{\text{ra}[\rho]^2+(\text{zo}-\text{za}[\rho])^2}\ ((\text{ra}'[\rho])^2+(\text{za}'[\rho])^2)}$$

$$+\ \frac{\text{ra}'[\rho]}{\sqrt{(\text{ra}'[\rho])^2+(\text{za}'[\rho])^2}}\sqrt{\left(1-\frac{\text{no}^2(\text{ra}[\rho]\text{ra}'[\rho]+(-\text{zo}+\text{za}[\rho])\ \text{za}'[\rho])^2}{\text{n}^2\ (\text{ra}[\rho]^2+(\text{zo}-\text{za}[\rho])^2)\ ((\text{ra}'[\rho])^2+(\text{za}'[\rho])^2)}\right)}$$

$$\varrho\text{r}[\rho\_] := \frac{\text{no(ra}'[\rho](\text{ra}[\rho]\text{ra}'[\rho]+(-\text{zo}+\text{za}[\rho])\ \text{za}'[\rho]))}{\text{n}\ \sqrt{\text{ra}[\rho]^2+(\text{zo}-\text{za}[\rho])^2}\ ((\text{ra}'[\rho])^2+(\text{za}'[\rho])^2)}$$

$$-\ \frac{\text{za}'[\rho]}{\sqrt{(\text{ra}'[\rho])^2+(\text{za}'[\rho])^2}}\sqrt{\left(1-\frac{\text{no}^2(\text{ra}[\rho]\text{ra}'[\rho]+(-\text{zo}+\text{za}[\rho])\ \text{za}'[\rho])^2}{\text{n}^2\ (\text{ra}[\rho]^2+(\text{zo}-\text{za}[\rho])^2)\ ((\text{ra}'[\rho])^2+(\text{za}'[\rho])^2)}\right)}$$

(** Homotopy for Eikonal **)

$\theta 1[\rho\_] := \text{ArcTan}[(\text{ra}[\rho])/(-\text{zo}+\text{za}[\rho])];$

$\theta 2[\rho\_] := \text{ArcTan}[(\text{ra}[\rho])/(-\text{za}[\rho] + \text{zi})];$

$\text{c1}[\rho\_] := \text{Sec}[\theta 1[\rho]](\text{za}[\rho] - \text{zo});$

(** Setting variable's **)

```
c = 1;
v = c/n;
```

```
no = 1;
n = 1.5;

ρmax = 16;
density = 4;
zo = −40;
zi = 40;
zin = n zi;


(** Plot Section **)

 SurfaceV:=ParametricPlot[{{za[ρ],ra[ρ]},{ 1.1*zi,−ra[ρ]}},{ρ,−ρmax,ρmax},
    PlotStyle →{Red,Transparent},AspectRatio→Automatic,AxesLabel  →{r,z}]

OvalSurfaceV:=ParametricPlot[{za[ρ],ra[ρ]},{ρ,− ρmax−250,ρmax+250},
  PlotStyle→Directive[{Black,Dashed}]]

LensV:=Graphics[{EdgeForm[Directive[{Black,Thin}]],
Directive[ Gray,Opacity[0.3]],
  Polygon[Join[#[[1,1]],List[{0,0}],#[[2,1]],#[[3, 1]],List[{0,0}],#[[4,1]]]&
[Cases[SurfaceV,_Line,∞]]]}]

 WavesV := Table[{
            ParametricPlot[{
               Piecewise[{
                     {Cos[θ1[ρ]]tt+zo,tt<=c1[ρ]},
                     {v Cos[θ2[ρ]](tt−c1[ρ])+za[ρ],c1[ρ]<tt<c2[ρ]}
                    }],

               Piecewise[{
                     {Sin[θ1[ρ]]tt,tt<=c1[ρ]},
                     {v Sin[θ2[ρ]](tt−c1[ρ])+ra[ρ],c1[ρ]<tt<c2[ρ]}
                    ]},

            {ρ,−ρmax,ρmax},
            PlotRange→All,
            PlotStyle →{Blue, Thickness[0.002]}]},
            {tt,0,− zo + n zi + 0.25,density}];

  RaysV := Table[{
         ParametricPlot[{
            Piecewise[{
                 {Cos[θ1[ρ]]tt+zo,tt<=c1[ρ]},
                 {v Cos[θ2[ρ]](tt−c1[ρ])+za[ρ],c1[ρ]<tt<c2[ρ]}
                 }],

            Piecewise[{
                 {Sin[θ1[ρ]]tt,tt<=c1[ρ]},
                 {v Sin[θ2[ρ]](tt−c1[ρ])+ra[ρ],c1[ρ]<tt<c2[ρ]}}]},

         {tt,0,−zo + n zi+0.25},
         PlotRange→All,PlotStyle →{Lighter[Purple], Thickness[0.002]}]},
         {ρ,−ρmax,ρmax,density}];

          Show[OvalSurfaceV,SurfaceV,LensV,WavesV,RaysV,
             PlotRange→All,ImageSize→{640,Automatic}]
```

12.2.2 Case 2: Real infinity object—real finite image

"

(** Definition of set variables **)

$$Ka = -\frac{no^2}{n^2};$$

$$coa = \frac{n}{n\ zi - no\ zi};$$

$$c1a = 0;$$

$$b1a = 0;$$

$$za[\rho\_] := \frac{\rho^2\ (coa + c1a\ \rho^2)}{1 + b1a\ \rho^2 + \sqrt{1 + (2\ b1a - coa^2\ Ka)\ \rho^2}}$$

$$ra[\rho\_] := \text{Piecewise}[\{\{\sqrt{\rho^2 - za[\rho]^2}, \rho \geq 0\}, \{-\sqrt{\rho^2 - za[\rho]^2}, \rho < 0\}\}]$$

$$\varrho z[\rho\_] := \frac{no\ za'[\rho]^2}{n\ (ra'[\rho]^2 + za'[\rho]^2)} + \frac{ra'[\rho]}{\sqrt{ra'[\rho]^2 + za'[\rho]^2}} \sqrt{\left(\frac{n^2\ ra'[\rho]^2 + (n^2 - no^2)\ za'[\rho]^2}{n^2\ (ra'[\rho]^2 + za'[\rho]^2)}\right)}$$

$$\varrho r[\rho\_] := \frac{za'[\rho]}{n\ (ra'[\rho]^2 + za'[\rho]^2)} \left(no\ ra'[\rho] - n \sqrt{ra'[\rho]^2 + za'[\rho]^2} \sqrt{\left(\frac{n^2 ra'[\rho]^2 + (n^2 - no^2) za'[\rho]^2}{n^2 (ra'[\rho]^2 + za'[\rho]^2)}\right)}\right)$$

(** Homotopy for Eikonal **)

$$\theta 1[\rho\_] := \text{ArcTan}[\text{Limit}[(ra[\rho])/(-zo + za[\rho]), zo \to -\text{Infinity}]]$$

$$\theta 2[\rho\_] := \text{ArcTan}[(ra[\rho])/(\ zi - za[\rho]\ )];$$

$$c1[\rho\_] := -Zo + za[\rho];$$

```
(** Setting variable's **)

c = 1;
v = c/n;

no = 1;
n = 1.5;

ρmax = 16;
density = 4;
(* zo --> -Infinity *)
Zo = -40;
zi = 40;
zin = n zi;


(** Plot Section **)


 SurfaceV:=ParametricPlot[{{za[ρ],ra[ρ]},{ 1.1*zi,-ra[ρ]}},{ρ,-ρmax,ρmax},
    PlotStyle →{Red,Transparent},AspectRatio→Automatic,AxesLabel  →{r,z}]

OvalSurfaceV:=ParametricPlot[{{za[ρ],ra[ρ]}},{ρ,- ρmax-250,ρmax+250},
  PlotStyle→Directive[{Black,Dashed}]]

LensV:=Graphics[{EdgeForm[Directive[{Black,Thin}]],
Directive[ Gray,Opacity[0.3]],
  Polygon[Join[#[[1,1]],List[{0,0}],#[[2,1]],#[[3, 1]],List[{0,0}],#[[4,1]]]&
[Cases[SurfaceV,_Line,∞]]]}]


 WavesV := Table[{
            ParametricPlot[{
               Piecewise[{
                     {Cos[θ1[ρ]]tt+Zo,tt<=c1[ρ]},
                     {v Cos[θ2[ρ]](tt-c1[ρ])+za[ρ],c1[ρ]<tt<c2[ρ]}}],

               Piecewise[{
                     {ra[ρ],tt<=c1[ρ]},
                     {v Sin[θ2[ρ]](tt-c1[ρ])+ra[ρ],c1[ρ]<tt<c2[ρ]}}]},

            {ρ,-ρmax,ρmax},
            PlotRange→All,
            PlotStyle →{Blue, Thickness[0.002]}]},
            {tt,0,-Zo + n zi+0.25,density}];

  RaysV := Table[{
         ParametricPlot[{
            Piecewise[{
                 {Cos[θ1[ρ]]tt+Zo,tt<=c1[ρ]},
                 {v Cos[θ2[ρ]](tt-c1[ρ])+za[ρ],c1[ρ]<tt<c2[ρ]}}],
```

```
Piecewise[{
     {ra[ρ],tt<=c1[ρ]},
     {v Sin[θ2[ρ]](tt−c1[ρ])+ra[ρ],c1[ρ]<tt<c2[ρ]}
    }]},

{tt,0,−Zo+ n zi+0.25},
PlotRange→All,PlotStyle →{Lighter[Purple], Thickness[0.002]}]},
{ρ,−ρmax,ρmax,density}];


Show[OvalSurfaceV,SurfaceV,LensV,WavesV,RaysV,
   PlotRange→All,ImageSize→{640,Automatic}]
```

12.2.3 Case 3: Real infinity object—virtual finite image

"

(** Definition of set variables **)

$$\text{Ka} = -\frac{\text{no}^2}{\text{n}^2};$$

$$\text{coa} = \frac{\text{n}}{\text{n zi}-\text{no zi}};$$

$$\text{c1a} = 0;$$

$$\text{b1a} = 0;$$

$$\text{za}[\rho\_] := \frac{\rho^2\ (\text{coa}+\text{c1a}\ \rho^2)}{1+\text{b1a}\ \rho^2+\sqrt{1+(2\ \text{b1a}-\text{coa}^2\ \text{Ka})\ \rho^2}}$$

$$\text{ra}[\rho\_] := \text{Piecewise}[\{\{\sqrt{\rho^2-\text{za}[\rho]^2},\rho\geq 0\},\{-\sqrt{\rho^2-\text{za}[\rho]^2},\rho<0\}\}]$$

$$\varrho\text{z}[\rho\_] := \frac{\text{no za}'[\rho]^2}{\text{n}\ (\text{ra}'[\rho]^2+\text{za}'[\rho]^2)} + \frac{\text{ra}'[\rho]}{\sqrt{\text{ra}'[\rho]^2+\text{za}'[\rho]^2}}$$
$$\sqrt{\left(\frac{\text{n}^2\ \text{ra}'[\rho]^2+(\text{n}^2-\text{no}^2)\ \text{za}'[\rho]^2}{\text{n}^2\ (\text{ra}'[\rho]^2+\text{za}'[\rho]^2)}\right)}$$

$$\varrho\text{r}[\rho\_] := \frac{\text{za}'[\rho]}{\text{n}\ (\text{ra}'[\rho]^2+\text{za}'[\rho]^2)}\Bigg(\text{no ra}'[\rho]-\text{n}$$
$$\sqrt{\text{ra}'[\rho]^2+\text{za}'[\rho]^2}\sqrt{\left(\frac{\text{n}^2\text{ra}'[\rho]^2+(\text{n}^2-\text{no}^2)\text{za}'[\rho]^2}{\text{n}^2(\text{ra}'[\rho]^2+\text{za}'[\rho]^2)}\right)}\Bigg)$$

```
(** Homotopy for Eikonal **)

θ1[ρ_] := ArcTan[Limit[(ra[ρ])/(−zo+za[ρ]),zo→−Infinity]]

θ2[ρ_] := ArcTan[(ra[ρ])/(zi−za[ρ] )];

c1[ρ_] := (−Zo+za[ρ];


(** Setting variable's **)

c = 1;
v = c/n;

no = 1;
n = 1.5;

ρmax = 16;
density = 4;
(* zo −> −Infinity *)
Zo = −40;
zi = 40;
zin = n zi;


(** Plot Section **)

SurfaceV:=ParametricPlot[{{za[ρ],ra[ρ]},{ 1.1*zi,−9ra[ρ]}},{ρ,−ρmax,ρmax},
   PlotStyle →{Red,Transparent},AspectRatio→Automatic,AxesLabel →{r,z}]

OvalSurfaceV:=ParametricPlot[{{za[ρ],ra[ρ]}},{ρ,− ρmax−250,ρmax+250},
  PlotStyle→Directive[{Black,Dashed}]]

LensV:=Graphics[{EdgeForm[Directive[{Black,Thin}]],
Directive[ Gray,Opacity[0.3]],
  Polygon[Join[#[[1,1]],List[{0,0}],#[[2,1]],#[[3, 1]],List[{0,0}],#[[4,1]]]&
[Cases[SurfaceV,_Line,∞]]]}]


WavesV := Table[{
           ParametricPlot[{
              Piecewise[{
                    {Cos[θ1[ρ]]tt+Zo,tt<=c1[ρ]},
                    {v Cos[θ2[ρ]](tt−c1[ρ])+za[ρ],c1[ρ]<tt<c2[ρ]}}],

              Piecewise[{
                    {Sin[θ1[ρ]]tt+ra[ρ],tt<=c1[ρ]},
                    {v Sin[θ2[ρ]](tt−c1[ρ])+ra[ρ],c1[ρ]<tt<c2[ρ]}
                    }]},

           {ρ,−ρmax,ρmax},
           PlotRange→All,
           PlotStyle →{Blue, Thickness[0.002]}]},
```

```
        {tt,0,−Zo+n zi+0.25,density}];

RaysV := Table[{
      ParametricPlot[{
         Piecewise[{
             {Cos[θ1[ρ]]tt+Zo,tt<=c1[ρ]},
             {v Cos[θ2[ρ]](tt−c1[ρ])+za[ρ],c1[ρ]<tt<c2[ρ]}
           }],

         Piecewise[{
             {Sin[θ1[ρ]]tt+ra[ρ],tt<=c1[ρ]},
             {v Sin[θ2[ρ]](tt−c1[ρ])+ra[ρ],c1[ρ]<tt<c2[ρ]}
            }]},

      {tt,0,−Zo+ n zi+0.25},
      PlotRange→All,PlotStyle →{Lighter[Purple], Thickness[0.002]}]},
      {ρ,−ρmax,ρmax,density}];


Show[OvalSurfaceV,SurfaceV,LensV,WavesV,RaysV,
    PlotRange→All,ImageSize→{640,Automatic}]
```

12.2.4 Case 4: Real finite object—virtual finite image

```
" ClearAll["Global'*"]
 (** Definition of set variables **)
```

$$\text{Ka} = \frac{(\text{n}^2\ \text{zin}-\text{no}^2\ \text{zo})^2}{\text{n no (n zin}-\text{no zo) (n zo}-\text{no zin)}};$$

$$\text{coa} = \frac{\text{n zo}-\text{no zin}}{\text{zin zo(n}-\text{no)}};$$

$$\text{c1a} = \frac{(\text{n}-\text{no})(\text{n}+\text{no})^2}{4\ \text{n no zin zo(n zin}-\text{no zo)}};$$

$$\text{b1a} = \frac{(\text{n}+\text{no})(\text{n}^2\text{zin}-\text{no}^2\text{zo})}{2\ \text{n no zin zo(n zin}-\text{no zo)}};$$

$$\text{za}[\rho\_] := \frac{\rho^2\ (\text{coa}+\text{c1a}\ \rho^2)}{1+\text{b1a}\ \rho^2+\sqrt{1+(2\ \text{b1a}-\text{coa}^2\ \text{Ka})\ \rho^2}}$$

$$\text{ra}[\rho\_] := \text{Piecewise}[\{\{\sqrt{\rho^2-\text{za}[\rho]^2},\rho\geq 0\},\{-\sqrt{\rho^2-\text{za}[\rho]^2},\rho<0\}\}]$$

$$\varrho z[\rho\_] := \frac{no(za'[\rho]\ (ra[\rho]ra'[\rho]+(-zo+za[\rho])\ za'[\rho]))}{n\ \sqrt{ra[\rho]^2+(zo-za[\rho])^2}\ ((ra'[\rho])^2+(za'[\rho])^2)}$$

$$+\ \frac{ra'[\rho]}{\sqrt{(ra'[\rho])^2+(za'[\rho])^2}}\sqrt{\left(1-\frac{no^2(ra[\rho]ra'[\rho]+(-zo+za[\rho])\ za'[\rho])^2}{n^2\ (ra[\rho]^2+(zo-za[\rho])^2)\ ((ra'[\rho])^2+(za'[\rho])^2)}\right)}$$

$$\varrho r[\rho\_]:=\frac{no(ra'[\rho](ra[\rho]ra'[\rho]+(-zo+za[\rho])\ za'[\rho]))}{n\ \sqrt{ra[\rho]^2+(zo-za[\rho])^2}\ ((ra'[\rho])^2+(za'[\rho])^2)}$$

$$-\ \frac{za'[\rho]}{\sqrt{(ra'[\rho])^2+(za'[\rho])^2}}\sqrt{\left(1-\frac{no^2(ra[\rho]ra'[\rho]+(-zo+za[\rho])\ za'[\rho])^2}{n^2\ (ra[\rho]^2+(zo-za[\rho])^2)\ ((ra'[\rho])^2+(za'[\rho])^2)}\right)}$$

```
(** Homotopy for Eikonal **)

θ1[ρ_] := ArcTan[(ra[ρ])/(−zo+za[ρ])];

θ2[ρ_] := ArcTan[ra[ρ]/(zi−za[ρ] )];

c1[ρ_] := Sec[θ1[ρ]](−zo+za[ρ]);

(** Setting variable's **)

c = 1;
v = c/n;

no = 1;
n = 1.5;

ρmax =12;
density =3;
zo  =  − 40;
zi  =  −15;
zin  =  n zi;

(** Plot Section **)

 SurfaceV:=ParametricPlot[{{za[ρ],ra[ρ]},{ −1.1*zi,−5 ra[ρ]}},{ρ,−ρmax,
ρmax},
      PlotStyle →{Red,Transparent},AspectRatio→Automatic,AxesLabel  →{r,z}]

 OvalSurfaceV:=ParametricPlot[{{za[ρ],ra[ρ]}},{ρ,− ρmax−150,ρmax+150},
```

```
PlotStyle→Directive[{Black,Dashed}]]

LensV:=Graphics[{EdgeForm[Directive[{Black,Thin}]],
Directive[ Gray,Opacity[0.3]],
  Polygon[Join[#[[1,1]],List[{0,0}],#[[2,1]],#[[3, 1]],List[{0,0}],#[[4,1]]]&
[Cases[SurfaceV,_Line,∞]]]}]


WavesV := Table[{
            ParametricPlot[{
               Piecewise[{
                     {Cos[θ1[ρ]]tt+zo,tt<=c1[ρ]},
                     {v Cos[θ2[ρ]](tt−c1[ρ])+za[ρ],c1[ρ]<tt<c2[ρ]},
                     }],

               Piecewise[{
                     {Sin[θ1[ρ]]tt,tt<=c1[ρ]},
                     {v Sin[θ2[ρ]](tt−c1[ρ])+ra[ρ],c1[ρ]<tt<c2[ρ]},
                     }]},

            {ρ,−ρmax,ρmax},
            PlotRange→All,
            PlotStyle →{Blue, Thickness[0.002]}]},
            {tt,0,−zo− n zi+0.25,density}];

RaysV := Table[{
         ParametricPlot[{
            Piecewise[{
                {Cos[θ1[ρ]]tt+zo,tt<=c1[ρ]},
                {v Cos[θ2[ρ]](tt−c1[ρ])+za[ρ],c1[ρ]<tt<c2[ρ]},
               }],

            Piecewise[{
                {Sin[θ1[ρ]]tt,tt<=c1[ρ]},
                {v Sin[θ2[ρ]](tt−c1[ρ])+ra[ρ],c1[ρ]<tt<c2[ρ]},
               }]},

         {tt,0,−zo − n zi+0.25},
         PlotRange→All,PlotStyle →{Lighter[Purple], Thickness[0.002]}]},
         {ρ,−ρmax,ρmax,density}];


Show[OvalSurfaceV,SurfaceV,LensV,WavesV,RaysV,
   PlotRange→All,ImageSize→{640,Automatic}]
```

12.2.5 Case 5: Real finite object—real infinite image

" ClearAll["Global'*"]

(** Definition of set variables **)

$$Ka = -\frac{n^2}{no^2};$$

$$coa = -\frac{no}{n\ zo - no\ zo};$$

$$c1a = 0;$$

$$b1a = 0;$$

$$za[\rho\_] := \frac{\rho^2\ (coa + c1a\ \rho^2)}{1 + b1a\ \rho^2 + \sqrt{1 + (2\ b1a - coa^2\ Ka)\ \rho^2}}$$

$$ra[\rho\_] := \text{Piecewise}[\{\{\sqrt{\rho^2 - za[\rho]^2}, \rho \geq 0\}, \{-\sqrt{\rho^2 - za[\rho]^2}, \rho < 0\}\}]$$

$$\varrho z[\rho\_] := \frac{no(za'[\rho]\ (ra[\rho]ra'[\rho] + (-zo + za[\rho])\ za'[\rho]))}{n\ \sqrt{ra[\rho]^2 + (zo - za[\rho])^2}\ ((ra'[\rho])^2 + (za'[\rho])^2)}$$

$$+ \frac{ra'[\rho]}{\sqrt{(ra'[\rho])^2 + (za'[\rho])^2}} \sqrt{\left(1 - \frac{no^2(ra[\rho]ra'[\rho] + (-zo + za[\rho])\ za'[\rho])^2}{n^2\ (ra[\rho]^2 + (zo - za[\rho])^2)\ ((ra'[\rho])^2 + (za'[\rho])^2)}\right)}$$

$$\varrho r[\rho\_] := \frac{no(ra'[\rho](ra[\rho]ra'[\rho] + (-zo + za[\rho])\ za'[\rho]))}{n\ \sqrt{ra[\rho]^2 + (zo - za[\rho])^2}\ ((ra'[\rho])^2 + (za'[\rho])^2)}$$

$$- \frac{za'[\rho]}{\sqrt{(ra'[\rho])^2 + (za'[\rho])^2}} \sqrt{\left(1 - \frac{no^2(ra[\rho]ra'[\rho] + (-zo + za[\rho])\ za'[\rho])^2}{n^2\ (ra[\rho]^2 + (zo - za[\rho])^2)\ ((ra'[\rho])^2 + (za'[\rho])^2)}\right)}$$

(** Homotopy for Eikonal **)

$$\theta 1[\rho\_] := \text{ArcTan}[(ra[\rho])/(-zo + za[\rho])];$$

$$c1[\rho\_] := \text{Sec}[\theta 1[\rho]](-zo + za[\rho]);$$

(** Setting variable's **)

```
c = 1;
v = c/n;

no = 1;
n = 1.5;
```

```
ρmax = 24;
density = 4;
zo = −40;
(* zi −> Infinity*)
Zi = 25;


(** Plot Section **)


 SurfaceV:=ParametricPlot[{{za[ρ],ra[ρ]},{ −1.1*zi, −ra[ρ]}},{ρ,−ρmax,ρmax},
    PlotStyle →{Red,Transparent},AspectRatio→Automatic,AxesLabel  →{r,z}]

OvalSurfaceV:=ParametricPlot[{{za[ρ],ra[ρ]}},{ρ,− ρmax−30,ρmax+30},
  PlotStyle→Directive[{Black,Dashed}]]

LensV:=Graphics[{EdgeForm[Directive[{Black,Thin}]],
Directive[ Gray,Opacity[0.3]],
  Polygon[Join[#[[1,1]],List[{0,0}],#[[2,1]],#[[3, 1]],List[{0,0}],#[[4,1]]]&
[Cases[SurfaceV,_Line,∞]]]}]


 WavesV  := Table[{
             ParametricPlot[{
                Piecewise[{
                      {Cos[θ1[ρ]]tt+zo,tt<=c1[ρ]},
                      {v (tt−c1[ρ])+za[ρ],c1[ρ]<tt<c2[ρ]}
                      }],

                Piecewise[{
                      {Sin[θ1[ρ]]tt,tt<=c1[ρ]},
                      {ra[ρ],c1[ρ]<tt<c2[ρ]}
                      }]},

             {ρ,−ρmax,ρmax},
             PlotRange→All,
             PlotStyle →{Blue, Thickness[0.002]}]},
             {tt,0,−zo  + n Zi + 0.25,density}];

  RaysV := Table[{
          ParametricPlot[{
            Piecewise[{
                      {Cos[θ1[ρ]]tt+zo,tt<=c1[ρ]},
                      {v (tt−c1[ρ])+za[ρ],c1[ρ]<tt<c2[ρ]}
                      }],

                Piecewise[{
                      {Sin[θ1[ρ]]tt,tt<=c1[ρ]},
                      {ra[ρ],c1[ρ]<tt<c2[ρ]}
                      }]},

          {tt,0,−zo +  n  Zi + 0.25},
          PlotRange→All,PlotStyle →{Lighter[Purple], Thickness[0.002]}]},
          {ρ,−ρmax,ρmax,density}];

          Show[OvalSurfaceV,SurfaceV,LensV,WavesV,RaysV,
              PlotRange→All,ImageSize→{640,Automatic}]
```

12.2.6 Case 6: Virtual finite object—real infinite image

```
ClearAll["Global`*"]
(** Definition of set variables **)

Ka = -n^2/no^2;
coa = -no/(n zo-no zo);

c1a = 0;

b1a = 0;

za[ρ_] := (ρ^2 (coa+c1a ρ^2))/(1+b1a ρ^2+Sqrt[1+(2 b1a-coa^2 Ka) ρ^2])

ra[ρ_] := Piecewise[{{Sqrt[ρ^2-za[ρ]^2],ρ≥0},{-Sqrt[ρ^2-za[ρ]^2],ρ<0}}]

ϱz[ρ_] := (no(za'[ρ] (ra[ρ]ra'[ρ]+(-zo+za[ρ]) za'[ρ])))/(n Sqrt[ra[ρ]^2+(zo-za[ρ])^2] ((ra'[ρ])^2+(za'[ρ])^2))
    + ra'[ρ]/Sqrt[(ra'[ρ])^2+(za'[ρ])^2] Sqrt[(1-(no^2(ra[ρ]ra'[ρ]+(-zo+za[ρ]) za'[ρ])^2)/(n^2 (ra[ρ]^2+(zo-za[ρ])^2) ((ra'[ρ])^2+(za'[ρ])^2)))]

ϱr[ρ_]:=(no(ra'[ρ](ra[ρ]ra'[ρ]+(-zo+za[ρ]) za'[ρ])))/(n Sqrt[ra[ρ]^2+(zo-za[ρ])^2] ((ra'[ρ])^2+(za'[ρ])^2))
    - za'[ρ]/Sqrt[(ra'[ρ])^2+(za'[ρ])^2] Sqrt[(1-(no^2(ra[ρ]ra'[ρ]+(-zo+za[ρ]) za'[ρ])^2)/(n^2 (ra[ρ]^2+(zo-za[ρ])^2) ((ra'[ρ])^2+(za'[ρ])^2)))]
```

```
(** Homotopy for Eikonal **)

θ1[ρ_] := ArcTan[(ra[ρ])/(−zo+za[ρ])];

c1[ρ_] := (zo + n Zi) + Sec[θ1[ρ]](−zo + za[ρ]);

(** Setting variable's **)

c = 1;
v = c/n;

no = 1;
n = 1.5;

ρmax = 16;
density = 4;
 zo = 25;
(* zi −> Infinity*)
Zi = 40;

(** Plot Section **)

 SurfaceV:=ParametricPlot[{{za[ρ],ra[ρ]},{ 1.1*zo, −ra[ρ]}},{ρ,−ρmax,ρmax},
    PlotStyle →{Red,Transparent},AspectRatio→Automatic,AxesLabel →{r,z}]

OvalSurfaceV:=ParametricPlot[{{za[ρ],ra[ρ]}},{ρ,− ρmax−30,ρmax+30},
  PlotStyle→Directive[{Black,Dashed}]]

LensV:=Graphics[{EdgeForm[Directive[{Black,Thin}]],
Directive[ Gray,Opacity[0.3]],
  Polygon[Join[#[[1,1]],List[{0,0}],#[[2,1]],#[[3, 1]],List[{0,0}],#[[4,1]]]&
[Cases[SurfaceV,_Line,∞]]]}]

 WavesV := Table[{
              ParametricPlot[{
                 Piecewise[{
                   {Cos[θ1[ρ]](tt−( zo+n Zi)) + zo,tt<=c1[ρ]},
                   {v (tt−c1[ρ])+za[ρ],c1[ρ]<tt<c2[ρ]}
                   }],

              Piecewise[{
                   {Sin[θ1[ρ]](tt−(zo + n Zi)),tt<=c1[ρ]},
                   { ra[ρ],c1[ρ]<tt<c2[ρ]}
                   }]},

              {ρ,−ρmax,ρmax},
              PlotRange→All,
              PlotStyle →{Blue, Thickness[0.002]}]},
```

```
        {tt,0, zo +  n Zi + 0.25,density}];

RaysV := Table[{
      ParametricPlot[{
         Piecewise[{
            {Cos[θ1[ρ]](tt−( zo+n Zi)) + zo,tt<=c1[ρ]},
            {v (tt−c1[ρ])+za[ρ],c1[ρ]<tt<c2[ρ]}
            }],

         Piecewise[{
            {Sin[θ1[ρ]](tt−(zo + n Zi)),tt<=c1[ρ]},
            { ra[ρ],c1[ρ]<tt<c2[ρ]}
            }]},

      {tt,0, n zo +n Zi + 0.25},
      PlotRange→All,PlotStyle →{Lighter[Purple], Thickness[0.002]}]},
      {ρ,−ρmax,ρmax,density}];


Show[OvalSurfaceV,SurfaceV,LensV,WavesV,RaysV,
   PlotRange→All,ImageSize→{640,Automatic}]
```

12.2.7 Case 7: Virtual finite object—real finite image

```
" ClearAll["Global`*"]
  (** Definition of set variables **)
```

$$\text{Ka} = \frac{(\text{n}^2\ \text{zin}-\text{no}^2\ \text{zo})^2}{\text{n no (n zin}-\text{no zo) (n zo}-\text{no zin)}};$$

$$\text{coa} = \frac{\text{n zo}-\text{no zin}}{\text{zin zo(n}-\text{no)}};$$

$$\text{c1a} = \frac{(\text{n}-\text{no})(\text{n}+\text{no})^2}{4\ \text{n no zin zo(n zin}-\text{no zo)}};$$

$$\text{b1a} = \frac{(\text{n}+\text{no})(\text{n}^2\text{zin}-\text{no}^2\text{zo})}{2\ \text{n no zin zo(n zin}-\text{no zo)}};$$

$$\text{za}[\rho\_] := \frac{\rho^2\ (\text{coa}+\text{c1a}\ \rho^2)}{1+\text{b1a}\ \rho^2+\sqrt{1+(2\ \text{b1a}-\text{coa}^2\ \text{Ka})\ \rho^2}}$$

$$\text{ra}[\rho\_] := \text{Piecewise}[\{\{\sqrt{\rho^2-\text{za}[\rho]^2},\rho\geq 0\},\{-\sqrt{\rho^2-\text{za}[\rho]^2},\rho<0\}\}]$$

$$\varrho z[\rho\_] := \frac{no(za'[\rho]\ (ra[\rho]ra'[\rho]+(-zo+za[\rho])\ za'[\rho]))}{n\ \sqrt{ra[\rho]^2+(zo-za[\rho])^2}\ ((ra'[\rho])^2+(za'[\rho])^2)} + \frac{ra'[\rho]}{\sqrt{(ra'[\rho])^2+(za'[\rho])^2}}\sqrt{\left(1-\frac{no^2(ra[\rho]ra'[\rho]+(-zo+za[\rho])\ za'[\rho])^2}{n^2\ (ra[\rho]^2+(zo-za[\rho])^2)\ ((ra'[\rho])^2+(za'[\rho])^2)}\right)}$$

$$\varrho r[\rho\_] := \frac{no(ra'[\rho](ra[\rho]ra'[\rho]+(-zo+za[\rho])\ za'[\rho]))}{n\ \sqrt{ra[\rho]^2+(zo-za[\rho])^2}\ ((ra'[\rho])^2+(za'[\rho])^2)} - \frac{za'[\rho]}{\sqrt{(ra'[\rho])^2+(za'[\rho])^2}}\sqrt{\left(1-\frac{no^2(ra[\rho]ra'[\rho]+(-zo+za[\rho])\ za'[\rho])^2}{n^2\ (ra[\rho]^2+(zo-za[\rho])^2)\ ((ra'[\rho])^2+(za'[\rho])^2)}\right)}$$

```
(** Homotopy for Eikonal **)

θ1[ρ_]  := ArcTan[ra[ρ]/(−zo+za[ρ])];

θ2[ρ_] := ArcTan[ra[ρ]/(zi−za[ρ])];

c1[ρ_] := (zo − n  zi) + Sec[θ1[ρ]](−zo+za[ρ]);

(** Setting variable's **)

c = 1;
v = c/n;

no = 1;
n = 1.5;

ρmax = 12;
density = 2;
zo = 40;
zi = 25;

(** Plot Section **)

SurfaceV:=ParametricPlot[{{za[ρ],ra[ρ]},{ 1.1 * zi, −ra[ρ]}},{ρ,−ρmax,ρmax},
   PlotStyle →{Red,Transparent},AspectRatio→Automatic,AxesLabel  →{r,z}]

OvalSurfaceV:=ParametricPlot[{{za[ρ],ra[ρ]}},{ρ,− ρmax−250,ρmax+250},
  PlotStyle→Directive[{Black,Dashed}]]

LensV:=Graphics[{EdgeForm[Directive[{Black,Thin}]],
Directive[ Gray,Opacity[0.3]],
```

```
  Polygon[Join[#[[1,1]],List[{0,0}],#[[2,1]],#[[3, 1]],List[{0,0}],#[[4,1]]]&
[Cases[SurfaceV,_Line,∞]]]}]

 WavesV := Table[{
           ParametricPlot[{
              Piecewise[{
                    {Cos[θ1[ρ]](tt−(zo + n zi))+zo,tt<=c1[ρ]},
                    {v Cos[θ2[ρ]](tt−c1[ρ])+za[ρ],c1[ρ]<tt<c2[ρ]}
                    }],

              Piecewise[{
                    {Sin[θ1[ρ]](tt−(zo + n zi)),tt<=c1[ρ]},
                    {v Sin[θ2[ρ]](tt−c1[ρ])+ra[ρ],c1[ρ]<tt<c2[ρ]}
                    }]},

           {ρ,−ρmax,ρmax},
           PlotRange→All,
           PlotStyle →{Blue, Thickness[0.002]}]},
           {tt,0,zo+zi+10,density}];

 RaysV := Table[{
        ParametricPlot[{
          Piecewise[{
                    {Cos[θ1[ρ]](tt−(zo + n zi))+zo,tt<=c1[ρ]},
                    {v Cos[θ2[ρ]](tt−c1[ρ])+za[ρ],c1[ρ]<tt<c2[ρ]}
                    }],

              Piecewise[{
                    {Sin[θ1[ρ]](tt−(zo + n zi)),tt<=c1[ρ]},
                    {v Sin[θ2[ρ]](tt−c1[ρ])+ra[ρ],c1[ρ]<tt<c2[ρ]}
                    }]},

        {tt,0,zo+zi+10},
        PlotRange→All,PlotStyle →{Lighter[Purple], Thickness[0.002]}]},
        {ρ,−ρmax,ρmax,density}];

 Show[OvalSurfaceV,SurfaceV,LensV,WavesV,RaysV,
     PlotRange→All,ImageSize→{640,Automatic}]
```

12.2.8 Case 8: Virtual finite object—virtual finite image

```
" ClearAll["Global'*"]
(** Definition of set variables **)
```

$$Ka = \frac{(n^2\ zin - no^2\ zo)^2}{n\ no\ (n\ zin - no\ zo)\ (n\ zo - no\ zin)};$$

$$coa = \frac{n\ zo - no\ zin}{zin\ zo(n - no)};$$

$$c1a = \frac{(n - no)(n + no)^2}{4\ n\ no\ zin\ zo(n\ zin - no\ zo)};$$

$$b1a = \frac{(n + no)(n^2 zin - no^2 zo)}{2\ n\ no\ zin\ zo(n\ zin - no\ zo)};$$

$$za[\rho\_] := \frac{\rho^2\ (coa + c1a\ \rho^2)}{1 + b1a\ \rho^2 + \sqrt{1 + (2\ b1a - coa^2\ Ka)\ \rho^2}}$$

$$ra[\rho\_] := \text{Piecewise}[\{\{\sqrt{\rho^2 - za[\rho]^2}, \rho \geq 0\}, \{-\sqrt{\rho^2 - za[\rho]^2}, \rho < 0\}\}]$$

$$\varrho z[\rho\_] := \frac{no(za'[\rho]\ (ra[\rho]ra'[\rho] + (-zo + za[\rho])\ za'[\rho]))}{n\ \sqrt{ra[\rho]^2 + (zo - za[\rho])^2}\ ((ra'[\rho])^2 + (za'[\rho])^2)} + \frac{ra'[\rho]}{\sqrt{(ra'[\rho])^2 + (za'[\rho])^2}} \sqrt{\left(1 - \frac{no^2 (ra[\rho]ra'[\rho] + (-zo + za[\rho])\ za'[\rho])^2}{n^2\ (ra[\rho]^2 + (zo - za[\rho])^2)\ ((ra'[\rho])^2 + (za'[\rho])^2)}\right)}$$

$$\varrho r[\rho\_] := \frac{no(ra'[\rho](ra[\rho]ra'[\rho] + (-zo + za[\rho])\ za'[\rho]))}{n\ \sqrt{ra[\rho]^2 + (zo - za[\rho])^2}\ ((ra'[\rho])^2 + (za'[\rho])^2)} - \frac{za'[\rho]}{\sqrt{(ra'[\rho])^2 + (za'[\rho])^2}} \sqrt{\left(1 - \frac{no^2 (ra[\rho]ra'[\rho] + (-zo + za[\rho])\ za'[\rho])^2}{n^2\ (ra[\rho]^2 + (zo - za[\rho])^2)\ ((ra'[\rho])^2 + (za'[\rho])^2)}\right)}$$

```
(** Homotopy for Eikonal **)

θ1[ρ_] := ArcTan[(ra[ρ])/(−zo+za[ρ])];

θ2[ρ_] := ArcTan[ra[ρ]/(zi−za[ρ] )];

c1[ρ_] := (zo − n zi) + Sec[θ1[ρ]](−zo+za[ρ]);

(** Setting variable's **)

c = 1;
v = c/n;
```

```
no = 1;
n = 1.5;

ρmax = 16;
density = 4;
zo = 25;
zi = −40;
zin = n zi;


(** Plot Section **)


 SurfaceV:=ParametricPlot[{{za[ρ],ra[ρ]},{ −n* zi+n, −3.5ra[ρ]}},{ρ,−ρmax,
ρmax},
    PlotStyle →{Red,Transparent},AspectRatio→Automatic,AxesLabel  →{r,z}]

 OvalSurfaceV:=ParametricPlot[{{za[ρ],ra[ρ]}},{ρ,− ρmax−250,ρmax+250},
   PlotStyle→Directive[{Black,Dashed}]]

 LensV:=Graphics[{EdgeForm[Directive[{Black,Thin}]],
 Directive[ Gray,Opacity[0.3]],
   Polygon[Join[#[[1,1]],List[{0,0}],#[[2,1]],#[[3, 1]],List[{0,0}],#[[4,1]]]&
[Cases[SurfaceV,_Line,∞]]]}]


  WavesV := Table[{
           ParametricPlot[{
              Piecewise[{
                     {Cos[θ1[ρ]](tt−(zo + nτ))+zo,tt<=c1[ρ]},
                     {v Cos[θ2[ρ]](tt−c1[ρ])+za[ρ],c1[ρ]<tt<c2[ρ]}
                     }],

              Piecewise[{
                     {Sin[θ1[ρ]](tt−(zo + nτ)),tt<=c1[ρ]},
                     {v Sin[θ2[ρ]](tt−c1[ρ])+ra[ρ],c1[ρ]<tt<c2[ρ]}
                     }]},

           {ρ,−ρmax,ρmax},
           PlotRange→All,
           PlotStyle →{Blue, Thickness[0.002]}]},
           {tt,0,4 zo − zi + 0.25,density}];

  RaysV := Table[{
         ParametricPlot[{
           Piecewise[{
                     {Cos[θ1[ρ]](tt−(zo + nτ))+zo,tt<=c1[ρ]},
                     {v Cos[θ2[ρ]](tt−c1[ρ])+za[ρ],c1[ρ]<tt<c2[ρ]}
                     }],

              Piecewise[{
                     {Sin[θ1[ρ]](tt−(zo + nτ)),tt<=c1[ρ]},
                     {v Sin[θ2[ρ]](tt−c1[ρ])+ra[ρ],c1[ρ]<tt<c2[ρ]}
```

```
            }]},
    {tt,0,4*zo -  zi + 0.25},
    PlotRange→All,PlotStyle →{Lighter[Purple], Thickness[0.002]}]},
    {ρ,−ρmax,ρmax,density}];

Show[OvalSurfaceV,SurfaceV,LensV,WavesV,RaysV,
   PlotRange→All,ImageSize→{640,Automatic}]
```

12.2.9 Case 9: Real infinite object—real infinite image

```
(** Definition of set variables **)
```

$$Ka = -\frac{no^2}{n^2};$$

coa = 0;

c1a = 0;

b1a = 0;

$$za[\rho\_] := \frac{\rho^2\ (coa+c1a\ \rho^2)}{1+b1a\ \rho^2+\sqrt{1+(2\ b1a-coa^2\ Ka)\ \rho^2}}$$

$$ra[\rho\_] := \text{Piecewise}[\{\{\sqrt{\rho^2-za[\rho]^2},\rho\geq 0\},\{-\sqrt{\rho^2-za[\rho]^2},\rho<0\}\}]$$

$$\varrho z[\rho\_] := \frac{no\ za'[\rho]^2}{n\ (ra'[\rho]^2+za'[\rho]^2)} + \frac{ra'[\rho]}{\sqrt{ra'[\rho]^2+za'[\rho]^2}} \sqrt{\left(\frac{n^2\ ra'[\rho]^2+(n^2-no^2)\ za'[\rho]^2}{n^2\ (ra'[\rho]^2+za'[\rho]^2)}\right)}$$

$$\varrho r[\rho\_] := \frac{za'[\rho]}{n\ (ra'[\rho]^2+za'[\rho]^2)}\left(no ra'[\rho]-n \sqrt{ra'[\rho]^2+za'[\rho]^2}\sqrt{\left(\frac{n^2 ra'[\rho]^2+(n^2-no^2) za'[\rho]^2}{n^2(ra'[\rho]^2+za'[\rho]^2)}\right)}\right)$$

```
(** Homotopy for Eikonal **)
```

```
θ1[ρ_] := ArcTan[Limit[(ra[ρ])/(−zo+za[ρ]),zo→−Infinity]];

θ2[ρ_] :=  ArcTan[Limit[(ra[ρ])/(zi−za[ρ]),zi→Infinity]];

c1[ρ_] :=za[ρ];

(** Setting variable's **)

c = 1;
v = c/n;

no = 1;
n = 1.5;

ρmax = 25;
density = 5;
(* zo −> −Infinity *)
Zo = −50;
(* zi −> Infinity*)
Zi = 60;


(** Plot Section **)

SurfaceV := ParametricPlot[{{za[ρ],ra[ρ]},{zb[ρ],rb[ρ]}},{ρ,−ρmax−1,ρmax+1},
          PlotStyle →{Black,Red}, AspectRatio→Automatic, AxesLabel →{r,z}]

OvalSurfaceV := ParametricPlot[{{za[ρ],ra[ρ]},{zb[ρ],rb[ρ]}},
           {ρ,−ρmax−500,ρmax+500},
           PlotStyle →{Directive[{Black,Dashed}],Directive[{Red,Dashed}]}]

LensV := Graphics[{EdgeForm[Directive[{Black,Thin}]],
         Directive[Gray,Opacity[0.3]],
         Polygon[Join[#[[1,1]],#[[2,1]],Reverse[#[[4,1]]],Reverse[#[[3,1]]]]&
         [Cases[SurfaceV,_Line,∞]]]}];

 WavesV:=Table[{
ParametricPlot[{
Piecewise[{
{−tt,tt<=c1[ρ]},
{v (tt−c1[ρ])−za[ρ],
c1[ρ]<tt}}],
Piecewise[{
{ra[ρ],tt<=c1[ρ]},
{ra[ρ],c1[ρ]<tt}}]},
{ρ,−ρmax,ρmax},

PlotRange→All,
PlotStyle →{Blue,  Thickness[0.002]}]},{tt,0,−Zo+( n  Zi),density}];

RaysV:=Table[{
```

```
ParametricPlot[{
Piecewise[{
{-tt,tt<=c1[ρ]},
{v (tt-c1[ρ])-za[ρ],c1[ρ]<tt}}],
Piecewise[{
{ra[ρ],tt<=c1[ρ]},
{ra[ρ],c1[ρ]<tt}}] },
{tt,0,-Zo+( n Zi)},PlotRange→All,

Show[OvalSurfaceV,SurfaceV,LensV,WavesV,RaysV,
   PlotRange→All,ImageSize→{640,Automatic}]
```

12.3 Programs for chapter 8

12.3.1 Case 1: Real finite object—real finite image

```
" ClearAll["Global`*"]
(** Definition of set variables **)
```

$$Ka = \frac{(n^2\ zin-no^2\ zo)^2}{n\ no\ (n\ zin-no\ zo)\ (n\ zo-no\ zin)};$$

$$coa = \frac{n\ zo-no\ zin}{zin\ zo(n-no)};$$

$$c1a = \frac{(n-no)(n+no)^2}{4\ n\ no\ zin\ zo(n\ zin-no\ zo)};$$

$$b1a = \frac{(n+no)(n^2zin-no^2zo)}{2\ n\ no\ zin\ zo(n\ zin-no\ zo)};$$

$$za[\rho\_] := \frac{\rho^2\ (coa+c1a\ \rho^2)}{1+b1a\ \rho^2+\sqrt{1+(2\ b1a-coa^2\ Ka)\ \rho^2}}$$

$$ra[\rho\_] := \mathrm{Piecewise}[\{\{\sqrt{\rho^2-za[\rho]^2},\rho\geq 0\},\{-\sqrt{\rho^2-za[\rho]^2},\rho<0\}\}]$$

$$f[\rho\_] := -no\ zo + n\ \tau + ni\ zi + \mathrm{Sign}[zo]\ \sqrt{ra[\rho]^2+(zo-za[\rho])^2}$$

$$z[\rho\_] := za[\rho] - \tau - zi$$

$$\varrho z[\rho\_] := \frac{no(za'[\rho]\ (ra[\rho]ra'[\rho]+(-zo+za[\rho])\ za'[\rho]))}{n\ \sqrt{ra[\rho]^2+(zo-za[\rho])^2}\ ((ra'[\rho])^2+(za'[\rho])^2)}$$

$$+\frac{ra'[\rho]}{\sqrt{(ra'[\rho])^2+(za'[\rho])^2}}\sqrt{\left(1-\frac{no^2(ra[\rho]ra'[\rho]+(-zo+za[\rho])\ za'[\rho])^2}{n^2\ (ra[\rho]^2+(zo-za[\rho])^2)\ ((ra'[\rho])^2+(za'[\rho])^2)}\right)}$$

$$\varrho r[\rho\_]:=\frac{no(ra'[\rho](ra[\rho]ra'[\rho]+(-zo+za[\rho])\ za'[\rho]))}{n\ \sqrt{ra[\rho]^2+(zo-za[\rho])^2}\ ((ra'[\rho])^2+(za'[\rho])^2)}$$

$$-\frac{za'[\rho]}{\sqrt{(ra'[\rho])^2+(za'[\rho])^2}}\sqrt{\left(1-\frac{no^2(ra[\rho]ra'[\rho]+(-zo+za[\rho])\ za'[\rho])^2}{n^2\ (ra[\rho]^2+(zo-za[\rho])^2)\ ((ra'[\rho])^2+(za'[\rho])^2)}\right)}$$

$$\vartheta[\rho\_] := \frac{1}{n^2-1}\Bigg(f[\rho]\ n+ni^2\ (ra[\rho]\ \varrho r[\rho]+z[\rho]\ \varrho z[\rho])+\mathbf{Sign}[n]\mathbf{Sign}[zi]$$

$$\sqrt{\left(ni^2\ (f[\rho]^2+n^2\ (ra[\rho]^2+z[\rho]^2)-ni^2\ (z[\rho]\ \varrho r[\rho]-ra[\rho]\ \varrho z[\rho])^2\right.}$$

$$\left.+\ 2\ f[\rho]\ n\ (ra[\rho]\ \varrho r[\rho]+z[\rho]\ \varrho z[\rho]))\right)\Bigg)$$

```
zb[ρ_] := za[ρ] + ϑ[ρ] ϱz[ρ];

rb[ρ_] := ra[ρ] + ϑ[ρ] ϱr[ρ];


(** Homotopy for Eikonal **)


θ1[ρ_] := ArcTan[(ra[ρ])/(−zo+za[ρ])];

θ2[ρ_] := ArcTan[(rb[ρ]−ra[ρ])/(zb[ρ]−za[ρ] )];

θ3[ρ_] := ArcTan[rb[ρ]/(zb[ρ]−τ−zi)];

c1[ρ_] := Sec[θ1[ρ]](−zo+za[ρ]);

c2[ρ_] := (Sec[θ2[ρ]](zb[ρ]−za[ ρ])/v)+c1[ρ];


(** Setting variable's **)

c = 1;
v = c/n;

no = 1;
```

```
ni = no;
n = 1.5;

ρmax = 16;
density = 4;
τ = 12;
zo = −40;
zi = 40;
zin = n zi;


(** Plot Section **)

SurfaceV := ParametricPlot[{{za[ρ],ra[ρ]},{zb[ρ],rb[ρ]}},{ρ,−ρmax−1,ρmax+1},
        PlotStyle →{Black,Red}, AspectRatio→Automatic, AxesLabel →{r,z}]

OvalSurfaceV := ParametricPlot[{{za[ρ],ra[ρ]},{zb[ρ],rb[ρ]}},
          {ρ,−ρmax−500,ρmax+500},
          PlotStyle →{Directive[{Black,Dashed}],Directive[{Red,Dashed}]}]

LensV := Graphics[{EdgeForm[Directive[{Black,Thin}]],
         Directive[Gray,Opacity[0.3]],
         Polygon[Join[#[[1,1]],#[[2,1]],Reverse[#[[4,1]]],Reverse[#[[3,1]]]]&
         [Cases[SurfaceV,_Line,∞]]]}];

 WavesV := Table[{
          ParametricPlot[{
             Piecewise[{
                   {Cos[θ1[ρ]]tt+zo,tt<=c1[ρ]},
                   {v Cos[θ2[ρ]](tt−c1[ρ])+za[ρ],c1[ρ]<tt<c2[ρ]},
                   {zb[ρ]+(tt−c2[ρ])Cos[θ3[ρ]],tt>=c2[ρ]}}],

             Piecewise[{
                   {Sin[θ1[ρ]]tt,tt<=c1[ρ]},
                   {v Sin[θ2[ρ]](tt−c1[ρ])+ra[ρ],c1[ρ]<tt<c2[ρ]},
                   {rb[ρ]+(tt−c2[ρ])Sin[θ3[ρ]],tt>=c2[ρ]}}]},
          {ρ,−ρmax,ρmax},
          PlotRange→All,
          PlotStyle →{Blue, Thickness[0.002]}]},
          {tt,0,−zo+nτ+zi+0.25,density}];

  RaysV := Table[{
        ParametricPlot[{
          Piecewise[{
               {Cos[θ1[ρ]]tt+zo,tt<=c1[ρ]},
               {v Cos[θ2[ρ]](tt−c1[ρ])+za[ρ],c1[ρ]<tt<c2[ρ]},
               {zb[ρ]+(tt−c2[ρ])Cos[θ3[ρ]],tt>=c2[ρ]}}],

          Piecewise[{
               {Sin[θ1[ρ]]tt,tt<=c1[ρ]},
               {v Sin[θ2[ρ]](tt−c1[ρ])+ra[ρ],c1[ρ]<tt<c2[ρ]},
               {rb[ρ]+(tt−c2[ρ])Sin[θ3[ρ]],tt>=c2[ρ]}}]},

       {tt,0,−zo+ n τ + zi+0.25},
```

```
        PlotRange→All,PlotStyle →{Lighter[Purple], Thickness[0.002]}]},
        {ρ,−ρmax,ρmax,density}];

Show[OvalSurfaceV,SurfaceV,LensV,WavesV,RaysV,
    PlotRange→All,ImageSize→{640,Automatic}]
```

12.3.2 Case 2: Real infinity object—real finite image

"

(** Definition of set variables **)

$$Ka = -\frac{no^2}{n^2};$$

$$coa = \frac{n}{n\ zi-no\ zi};$$

$$c1a = 0;$$

$$b1a = 0;$$

$$za[\rho\_] := \frac{\rho^2\ (coa+c1a\ \rho^2)}{1+b1a\ \rho^2+\sqrt{1+(2\ b1a-coa^2\ Ka)\ \rho^2}}$$

$$ra[\rho\_] := \text{Piecewise}[\{\{\sqrt{\rho^2-za[\rho]^2},\rho\geq 0\},\{-\sqrt{\rho^2-za[\rho]^2},\rho<0\}\}]$$

$$f[\rho\_] := n\ \tau + ni\ zi - za[\rho]$$

$$z[\rho\_] := za[\rho] - \tau - zi$$

$$\varrho z[\rho\_] := \frac{no\ za'[\rho]^2}{n\ (ra'[\rho]^2+za'[\rho]^2)} + \frac{ra'[\rho]}{\sqrt{ra'[\rho]^2+za'[\rho]^2}} \sqrt{\left(\frac{n^2\ ra'[\rho]^2+(n^2-no^2)\ za'[\rho]^2}{n^2\ (ra'[\rho]^2+za'[\rho]^2)}\right)}$$

$$\varrho r[\rho\_] := \frac{za'[\rho]}{n\ (ra'[\rho]^2+za'[\rho]^2)}\left(no\ ra'[\rho]-n \sqrt{ra'[\rho]^2+za'[\rho]^2}\sqrt{\left(\frac{n^2 ra'[\rho]^2+(n^2-no^2) za'[\rho]^2}{n^2(ra'[\rho]^2+za'[\rho]^2)}\right)}\right)$$

$$\vartheta[\rho\_] := \frac{1}{-1+n^2}\Big(n\ f[\rho]+ni^2\ (ra[\rho]\ \varrho r[\rho]+z[\rho]\ \varrho z[\rho])+\text{Sign}[n]\text{Sign}[zi]$$

```
√(ni² (f[ρ]²+n² (ra[ρ]²+z[ρ]²)−ni² (z[ρ] ϱr[ρ]−ra[ρ] ϱz[ρ])²
   + 2 n f[ρ] (ra[ρ] ϱr[ρ]+z[ρ] ϱz[ρ]))))

zb[ρ_] := za[ρ] + ϑ[ρ] ϱz[ρ];

rb[ρ_] := ra[ρ] + ϑ[ρ] ϱr[ρ];


(** Homotopy for Eikonal **)

θ1[ρ_] := ArcTan[Limit[(ra[ρ])/(−zo+za[ρ]),zo→−Infinity]]

θ2[ρ_] := ArcTan[(rb[ρ]−ra[ρ])/(zb[ρ]−za[ρ])];

θ3[ρ_] := ArcTan[rb[ρ]/(zb[ρ]−τ−zi)];

c1[ρ_] := (−Zo+za[ρ]);

c2[ρ_] := (Sec[θ2[ρ]](zb[ρ]−za[ ρ])/v)+c1[ρ];


(** Setting variable's **)

c = 1;
v = c/n;

no = 1;
ni = no;
n = 1.5;

ρmax = 16;
density = 4;
τ = 12;
(* zo −> −Infinity *)
Zo = −40;
zi = 40;
zin = n zi;


(** Plot Section **)

SurfaceV := ParametricPlot[{{za[ρ],ra[ρ]},{zb[ρ],rb[ρ]}},{ρ,−ρmax−1,ρmax+1},
        PlotStyle →{Black,Red}, AspectRatio→Automatic, AxesLabel →{r,z}]

OvalSurfaceV := ParametricPlot[{{za[ρ],ra[ρ]},{zb[ρ],rb[ρ]}},
          {ρ,−ρmax−500,ρmax+500},
          PlotStyle →{Directive[{Black,Dashed}],Directive[{Red,Dashed}]}]

LensV := Graphics[{EdgeForm[Directive[{Black,Thin}]],
         Directive[Gray,Opacity[0.3]],
```

```
        Polygon[Join[#[[1,1]],#[[2,1]],Reverse[#[[4,1]]],Reverse[#[[3,1]]]]&
        [Cases[SurfaceV,_Line,∞]]]}];

WavesV := Table[{
          ParametricPlot[{
            Piecewise[{
                    {Cos[θ1[ρ]]tt+Zo,tt<=c1[ρ]},
                    {v Cos[θ2[ρ]](tt−c1[ρ])+za[ρ],c1[ρ]<tt<c2[ρ]},
                    {zb[ρ]+(tt−c2[ρ])Cos[θ3[ρ]],tt>=c2[ρ]}}],

            Piecewise[{
                    {ra[ρ],tt<=c1[ρ]},
                    {v Sin[θ2[ρ]](tt−c1[ρ])+ra[ρ],c1[ρ]<tt<c2[ρ]},
                    {rb[ρ]+(tt−c2[ρ])Sin[θ3[ρ]],tt>=c2[ρ]}}]},
          {ρ,−ρmax,ρmax},
          PlotRange→All,
          PlotStyle →{Blue, Thickness[0.002]}]},
          {tt,0,−Zo+nτ+zi+0.25,density}];

RaysV := Table[{
        ParametricPlot[{
          Piecewise[{
                {Cos[θ1[ρ]]tt+Zo,tt<=c1[ρ]},
                {v Cos[θ2[ρ]](tt−c1[ρ])+za[ρ],c1[ρ]<tt<c2[ρ]},
                {zb[ρ]+(tt−c2[ρ])Cos[θ3[ρ]],tt>=c2[ρ]}}],

          Piecewise[{
                {ra[ρ],tt<=c1[ρ]},
                {v Sin[θ2[ρ]](tt−c1[ρ])+ra[ρ],c1[ρ]<tt<c2[ρ]},
                {rb[ρ]+(tt−c2[ρ])Sin[θ3[ρ]],tt>=c2[ρ]}}]},

        {tt,0,−Zo+ n τ + zi+0.25},
        PlotRange→All,PlotStyle →{Lighter[Purple], Thickness[0.002]}]},
        {ρ,−ρmax,ρmax,density}];


Show[OvalSurfaceV,SurfaceV,LensV,WavesV,RaysV,
    PlotRange→All,ImageSize→{640,Automatic}]
```

12.3.3 Case 3: Real infinity object—virtual finite image

(** Definition of set variables **)

$$\text{Ka} = -\frac{\text{no}^2}{\text{n}^2};$$

$$\text{coa} = \frac{\text{n}}{\text{n zi}-\text{no zi}};$$

$$\text{c1a} = 0;$$

$$\text{b1a} = 0;$$

$$\text{za}[\rho\_] := \frac{\rho^2\ (\text{coa}+\text{c1a}\ \rho^2)}{1+\text{b1a}\ \rho^2+\sqrt{1+(2\ \text{b1a}-\text{coa}^2\ \text{Ka})\ \rho^2}}$$

$$\text{ra}[\rho\_] := \text{Piecewise}[\{\{\sqrt{\rho^2-\text{za}[\rho]^2},\rho\geq 0\},\{-\sqrt{\rho^2-\text{za}[\rho]^2},\rho<0\}\}]$$

$$\text{f}[\rho\_] := \text{n}\ \tau+\text{ni zi}-\text{za}[\rho]$$

$$z[\rho\_] := \text{za}[\rho]-\tau-\text{zi}$$

$$\varrho\text{z}[\rho\_] := \frac{\text{no za}'[\rho]^2}{\text{n}\ (\text{ra}'[\rho]^2+\text{za}'[\rho]^2)}+\frac{\text{ra}'[\rho]}{\sqrt{\text{ra}'[\rho]^2+\text{za}'[\rho]^2}}$$
$$\sqrt{\left(\frac{\text{n}^2\ \text{ra}'[\rho]^2+(\text{n}^2-\text{no}^2)\ \text{za}'[\rho]^2}{\text{n}^2\ (\text{ra}'[\rho]^2+\text{za}'[\rho]^2)}\right)}$$

$$\varrho\text{r}[\rho\_] := \frac{\text{za}'[\rho]}{\text{n}\ (\text{ra}'[\rho]^2+\text{za}'[\rho]^2)}\left(\text{no ra}'[\rho]-\text{n}\right.$$
$$\left.\sqrt{\text{ra}'[\rho]^2+\text{za}'[\rho]^2}\sqrt{\left(\frac{\text{n}^2\text{ra}'[\rho]^2+(\text{n}^2-\text{no}^2)\text{za}'[\rho]^2}{\text{n}^2(\text{ra}'[\rho]^2+\text{za}'[\rho]^2)}\right)}\right)$$

$$\vartheta[\rho\_] := \frac{1}{-1+\text{n}^2}\Big(\text{n f}[\rho]+\text{ni}^2\ (\text{ra}[\rho]\ \varrho\text{r}[\rho]+z[\rho]\ \varrho\text{z}[\rho])-\text{Sign}[\text{n}]\text{Sign}[\text{zi}]$$
$$\sqrt{\Big(\text{ni}^2\ (\text{f}[\rho]^2+\text{n}^2\ (\text{ra}[\rho]^2+z[\rho]^2)-\text{ni}^2\ (z[\rho]\ \varrho\text{r}[\rho]-\text{ra}[\rho]\ \varrho\text{z}[\rho])^2}$$
$$+\,2\ \text{n f}[\rho]\ (\text{ra}[\rho]\ \varrho\text{r}[\rho]+z[\rho]\ \varrho\text{z}[\rho]))\Big)\Big)$$

$$\text{zb}[\rho\_] := \text{za}[\rho]+\vartheta[\rho]\ \varrho\text{z}[\rho];$$

$$\text{rb}[\rho\_] := \text{ra}[\rho]+\vartheta[\rho]\ \varrho\text{r}[\rho];$$

(** Homotopy for Eikonal **)

$$\theta1[\rho\_] := \text{ArcTan}[\text{Limit}[(\text{ra}[\rho])/(-\text{zo}+\text{za}[\rho]),\text{zo}\to-\text{Infinity}]]$$

$$\theta2[\rho\_] := \text{ArcTan}[(\text{rb}[\rho]-\text{ra}[\rho])/(\text{zb}[\rho]-\text{za}[\rho])];$$

$$\theta3[\rho\_] := \text{ArcTan}[\text{rb}[\rho]/(\text{zb}[\rho]-\tau-\text{zi})];$$

$$\text{c1}[\rho\_] := (-\text{Zo}+\text{za}[\rho]);$$

```
c2[ρ_] := (Sec[θ2[ρ]](zb[ρ]−za[ ρ])/v)+c1[ρ];


(** Setting variable's **)

c = 1;
v = c/n;

no = 1;
ni = no;
n = 1.5;

ρmax = 15;
density = 5;
τ = 20;
(* zo −> −Infinity *)
Zo = −25;
zi = −20;
zin = n zi;


(** Plot Section **)

SurfaceV := ParametricPlot[{{za[ρ],ra[ρ]},{zb[ρ],rb[ρ]}},{ρ,−ρmax−1,ρmax+1},
        PlotStyle →{Black,Red}, AspectRatio→Automatic, AxesLabel →{r,z}]

OvalSurfaceV := ParametricPlot[{{za[ρ],ra[ρ]},{zb[ρ],rb[ρ]}},
        {ρ,−ρmax−500,ρmax+500},
        PlotStyle →{Directive[{Black,Dashed}],Directive[{Red,Dashed}]}]

LensV := Graphics[{EdgeForm[Directive[{Black,Thin}]],
        Directive[Gray,Opacity[0.3]],
        Polygon[Join[#[[1,1]],#[[2,1]],Reverse[#[[4,1]]],Reverse[#[[3,1]]]]&
        [Cases[SurfaceV,_Line,∞]]]}];

 WavesV := Table[{
        ParametricPlot[{
            Piecewise[{
                {Cos[θ1[ρ]]tt+Zo,tt<=c1[ρ]},
                {v Cos[θ2[ρ]](tt−c1[ρ])+za[ρ],c1[ρ]<tt<c2[ρ]},
                {zb[ρ]+(tt−c2[ρ])Cos[θ3[ρ]],tt>=c2[ρ]}}],

            Piecewise[{
                {Sin[θ1[ρ]]tt+ra[ρ],tt<=c1[ρ]},
                {v Sin[θ2[ρ]](tt−c1[ρ])+ra[ρ],c1[ρ]<tt<c2[ρ]},
                {rb[ρ]+(tt−c2[ρ])Sin[θ3[ρ]],tt>=c2[ρ]}}]},
        {ρ,−ρmax,ρmax},
        PlotRange→All,
        PlotStyle →{Blue, Thickness[0.002]}]},
        {tt,0,−Zo+nτ−zi+0.25,density}];

 RaysV := Table[{
        ParametricPlot[{
            Piecewise[{
```

```
        {Cos[θ1[ρ]]tt+Zo,tt<=c1[ρ]},
        {v Cos[θ2[ρ]](tt−c1[ρ])+za[ρ],c1[ρ]<tt<c2[ρ]},
        {zb[ρ]+(tt−c2[ρ])Cos[θ3[ρ]],tt>=c2[ρ]}}],

    Piecewise[{
        {Sin[θ1[ρ]]tt+ra[ρ],tt<=c1[ρ]},
        {v Sin[θ2[ρ]](tt−c1[ρ])+ra[ρ],c1[ρ]<tt<c2[ρ]},
        {rb[ρ]+(tt−c2[ρ])Sin[θ3[ρ]],tt>=c2[ρ]}}]},

  {tt,0,−Zo+ n τ − zi+0.25},
  PlotRange→All,PlotStyle →{Lighter[Purple], Thickness[0.002]}]},
  {ρ,−ρmax,ρmax,density}];


Show[OvalSurfaceV,SurfaceV,LensV,WavesV,RaysV,
   PlotRange→All,ImageSize→{640,Automatic}]
```

12.3.4 Case 4: Real finite object—virtual finite image

```
" ClearAll["Global'*"]
 (** Definition of set variables **)
```

$$Ka = \frac{(n^2\ zin-no^2\ zo)^2}{n\ no\ (n\ zin-no\ zo)\ (n\ zo-no\ zin)};$$

$$coa = \frac{n\ zo-no\ zin}{zin\ zo(n-no)};$$

$$c1a = \frac{(n-no)(n+no)^2}{4\ n\ no\ zin\ zo(n\ zin-no\ zo)};$$

$$b1a = \frac{(n+no)(n^2 zin-no^2 zo)}{2\ n\ no\ zin\ zo(n\ zin-no\ zo)};$$

$$za[\rho\_] := \frac{\rho^2\ (coa+c1a\ \rho^2)}{1+b1a\ \rho^2+\sqrt{1+(2\ b1a-coa^2\ Ka)\ \rho^2}}$$

$$ra[\rho\_] := Piecewise[\{\{\sqrt{\rho^2-za[\rho]^2},\rho\geq 0\},\{-\sqrt{\rho^2-za[\rho]^2},\rho<0\}\}]$$

$$f[\rho\_] := -no\ zo+n\ \tau+ni\ zi+Sign[zo]\ \sqrt{ra[\rho]^2+(zo-za[\rho])^2}$$

```
z[ρ_] := za[ρ] − τ − zi
```

$$\varrho z[\rho\_] := \frac{no(za'[\rho]\ (ra[\rho]ra'[\rho]+(-zo+za[\rho])\ za'[\rho]))}{n\ \sqrt{ra[\rho]^2+(zo-za[\rho])^2}\ ((ra'[\rho])^2+(za'[\rho])^2)}$$

$$+\frac{ra'[\rho]}{\sqrt{(ra'[\rho])^2+(za'[\rho])^2}}\sqrt{\left(1-\frac{no^2(ra[\rho]ra'[\rho]+(-zo+za[\rho])\ za'[\rho])^2}{n^2\ (ra[\rho]^2+(zo-za[\rho])^2)\ ((ra'[\rho])^2+(za'[\rho])^2)}\right)}$$

$$\varrho r[\rho\_] := \frac{no(ra'[\rho](ra[\rho]ra'[\rho]+(-zo+za[\rho])\ za'[\rho]))}{n\ \sqrt{ra[\rho]^2+(zo-za[\rho])^2}\ ((ra'[\rho])^2+(za'[\rho])^2)}$$

$$-\frac{za'[\rho]}{\sqrt{(ra'[\rho])^2+(za'[\rho])^2}}\sqrt{\left(1-\frac{no^2(ra[\rho]ra'[\rho]+(-zo+za[\rho])\ za'[\rho])^2}{n^2\ (ra[\rho]^2+(zo-za[\rho])^2)\ ((ra'[\rho])^2+(za'[\rho])^2)}\right)}$$

$$\vartheta[\rho\_] := \frac{1}{n^2-1}\left(f[\rho]\ n+ni^2\ (ra[\rho]\ \varrho r[\rho]+z[\rho]\ \varrho z[\rho])-\mathbf{Sign[n]Sign[zi]}\right.$$

$$\sqrt{\left(ni^2\ (f[\rho]^2+n^2\ (ra[\rho]^2+z[\rho]^2)-ni^2\ (z[\rho]\ \varrho r[\rho]-ra[\rho]\ \varrho z[\rho])^2\right.}$$

$$\left.\left.+\ 2\ f[\rho]\ n\ (ra[\rho]\ \varrho r[\rho]+z[\rho]\ \varrho z[\rho]))\right)\right)$$

```
zb[ρ_] := za[ρ] + ϑ[ρ] ϱz[ρ];

rb[ρ_] := ra[ρ] + ϑ[ρ] ϱr[ρ];


(** Homotopy for Eikonal **)


θ1[ρ_] := ArcTan[(ra[ρ])/(−zo+za[ρ])];

θ2[ρ_] := ArcTan[(rb[ρ]−ra[ρ])/(zb[ρ]−za[ρ] )];

θ3[ρ_] := ArcTan[rb[ρ]/(zb[ρ]−τ−zi)];

c1[ρ_] := Sec[θ1[ρ]](−zo+za[ρ]);

c2[ρ_] := (Sec[θ2[ρ]](zb[ρ]−za[ ρ])/v)+c1[ρ];


(** Setting variable's **)

c = 1;
v = c/n;
```

```
no = 1;
ni = no;
n = 1.5;

ρmax = 5;
density = 1;
τ = 10;
zo = −15;
zi = −5;
zin = n zi;


(** Plot Section **)

SurfaceV := ParametricPlot[{{za[ρ],ra[ρ]},{zb[ρ],rb[ρ]}},{ρ,−ρmax−1,ρmax+1},
        PlotStyle →{Black,Red}, AspectRatio→Automatic, AxesLabel →{r,z}]

OvalSurfaceV := ParametricPlot[{{za[ρ],ra[ρ]},{zb[ρ],rb[ρ]}},
          {ρ,−ρmax−500,ρmax+500},
          PlotStyle →{Directive[{Black,Dashed}],Directive[{Red,Dashed}]}]

LensV := Graphics[{EdgeForm[Directive[{Black,Thin}]],
        Directive[Gray,Opacity[0.3]],
        Polygon[Join[#[[1,1]],#[[2,1]],Reverse[#[[4,1]]],Reverse[#[[3,1]]]]&
        [Cases[SurfaceV,_Line,∞]]]}];

 WavesV := Table[{
          ParametricPlot[{
            Piecewise[{
                {Cos[θ1[ρ]]tt+zo,tt<=c1[ρ]},
                {v Cos[θ2[ρ]](tt−c1[ρ])+za[ρ],c1[ρ]<tt<c2[ρ]},
                {zb[ρ]+(tt−c2[ρ])Cos[θ3[ρ]],tt>=c2[ρ]}}],

            Piecewise[{
                {Sin[θ1[ρ]]tt,tt<=c1[ρ]},
                {v Sin[θ2[ρ]](tt−c1[ρ])+ra[ρ],c1[ρ]<tt<c2[ρ]},
                {rb[ρ]+(tt−c2[ρ])Sin[θ3[ρ]],tt>=c2[ρ]}}]},
          {ρ,−ρmax,ρmax},
          PlotRange→All,
          PlotStyle →{Blue, Thickness[0.002]}]},
          {tt,0,−zo+nτ−zi+0.25,density}];

  RaysV := Table[{
        ParametricPlot[{
          Piecewise[{
              {Cos[θ1[ρ]]tt+zo,tt<=c1[ρ]},
              {v Cos[θ2[ρ]](tt−c1[ρ])+za[ρ],c1[ρ]<tt<c2[ρ]},
              {zb[ρ]+(tt−c2[ρ])Cos[θ3[ρ]],tt>=c2[ρ]}}],

          Piecewise[{
              {Sin[θ1[ρ]]tt,tt<=c1[ρ]},
              {v Sin[θ2[ρ]](tt−c1[ρ])+ra[ρ],c1[ρ]<tt<c2[ρ]},
              {rb[ρ]+(tt−c2[ρ])Sin[θ3[ρ]],tt>=c2[ρ]}}]},
```

```
{tt,0,−zo+ n τ − zi+0.25},
PlotRange→All,PlotStyle →{Lighter[Purple], Thickness[0.002]}]},
{ρ,−ρmax,ρmax,density}];
```

```
Show[OvalSurfaceV,SurfaceV,LensV,WavesV,RaysV,
  PlotRange→All,ImageSize→{640,Automatic}]
```

12.3.5 Case 5: Real finite object—real infinite image

```
" ClearAll["Global`*"]
(** Definition of set variables **)
```

$$\mathrm{Ka} = -\frac{\mathrm{n}^2}{\mathrm{no}^2};$$

$$\mathrm{coa} = -\frac{\mathrm{no}}{\mathrm{n\ zo-no\ zo}};$$

$$\mathrm{coa} = 0;$$

$$\mathrm{c1a} = 0;$$

$$\mathrm{b1a} = \frac{(\mathrm{n+no})(\mathrm{n}^2\mathrm{zin}-\mathrm{no}^2\mathrm{zo})}{2\ \mathrm{n\ no\ zin\ zo}(\mathrm{n\ zin-no\ zo})};$$

$$\mathrm{za}[\rho\_] := \frac{\rho^2\ (\mathrm{coa}+\mathrm{c1a}\ \rho^2)}{1+\mathrm{b1a}\ \rho^2+\sqrt{1+(2\ \mathrm{b1a}-\mathrm{coa}^2\ \mathrm{Ka})\ \rho^2}}$$

$$\mathrm{ra}[\rho\_] := \mathrm{Piecewise}[\{\{\sqrt{\rho^2-\mathrm{za}[\rho]^2},\rho\geq 0\},\{-\sqrt{\rho^2-\mathrm{za}[\rho]^2},\rho<0\}\}]$$

$$\varrho\mathrm{z}[\rho\_] := \frac{\mathrm{no}(\mathrm{za}'[\rho]\ (\mathrm{ra}[\rho]\mathrm{ra}'[\rho]+(-\mathrm{zo}+\mathrm{za}[\rho])\ \mathrm{za}'[\rho]))}{\mathrm{n}\ \sqrt{\mathrm{ra}[\rho]^2+(\mathrm{zo}-\mathrm{za}[\rho])^2}\ ((\mathrm{ra}'[\rho])^2+(\mathrm{za}'[\rho])^2)}$$

$$+\frac{\mathrm{ra}'[\rho]}{\sqrt{(\mathrm{ra}'[\rho])^2+(\mathrm{za}'[\rho])^2}}\sqrt{\left(1-\frac{\mathrm{no}^2(\mathrm{ra}[\rho]\mathrm{ra}'[\rho]+(-\mathrm{zo}+\mathrm{za}[\rho])\ \mathrm{za}'[\rho])^2}{\mathrm{n}^2\ (\mathrm{ra}[\rho]^2+(\mathrm{zo}-\mathrm{za}[\rho])^2)\ ((\mathrm{ra}'[\rho])^2+(\mathrm{za}'[\rho])^2)}\right)}$$

$$\varrho\mathrm{r}[\rho\_]:=\frac{\mathrm{no}(\mathrm{ra}'[\rho](\mathrm{ra}[\rho]\mathrm{ra}'[\rho]+(-\mathrm{zo}+\mathrm{za}[\rho])\ \mathrm{za}'[\rho]))}{\mathrm{n}\ \sqrt{\mathrm{ra}[\rho]^2+(\mathrm{zo}-\mathrm{za}[\rho])^2}\ ((\mathrm{ra}'[\rho])^2+(\mathrm{za}'[\rho])^2)}$$

$$- \frac{za'[\rho]}{\sqrt{(ra'[\rho])^2+(za'[\rho])^2}} \sqrt{\left(1-\frac{no^2(ra[\rho]ra'[\rho]+(-zo+za[\rho])\ za'[\rho])^2}{n^2\ (ra[\rho]^2+(zo-za[\rho])^2)\ ((ra'[\rho])^2+(za'[\rho])^2)}\right)}$$

```
ϑ[ρ_] := ((−1+n)τ −zo+za[ρ] − Sign[n]Sign[Zi]Sqrt[ra[ρ]^2 + (za[ρ]−zo)^2]
)/(n−ϱz[ρ]);
zb[ρ_] := za[ρ] + ϑ[ρ] ϱz[ρ];

rb[ρ_] := ra[ρ] + ϑ[ρ] ϱr[ρ];


(** Homotopy for Eikonal **)

θ1[ρ_] := ArcTan[(ra[ρ])/(−zo+za[ρ])];

θ2[ρ_] := ArcTan[(rb[ρ]−ra[ρ])/(zb[ρ]−za[ρ] )];

θ3[ρ_] := ArcTan[Limit[rb[ρ]/(zb[ρ]−τ−zi),zi→Infinity]];

c1[ρ_] := Sec[θ1[ρ]](−zo+za[ρ]);

c2[ρ_] := (Sec[θ2[ρ]](zb[ρ]−za[ρ])/v)+c1[ρ];

(** Setting variable's **)

c = 1;
v = c/n;

no = 1;
ni = no;
n = 1.5;

ρmax = 45;
density = 5;
τ = 30;
zo = −55;
(* zi −> Infinity*)
Zi = 35;


(** Plot Section **)

SurfaceV := ParametricPlot[{{za[ρ],ra[ρ]},{zb[ρ],rb[ρ]}},{ρ,−ρmax−1,ρmax+1},
          PlotStyle →{Black,Red}, AspectRatio→Automatic, AxesLabel →{r,z}]

OvalSurfaceV := ParametricPlot[{{za[ρ],ra[ρ]},{zb[ρ],rb[ρ]}},
            {ρ,−ρmax−50,ρmax+50},
            PlotStyle →{Directive[{Black,Dashed}],Directive[{Red,Dashed}]}]

LensV := Graphics[{EdgeForm[Directive[{Black,Thin}]],
```

```
        Directive[Gray,Opacity[0.3]],
        Polygon[Join[#[[1,1]],#[[2,1]],Reverse[#[[4,1]]],Reverse[#[[3,1]]]]&
        [Cases[SurfaceV,_Line,∞]]]}];

WavesV := Table[{
          ParametricPlot[{
             Piecewise[{
                   {Cos[θ1[ρ]]tt+zo,tt<=c1[ρ]},
                   {v Cos[θ2[ρ]](tt−c1[ρ])+za[ρ],c1[ρ]<tt<c2[ρ]},
                   {zb[ρ]+(tt−c2[ρ])Cos[θ3[ρ]],tt>=c2[ρ]}}],

             Piecewise[{
                   {Sin[θ1[ρ]]tt,tt<=c1[ρ]},
                   {v Sin[θ2[ρ]](tt−c1[ρ])+ra[ρ],c1[ρ]<tt<c2[ρ]},
                   {rb[ρ]+(tt−c2[ρ])Sin[θ3[ρ]],tt>=c2[ρ]}}]},
          {ρ,−ρmax,ρmax},
          PlotRange→All,
          PlotStyle →{Blue, Thickness[0.002]}]},
          {tt,0,−zo + n τ + Zi + 0.25,density}];

RaysV := Table[{
       ParametricPlot[{
          Piecewise[{
               {Cos[θ1[ρ]]tt+zo,tt<=c1[ρ]},
               {v Cos[θ2[ρ]](tt−c1[ρ])+za[ρ],c1[ρ]<tt<c2[ρ]},
               {zb[ρ]+(tt−c2[ρ])Cos[θ3[ρ]],tt>=c2[ρ]}}],

          Piecewise[{
               {Sin[θ1[ρ]]tt,tt<=c1[ρ]},
               {v Sin[θ2[ρ]](tt−c1[ρ])+ra[ρ],c1[ρ]<tt<c2[ρ]},
               {rb[ρ]+(tt−c2[ρ])Sin[θ3[ρ]],tt>=c2[ρ]}}]},

       {tt,0,−zo + n τ + Zi + 0.25},
       PlotRange→All,PlotStyle →{Lighter[Purple], Thickness[0.002]}]},
       {ρ,−ρmax,ρmax,density}];


Show[OvalSurfaceV,SurfaceV,LensV,WavesV,RaysV,
   PlotRange→All,ImageSize→{640,Automatic}]
```

12.3.6 Case 6: Virtual finite object—real infinite image

```
" ClearAll["Global`*"]
(** Definition of set variables **)
```

$$Ka = -\frac{n^2}{no^2};$$

$$coa = -\frac{no}{n\ zo - no\ zo};$$

$$coa = 0;$$

$$c1a = 0;$$

$$b1a = \frac{(n+no)(n^2 zin - no^2 zo)}{2\ n\ no\ zin\ zo(n\ zin - no\ zo)};$$

$$za[\rho\_] := \frac{\rho^2\ (coa + c1a\ \rho^2)}{1 + b1a\ \rho^2 + \sqrt{1 + (2\ b1a - coa^2\ Ka)\ \rho^2}}$$

$$ra[\rho\_] := \mathrm{Piecewise}[\{\{\sqrt{\rho^2 - za[\rho]^2}, \rho \geq 0\}, \{-\sqrt{\rho^2 - za[\rho]^2}, \rho < 0\}\}]$$

$$\varrho z[\rho\_] := \frac{no(za'[\rho]\ (ra[\rho]ra'[\rho] + (-zo + za[\rho])\ za'[\rho]))}{n\ \sqrt{ra[\rho]^2 + (zo - za[\rho])^2}\ ((ra'[\rho])^2 + (za'[\rho])^2)} + \frac{ra'[\rho]}{\sqrt{(ra'[\rho])^2 + (za'[\rho])^2}} \sqrt{\left(1 - \frac{no^2(ra[\rho]ra'[\rho] + (-zo + za[\rho])\ za'[\rho])^2}{n^2\ (ra[\rho]^2 + (zo - za[\rho])^2)\ ((ra'[\rho])^2 + (za'[\rho])^2)}\right)}$$

$$\varrho r[\rho\_] := \frac{no(ra'[\rho](ra[\rho]ra'[\rho] + (-zo + za[\rho])\ za'[\rho]))}{n\ \sqrt{ra[\rho]^2 + (zo - za[\rho])^2}\ ((ra'[\rho])^2 + (za'[\rho])^2)} - \frac{za'[\rho]}{\sqrt{(ra'[\rho])^2 + (za'[\rho])^2}} \sqrt{\left(1 - \frac{no^2(ra[\rho]ra'[\rho] + (-zo + za[\rho])\ za'[\rho])^2}{n^2\ (ra[\rho]^2 + (zo - za[\rho])^2)\ ((ra'[\rho])^2 + (za'[\rho])^2)}\right)}$$

```
ϑ[ρ_] := ((-1+n)τ -zo+za[ρ] + Sign[n]Sign[zo]Sqrt[ra[ρ]^2 + (za[ρ]-zo)^2]
)/(n-ϱz[ρ]);
zb[ρ_] := za[ρ] + ϑ[ρ] ϱz[ρ];
rb[ρ_] := ra[ρ] + ϑ[ρ] ϱr[ρ];


(** Homotopy for Eikonal **)
θ1[ρ_] := ArcTan[(ra[ρ])/(-zo+za[ρ])];
θ2[ρ_] := ArcTan[(rb[ρ]-ra[ρ])/(zb[ρ]-za[ρ] )];
θ3[ρ_] := ArcTan[Limit[rb[ρ]/(zb[ρ]-τ-zi),zi->Infinity]];
```

```
c1[ρ_] := (n zo + n τ) + Sec[θ1[ρ]](−zo + za[ρ]);

c2[ρ_] := (Sec[θ2[ρ]](zb[ρ]−za[ρ])/v)+c1[ρ];

(** Setting variable's **)

c = 1;
v = c/n;

no = 1;
ni = no;
n = 1.5;

ρmax = 16;
density = 2;
τ = 1;
zo = 16;
(* zi −> Infinity*)
Zi = 16;


(** Plot Section **)

SurfaceV := ParametricPlot[{{za[ρ],ra[ρ]},{zb[ρ],rb[ρ]}},{ρ,−ρmax−1,ρmax+1},
        PlotStyle →{Black,Red}, AspectRatio→Automatic, AxesLabel →{r,z}]

OvalSurfaceV := ParametricPlot[{{za[ρ],ra[ρ]},{zb[ρ],rb[ρ]}},
          {ρ,−ρmax−5,ρmax+5},
          PlotStyle →{Directive[{Black,Dashed}],Directive[{Red,Dashed}]}]

LensV := Graphics[{EdgeForm[Directive[{Black,Thin}]],
         Directive[Gray,Opacity[0.3]],
         Polygon[Join[#[[1,1]],#[[2,1]],Reverse[#[[4,1]]],Reverse[#[[3,1]]]]&
         [Cases[SurfaceV,_Line,∞]]]}];

 WavesV := Table[{
          ParametricPlot[{
             Piecewise[{
                   {Cos[θ1[ρ]](tt−(n zo+n τ))+zo,tt<=c1[ρ]},
                   {v Cos[θ2[ρ]](tt−c1[ρ])+za[ρ],c1[ρ]<tt<c2[ρ]},
                   {zb[ρ]+(tt−c2[ρ])Cos[θ3[ρ]],tt>=c2[ρ]}}],

             Piecewise[{
                   {Sin[θ1[ρ]](tt−(n zo+n τ)),tt<=c1[ρ]},
                   {v Sin[θ2[ρ]](tt−c1[ρ])+ra[ρ],c1[ρ]<tt<c2[ρ]},
                   {rb[ρ]+(tt−c2[ρ])Sin[θ3[ρ]],tt>=c2[ρ]}}]},
          {ρ,−ρmax,ρmax},
          PlotRange→All,
          PlotStyle →{Blue, Thickness[0.002]}]},
          {tt,0, zo + n τ + n Zi + 0.25,density}];

 RaysV := Table[{
       ParametricPlot[{
```

```
    Piecewise[{
        {Cos[θ1[ρ]](tt−(n zo+n τ))+zo,tt<=c1[ρ]},
        {v Cos[θ2[ρ]](tt−c1[ρ])+za[ρ],c1[ρ]<tt<c2[ρ]},
        {zb[ρ]+(tt−c2[ρ])Cos[θ3[ρ]],tt>=c2[ρ]}}],

    Piecewise[{
        {Sin[θ1[ρ]](tt−(n zo+n τ)),tt<=c1[ρ]},
        {v Sin[θ2[ρ]](tt−c1[ρ])+ra[ρ],c1[ρ]<tt<c2[ρ]},
        {rb[ρ]+(tt−c2[ρ])Sin[θ3[ρ]],tt>=c2[ρ]}}]},

  {tt,0, zo + n τ + n Zi + 0.25},
  PlotRange→All,PlotStyle →{Lighter[Purple], Thickness[0.002]}]},
  {ρ,−ρmax,ρmax,density}];


Show[OvalSurfaceV,SurfaceV,LensV,WavesV,RaysV,
   PlotRange→All,ImageSize→{640,Automatic}]
```

12.3.7 Case 7: Virtual finite object—real finite image

```
" ClearAll["Global`*"]
(** Definition of set variables **)
```

$$Ka = \frac{(n^2\ zin - no^2\ zo)^2}{n\ no\ (n\ zin - no\ zo)\ (n\ zo - no\ zin)};$$

$$coa = \frac{n\ zo - no\ zin}{zin\ zo(n - no)};$$

$$c1a = \frac{(n-no)(n+no)^2}{4\ n\ no\ zin\ zo(n\ zin - no\ zo)};$$

$$b1a = \frac{(n+no)(n^2 zin - no^2 zo)}{2\ n\ no\ zin\ zo(n\ zin - no\ zo)};$$

$$za[\rho\_] := \frac{\rho^2\ (coa + c1a\ \rho^2)}{1 + b1a\ \rho^2 + \sqrt{1 + (2\ b1a - coa^2\ Ka)\ \rho^2}}$$

$$ra[\rho\_] := Piecewise[\{\{\sqrt{\rho^2 - za[\rho]^2}, \rho \geq 0\}, \{-\sqrt{\rho^2 - za[\rho]^2}, \rho < 0\}\}]$$

$$f[\rho\_] := -no\ zo + n\ \tau + ni\ zi + Sign[zo]\ \sqrt{ra[\rho]^2 + (zo - za[\rho])^2}$$

```
z[ρ_] := za[ρ] − τ − zi

ϱz[ρ_] := no(za'[ρ] (ra[ρ]ra'[ρ]+(−zo+za[ρ]) za'[ρ])) / (n Sqrt[ra[ρ]^2+(zo−za[ρ])^2] ((ra'[ρ])^2+(za'[ρ])^2))
   + ra'[ρ]/Sqrt[(ra'[ρ])^2+(za'[ρ])^2] √(1− no^2(ra[ρ]ra'[ρ]+(−zo+za[ρ]) za'[ρ])^2/(n^2 (ra[ρ]^2+(zo−za[ρ])^2) ((ra'[ρ])^2+(za'[ρ])^2)))

ϱr[ρ_]:= no(ra'[ρ](ra[ρ]ra'[ρ]+(−zo+za[ρ]) za'[ρ])) / (n Sqrt[ra[ρ]^2+(zo−za[ρ])^2] ((ra'[ρ])^2+(za'[ρ])^2))
   − za'[ρ]/Sqrt[(ra'[ρ])^2+(za'[ρ])^2] √(1− no^2(ra[ρ]ra'[ρ]+(−zo+za[ρ]) za'[ρ])^2/(n^2 (ra[ρ]^2+(zo−za[ρ])^2) ((ra'[ρ])^2+(za'[ρ])^2)))

ϑ[ρ_] := 1/(n^2−1) (f[ρ] n + ni^2 (ra[ρ] ϱr[ρ] + z[ρ] ϱz[ρ]) − Sign[n]Sign[zi]
    √(ni^2 (f[ρ]^2 + n^2 (ra[ρ]^2 + z[ρ]^2) − ni^2 (z[ρ] ϱr[ρ] − ra[ρ] ϱz[ρ])^2
     + 2 f[ρ] n (ra[ρ] ϱr[ρ] + z[ρ] ϱz[ρ]))))

zb[ρ_] := za[ρ] + ϑ[ρ] ϱz[ρ];

rb[ρ_] := ra[ρ] + ϑ[ρ] ϱr[ρ];


(** Homotopy for Eikonal **)


θ1[ρ_] := ArcTan[(ra[ρ])/(−zo+za[ρ])];

θ2[ρ_] := ArcTan[(rb[ρ]−ra[ρ])/(zb[ρ]−za[ρ] )];

θ3[ρ_] := ArcTan[rb[ρ]/(zb[ρ]−τ−zi)];

c1[ρ_] := (zo+zi) + Sec[θ1[ρ]](−zo+za[ρ]);

c2[ρ_] := (Sec[θ2[ρ]](zb[ρ]−za[ ρ])/v)+c1[ρ];

(** Setting variable's **)

c = 1;
```

```
v = c/n;

no = 1;
ni = no;
n = 1.5;

ρmax = 15;
density = 3;
τ = 20;
zo = 40;
zi = 25;
zin = n zi;


(** Plot Section **)

SurfaceV := ParametricPlot[{{za[ρ],ra[ρ]},{zb[ρ],rb[ρ]}},{ρ,−ρmax−1,ρmax+1},
        PlotStyle →{Black,Red}, AspectRatio→Automatic, AxesLabel →{r,z}]

OvalSurfaceV := ParametricPlot[{{za[ρ],ra[ρ]},{zb[ρ],rb[ρ]}},
            {ρ,−ρmax−100,ρmax+100},
            PlotStyle →{Directive[{Black,Dashed}],Directive[{Red,Dashed}]}]

LensV := Graphics[{EdgeForm[Directive[{Black,Thin}]],
          Directive[Gray,Opacity[0.3]],
          Polygon[Join[#[[1,1]],#[[2,1]],Reverse[#[[4,1]]],Reverse[#[[3,1]]]]&
          [Cases[SurfaceV,_Line,∞]]]}];

 WavesV := Table[{
            ParametricPlot[{
               Piecewise[{
                     {Cos[θ1[ρ]](tt−(zo+zi))+zo,tt<=c1[ρ]},
                     {v Cos[θ2[ρ]](tt−c1[ρ])+za[ρ],c1[ρ]<tt<c2[ρ]},
                     {zb[ρ]+(tt−c2[ρ])Cos[θ3[ρ]],tt>=c2[ρ]}}],

               Piecewise[{
                     {Sin[θ1[ρ]](tt−(zo+zi)),tt<=c1[ρ]},
                     {v Sin[θ2[ρ]](tt−c1[ρ])+ra[ρ],c1[ρ]<tt<c2[ρ]},
                     {rb[ρ]+(tt−c2[ρ])Sin[θ3[ρ]],tt>=c2[ρ]}}]},
            {ρ,−ρmax,ρmax},
            PlotRange→All,
            PlotStyle →{Blue, Thickness[0.002]}]},
            {tt,0,zo+n τ+zi+0.25,density}];

  RaysV := Table[{
          ParametricPlot[{
            Piecewise[{
                {Cos[θ1[ρ]](tt−(zo+zi))+zo,tt<=c1[ρ]},
                {v Cos[θ2[ρ]](tt−c1[ρ])+za[ρ],c1[ρ]<tt<c2[ρ]},
                {zb[ρ]+(tt−c2[ρ])Cos[θ3[ρ]],tt>=c2[ρ]}}],

            Piecewise[{
                {Sin[θ1[ρ]](tt−(zo+zi)),tt<=c1[ρ]},
                {v Sin[θ2[ρ]](tt−c1[ρ])+ra[ρ],c1[ρ]<tt<c2[ρ]},
```

```
        {rb[ρ]+(tt−c2[ρ])Sin[θ3[ρ]],tt>=c2[ρ]}}]},

    {tt,0,zo+ n τ+zi+0.25},
    PlotRange→All,PlotStyle →{Lighter[Purple], Thickness[0.002]}]},
    {ρ,−ρmax,ρmax,density}];


Show[OvalSurfaceV,SurfaceV,LensV,WavesV,RaysV,
   PlotRange→All,ImageSize→{640,Automatic}]
```

12.3.8 Case 8: Virtual finite object—virtual finite image

```
" ClearAll["Global`*"]
 (** Definition of set variables **)
```

$$\text{Ka} = \frac{(\text{n}^2\ \text{zin}-\text{no}^2\ \text{zo})^2}{\text{n no (n zin}-\text{no zo) (n zo}-\text{no zin)}};$$

$$\text{coa} = \frac{\text{n zo}-\text{no zin}}{\text{zin zo(n}-\text{no)}};$$

$$\text{c1a} = \frac{(\text{n}-\text{no})(\text{n}+\text{no})^2}{4\ \text{n no zin zo(n zin}-\text{no zo)}};$$

$$\text{b1a} = \frac{(\text{n}+\text{no})(\text{n}^2\text{zin}-\text{no}^2\text{zo})}{2\ \text{n no zin zo(n zin}-\text{no zo)}};$$

$$\text{za}[\rho\_] := \frac{\rho^2\ (\text{coa}+\text{c1a}\ \rho^2)}{1+\text{b1a}\ \rho^2+\sqrt{1+(2\ \text{b1a}-\text{coa}^2\ \text{Ka})\ \rho^2}}$$

$$\text{ra}[\rho\_] := \text{Piecewise}[\{\{\sqrt{\rho^2-\text{za}[\rho]^2},\rho\geq 0\},\{-\sqrt{\rho^2-\text{za}[\rho]^2},\rho<0\}\}]$$

$$\text{f}[\rho\_] := -\text{no zo}+\text{n}\ \tau+\text{ni zi}+\text{Sign}[\text{zo}]\ \sqrt{\text{ra}[\rho]^2+(\text{zo}-\text{za}[\rho])^2}$$

$$z[\rho\_] := \text{za}[\rho]-\tau-\text{zi}$$

$$\varrho\text{z}[\rho\_] := \frac{\text{no}(\text{za}'[\rho]\ (\text{ra}[\rho]\text{ra}'[\rho]+(-\text{zo}+\text{za}[\rho])\ \text{za}'[\rho]))}{\text{n}\ \sqrt{\text{ra}[\rho]^2+(\text{zo}-\text{za}[\rho])^2}\ ((\text{ra}'[\rho])^2+(\text{za}'[\rho])^2)}$$

$$+\frac{\mathrm{ra}'[\rho]}{\sqrt{(\mathrm{ra}'[\rho])^2+(\mathrm{za}'[\rho])^2}}\sqrt{\left(1-\frac{\mathrm{no}^2(\mathrm{ra}[\rho]\mathrm{ra}'[\rho]+(-\mathrm{zo}+\mathrm{za}[\rho])\ \mathrm{za}'[\rho])^2}{\mathrm{n}^2\ (\mathrm{ra}[\rho]^2+(\mathrm{zo}-\mathrm{za}[\rho])^2)\ ((\mathrm{ra}'[\rho])^2+(\mathrm{za}'[\rho])^2)}\right)}$$

$$\varrho\mathrm{r}[\rho\_]:=\frac{\mathrm{no}(\mathrm{ra}'[\rho](\mathrm{ra}[\rho]\mathrm{ra}'[\rho]+(-\mathrm{zo}+\mathrm{za}[\rho])\ \mathrm{za}'[\rho]))}{\mathrm{n}\ \sqrt{\mathrm{ra}[\rho]^2+(\mathrm{zo}-\mathrm{za}[\rho])^2}\ ((\mathrm{ra}'[\rho])^2+(\mathrm{za}'[\rho])^2)}$$

$$-\frac{\mathrm{za}'[\rho]}{\sqrt{(\mathrm{ra}'[\rho])^2+(\mathrm{za}'[\rho])^2}}\sqrt{\left(1-\frac{\mathrm{no}^2(\mathrm{ra}[\rho]\mathrm{ra}'[\rho]+(-\mathrm{zo}+\mathrm{za}[\rho])\ \mathrm{za}'[\rho])^2}{\mathrm{n}^2\ (\mathrm{ra}[\rho]^2+(\mathrm{zo}-\mathrm{za}[\rho])^2)\ ((\mathrm{ra}'[\rho])^2+(\mathrm{za}'[\rho])^2)}\right)}$$

$$\vartheta[\rho\_] := \frac{1}{\mathrm{n}^2-1}\left(\mathrm{f}[\rho]\ \mathrm{n}+\mathrm{ni}^2\ (\mathrm{ra}[\rho]\ \varrho\mathrm{r}[\rho]+z[\rho]\ \varrho\mathrm{z}[\rho])+\mathbf{Sign}[\mathrm{n}]\mathbf{Sign}[\mathrm{zi}]\right.$$

$$\left.\sqrt{\left(\mathrm{ni}^2\ (\mathrm{f}[\rho]^2+\mathrm{n}^2\ (\mathrm{ra}[\rho]^2+z[\rho]^2)-\mathrm{ni}^2\ (z[\rho]\ \varrho\mathrm{r}[\rho]-\mathrm{ra}[\rho]\ \varrho\mathrm{z}[\rho])^2 + 2\ \mathrm{f}[\rho]\ \mathrm{n}\ (\mathrm{ra}[\rho]\ \varrho\mathrm{r}[\rho]+z[\rho]\ \varrho\mathrm{z}[\rho]))\right)}\right)$$

```
zb[ρ_] := za[ρ] + ϑ[ρ] ϱz[ρ];

rb[ρ_] := ra[ρ] + ϑ[ρ] ϱr[ρ];


(** Homotopy for Eikonal **)


θ1[ρ_] := ArcTan[(ra[ρ])/(−zo+za[ρ])];

θ2[ρ_] := ArcTan[(rb[ρ]−ra[ρ])/(zb[ρ]−za[ρ] )];

θ3[ρ_] := ArcTan[rb[ρ]/(zb[ρ]−τ−zi)];

c1[ρ_] := (zo + nτ) + Sec[θ1[ρ]](−zo+za[ρ]);

c2[ρ_] := (Sec[θ2[ρ]](zb[ρ]−za[ ρ])/v)+c1[ρ];


(** Setting variable's **)

c = 1;
v = c/n;

no = 1;
ni = no;
n = 1.5;

ρmax = 30;
density = 3;
```

```
τ = 30;
zo = 50;
zi = −60;
zin = n zi;


(** Plot Section **)

SurfaceV := ParametricPlot[{{za[ρ],ra[ρ]},{zb[ρ],rb[ρ]}},{ρ,−ρmax−1,ρmax+1},
          PlotStyle →{Black,Red}, AspectRatio→Automatic, AxesLabel →{r,z}]

OvalSurfaceV := ParametricPlot[{{za[ρ],ra[ρ]},{zb[ρ],rb[ρ]}},
            {ρ,−ρmax−10,ρmax+10},
            PlotStyle →{Directive[{Black,Dashed}],Directive[{Red,Dashed}]}]

LensV := Graphics[{EdgeForm[Directive[{Black,Thin}]],
          Directive[Gray,Opacity[0.3]],
          Polygon[Join[#[[1,1]],#[[2,1]],Reverse[#[[4,1]]],Reverse[#[[3,1]]]]&
          [Cases[SurfaceV,_Line,∞]]]}];

 WavesV := Table[{
            ParametricPlot[{
               Piecewise[{
                     {Cos[θ1[ρ]](tt−(zo + nτ))+zo,tt<=c1[ρ]},
                     {v Cos[θ2[ρ]](tt−c1[ρ])+za[ρ],c1[ρ]<tt<c2[ρ]},
                     {zb[ρ]+(tt−c2[ρ])Cos[θ3[ρ]],tt>=c2[ρ]}}],

               Piecewise[{
                     {Sin[θ1[ρ]](tt−(zo + nτ)),tt<=c1[ρ]},
                     {v Sin[θ2[ρ]](tt−c1[ρ])+ra[ρ],c1[ρ]<tt<c2[ρ]},
                     {rb[ρ]+(tt−c2[ρ])Sin[θ3[ρ]],tt>=c2[ρ]}}]},
            {ρ,−ρmax,ρmax},
            PlotRange→All,
            PlotStyle →{Blue, Thickness[0.002]}]},
            {tt,0,zo+n τ − n zi+0.25,density}];

  RaysV := Table[{
         ParametricPlot[{
            Piecewise[{
                  {Cos[θ1[ρ]](tt−(zo + nτ))+zo,tt<=c1[ρ]},
                  {v Cos[θ2[ρ]](tt−c1[ρ])+za[ρ],c1[ρ]<tt<c2[ρ]},
                  {zb[ρ]+(tt−c2[ρ])Cos[θ3[ρ]],tt>=c2[ρ]}}],

            Piecewise[{
                  {Sin[θ1[ρ]](tt−(zo + nτ)),tt<=c1[ρ]},
                  {v Sin[θ2[ρ]](tt−c1[ρ])+ra[ρ],c1[ρ]<tt<c2[ρ]},
                  {rb[ρ]+(tt−c2[ρ])Sin[θ3[ρ]],tt>=c2[ρ]}}]},

         {tt,0,zo+ n τ − n zi+0.25},
         PlotRange→All,PlotStyle →{Lighter[Purple], Thickness[0.002]}]},
         {ρ,−ρmax,ρmax,density}];


 Show[OvalSurfaceV,SurfaceV,LensV,WavesV,RaysV,
 PlotRange→All,ImageSize→{640,Automatic}]
```

12.3.9 Case 9: Real infinite object—real infinite image

"

(** Definition of set variables **)

$$Ka = -\frac{no^2}{n^2};$$

coa = 0;

c1a = 0;

b1a = 0;

$$za[\rho\_] := \frac{\rho^2\ (coa + c1a\ \rho^2)}{1 + b1a\ \rho^2 + \sqrt{1 + (2\ b1a - coa^2\ Ka)\ \rho^2}}$$

$$ra[\rho\_] := \text{Piecewise}[\{\{\sqrt{\rho^2 - za[\rho]^2}, \rho \geq 0\}, \{-\sqrt{\rho^2 - za[\rho]^2}, \rho < 0\}\}]$$

$$\varrho z[\rho\_] := \frac{no\ za'[\rho]^2}{n\ (ra'[\rho]^2 + za'[\rho]^2)} + \frac{ra'[\rho]}{\sqrt{ra'[\rho]^2 + za'[\rho]^2}} \sqrt{\left(\frac{n^2\ ra'[\rho]^2 + (n^2 - no^2)\ za'[\rho]^2}{n^2\ (ra'[\rho]^2 + za'[\rho]^2)}\right)}$$

$$\varrho r[\rho\_] := \frac{za'[\rho]}{n\ (ra'[\rho]^2 + za'[\rho]^2)} \left(no\,ra'[\rho] - n \sqrt{ra'[\rho]^2 + za'[\rho]^2} \sqrt{\left(\frac{n^2 ra'[\rho]^2 + (n^2 - no^2) za'[\rho]^2}{n^2 (ra'[\rho]^2 + za'[\rho]^2)}\right)} \right)$$

$$\vartheta[\rho\_] := \left((-1+n)\tau - zo + za[\rho] + \text{Sign}[n]\text{Sign}[zo]\text{Sqrt}[ra[\rho]\hat{\ }2 + (za[\rho] - zo)\hat{\ }2] \right) / (n - \varrho z[\rho]);$$

$$zb[\rho\_] := za[\rho] + \vartheta[\rho]\ \varrho z[\rho];$$

$$rb[\rho\_] := ra[\rho] + \vartheta[\rho]\ \varrho r[\rho];$$

(** Homotopy for Eikonal **)

```
θ1[ρ_] := ArcTan[Limit[(ra[ρ])/(−zo+za[ρ]),zo→−Infinity]];

θ2[ρ_] := ArcTan[(rb[ρ]−ra[ρ])/(zb[ρ]−za[ρ])];

θ3[ρ_] := ArcTan[Limit[rb[ρ]/(zb[ρ]−τ−zi),zi→Infinity]];

c1[ρ_] := Sec[θ1[ρ]](−zo+za[ρ]);

c2[ρ_] := (Sec[θ2[ρ]](zb[ρ]−za[ρ])/v)+c1[ρ];


(** Setting variable's **)

c = 1;
v = c/n;

no = 1;
ni = no;
n = 1.5;

ρmax = 25;
density = 5;
τ = 25;
(*zo −> −Infinity*)
Zo = 50;
(* zi −> Infinity*)
Zi = 60;



(** Plot Section **)

SurfaceV := ParametricPlot[{{za[ρ],ra[ρ]},{zb[ρ],rb[ρ]}},{ρ,−ρmax−1,ρmax+1},
          PlotStyle →{Black,Red}, AspectRatio→Automatic, AxesLabel →{r,z}]

OvalSurfaceV := ParametricPlot[{{za[ρ],ra[ρ]},{zb[ρ],rb[ρ]}},
              {ρ,−ρmax−5,ρmax+5},
              PlotStyle →{Directive[{Black,Dashed}],Directive[{Red,Dashed}]}]

LensV := Graphics[{EdgeForm[Directive[{Black,Thin}]],
           Directive[Gray,Opacity[0.3]],
           Polygon[Join[#[[1,1]],#[[2,1]],Reverse[#[[4,1]]],Reverse[#[[3,1]]]]&
           [Cases[SurfaceV,_Line,∞]]]}];

 WavesV := Table[{
              ParametricPlot[{
                  Piecewise[{
                      {−tt,tt<=c1[ρ]},
                      {v (tt−c1[ρ])+za[ρ],c1[ρ]<tt<c2[ρ]},
                      {zb[ρ]+(tt−c2[ρ]),tt>=c2[ρ]}}],

                  Piecewise[{
```

```
                {ra[ρ],tt<=c1[ρ]},
                {v ra[ρ],c1[ρ]<tt<c2[ρ]},
                {rb[ρ],tt>=c2[ρ]}}]},

          {ρ,−ρmax,ρmax},
          PlotRange→All,
          PlotStyle →{Blue, Thickness[0.002]}]},
          {tt,0,−Zo+nτ + Zi+0.25,density}];

RaysV := Table[{
       ParametricPlot[{
          Piecewise[{
               {−tt,tt<=c1[ρ]},
               {v (tt−c1[ρ])+za[ρ],c1[ρ]<tt<c2[ρ]},
               {zb[ρ]+(tt−c2[ρ]),tt>=c2[ρ]}}],

          Piecewise[{
               {ra[ρ],tt<=c1[ρ]},
               {v ra[ρ],c1[ρ]<tt<c2[ρ]},
               {rb[ρ],tt>=c2[ρ]}}]},

       {tt,0,−Zo+ n τ + zi+0.25},
       PlotRange→All,PlotStyle →{Lighter[Purple], Thickness[0.002]}]},
       {ρ,−ρmax,ρmax,density}];


Show[OvalSurfaceV,SurfaceV,LensV,WavesV,RaysV,
    PlotRange→All,ImageSize→{640,Automatic}]
```

12.4 Programs for chapter 9

12.4.1 Case 1: Real finite object—real finite image

```
" ClearAll["Global`*"]
 (** Definition of set variables **)


za[ra_] := −(29−Sqrt[29^2−ra^2])


f[ra_] := −no zo + n τ + ni zi + Sign[zo] √(ra^2+(zo−za[ra])^2)


z[ra_] := za[ra] − τ − zi
```

$$\varrho z[ra\_] := \frac{no(za'[ra]\ (ra+(-zo+za[ra])\ za'[ra]))}{n\ \sqrt{ra^2+(zo-za[ra])^2}\ (1+(za'[ra])^2)} + \frac{1}{\sqrt{1+(za'[ra])^2}}\sqrt{\left(1-\frac{no^2(ra+(-zo+za[ra])\ za'[ra])^2}{n^2\ (ra^2+(zo-za[ra])^2)\ (1+(za'[ra])^2)}\right)}$$

$$\varrho r[ra\_] := \frac{no(ra+(-zo+za[ra])\ za'[ra])}{n\ \sqrt{ra^2+(zo-za[ra])^2}\ (1+(za'[ra])^2)} - \frac{za'[ra]}{\sqrt{1+(za'[ra])^2}}\sqrt{\left(1-\frac{no^2(ra+(-zo+za[ra])\ za'[ra])^2)}{n^2\ (ra^2+(zo-za[ra])^2)\ (1+(za'[ra])^2)}\right)}$$

$$\vartheta[ra\_] := \frac{1}{n^2-1}\left(f[ra]\ n + ni^2\ (ra\ \varrho r[ra] + z[ra]\ \varrho z[ra]) + \mathrm{Sign}[n]\mathrm{Sign}[zi] \sqrt{\left(ni^2\ (f[ra]^2 + n^2\ (ra^2 + z[ra]^2)) - ni^2\ (z[ra]\ \varrho r[ra] - ra\ \varrho z[ra])^2 + 2\ f[ra]\ n\ (ra\ \varrho r[ra] + z[ra]\ \varrho z[ra])\right)}\right)$$

```
zb[ra_] := za[ra] + ϑ[ra] ϱz[ra];

rb[ra_] := ra + ϑ[ra] ϱr[ra];


(** Homotopy for Eikonal **)

θ1[ra_] := ArcTan[(ra)/(−zo+za[ra])];

θ2[ra_] := ArcTan[(rb[ra]−ra)/(zb[ra]−za[ra] )];

θ3[ra_] := ArcTan[rb[ra]/(zb[ra]−τ−zi)];

c1[ra_] := Sec[θ1[ra]](−zo+za[ra]);

c2[ra_] := (Sec[θ2[ra]](zb[ra]−za[ ra])/v)+c1[ra];

(** Setting variable's **)

c = 1;
v = c/n;

no = 1;
```

```
ni = no;
n = 1.5;

rmax = 20;
density = 4;
τ = 29;
zo = −55;
zi = 30;


(** Plot Section **)

SurfaceV := ParametricPlot[{{za[ra],ra},{zb[ra],rb[ra]}},{ra,−rmax−1,rmax+1},
         PlotStyle →{Black,Red}, AspectRatio→Automatic, AxesLabel →{r,z}]


LensV := Graphics[{EdgeForm[Directive[{Black,Thin}]],
          Directive[Gray,Opacity[0.3]],
          Polygon[Join[#[[1,1]],#[[2,1]],Reverse[#[[4,1]]],Reverse[#[[3,1]]]]&
          [Cases[SurfaceV,_Line,∞]]]}];

 WavesV := Table[{
          ParametricPlot[{
             Piecewise[{
                    {Cos[θ1[ra]]tt+zo,tt<=c1[ra]},
                    {v Cos[θ2[ra]](tt−c1[ra])+za[ra],c1[ra]<tt<c2[ra]},
                    {zb[ra]+(tt−c2[ra])Cos[θ3[ra]],tt>=c2[ra]}}],

             Piecewise[{
                    {Sin[θ1[ra]]tt,tt<=c1[ra]},
                    {v Sin[θ2[ra]](tt−c1[ra])+ra,c1[ra]<tt<c2[ra]},
                    {rb[ra]+(tt−c2[ra])Sin[θ3[ra]],tt>=c2[ra]}}]},
          {ra,−rmax,rmax},
          PlotRange→All,
          PlotStyle →{Blue, Thickness[0.002]}]},
          {tt,0,−zo+nτ+zi,density}];

  RaysV := Table[{
        ParametricPlot[{
           Piecewise[{
                {Cos[θ1[ra]]tt+zo,tt<=c1[ra]},
                {v Cos[θ2[ra]](tt−c1[ra])+za[ra],c1[ra]<tt<c2[ra]},
                {zb[ra]+(tt−c2[ra])Cos[θ3[ra]],tt>=c2[ra]}}],

           Piecewise[{
                {Sin[θ1[ra]]tt,tt<=c1[ra]},
                {v Sin[θ2[ra]](tt−c1[ra])+ra,c1[ra]<tt<c2[ra]},
                {rb[ra]+(tt−c2[ra])Sin[θ3[ra]],tt>=c2[ra]}}]},

       {tt,0,−zo+ n τ + zi},
       PlotRange→All,PlotStyle →{Lighter[Purple], Thickness[0.002]}]},
       {ra,−rmax,rmax,density}];
        Show[SurfaceV,LensV,WavesV,RaysV,
            PlotRange→All,ImageSize→{640,Automatic}]
```

12.4.2 Case 2: Real infinity object—real finite image

"

```
(** Definition of set variables **)

za[ra_] := -(29-Sqrt[29^2-ra^2])

f[ra_] := n τ + ni zi - za[ra]

z[ra_] := za[ra] - τ - zi
```

$$\varrho z[ra\_] := \frac{1}{n(1+za'[ra]^2)^{3/2}}\sqrt{\frac{n^2+(-1+n^2)\ za'[ra]^2}{n^2\ (1+za'[ra]^2)}} + \frac{za'[ra]^2}{n\ (1+za'[ra]^2)^{3/2}}\left(\sqrt{1+za'[ra]^2} + n\sqrt{\frac{n^2+(-1+n^2)\ za'[ra]^2}{n^2\ (1+za'[ra]^2)}}\right)$$

$$\varrho r[ra\_] := \frac{za'[ra]}{n\ (1+za'[ra]^2)^{3/2}}\left(\sqrt{1+za'[ra]^2} - n\sqrt{\frac{n^2+(-1+n^2)\ za'[ra]^2}{n^2\ (1+za'[ra]^2)}} - n\ za'[ra]^2\sqrt{\frac{n^2+(-1+n^2)\ za'[ra]^2}{n^2\ (1+za'[ra]^2)}}\right);$$

$$\vartheta[ra\_] := \frac{1}{-1+n^2}\Big(n\ f[ra]+ni^2\ (ra\ \varrho r[ra]+z[ra]\ \varrho z[ra])+\mathrm{Sign}[n]\mathrm{Sign}[zi]\ \sqrt{\Big(ni^2\ (f[ra]^2+n^2\ (ra^2+z[ra]^2)-ni^2\ (z[ra]\ \varrho r[ra]-ra\ \varrho z[ra])^2 + 2\ n\ f[ra]\ (ra\ \varrho r[ra]+z[ra]\ \varrho z[ra]))\Big)}\Big)$$

$$zb[ra\_] := za[ra] + \vartheta[ra]\ \varrho z[ra];$$

```
rb[ra_] := ra + ϑ[ra] ϱr[ra];


(** Homotopy for Eikonal **)

θ1[ra_] := ArcTan[Limit[(ra)/(−zo+za[ra]),zo→−Infinity]]

θ2[ra_] := ArcTan[(rb[ra]−ra)/(zb[ra]−za[ra] )];

θ3[ra_] := ArcTan[rb[ra]/(zb[ra]−τ−zi)];

c1[ra_] := (−Zo+za[ra]);

c2[ra_] := (Sec[θ2[ra]](zb[ra]−za[ ra])/v)+c1[ra];


(** Setting variable's **)

c = 1;
v = c/n;

no = 1;
ni = no;
n = 1.5;

rmax = 26;
density = 4;
τ = 29;
(* zo −> −Infinity *)
Zo = −55;
zi = 30;


(** Plot Section **)

SurfaceV := ParametricPlot[{{za[ra],ra},{zb[ra],rb[ra]}},{ra,−rmax−1,rmax+1},
         PlotStyle →{Black,Red}, AspectRatio→Automatic, AxesLabel →{r,z}]

LensV := Graphics[{EdgeForm[Directive[{Black,Thin}]],
          Directive[Gray,Opacity[0.3]],
          Polygon[Join[#[[1,1]],#[[2,1]],Reverse[#[[4,1]]],Reverse[#[[3,1]]]]&
          [Cases[SurfaceV,_Line,∞]]]}];

 WavesV := Table[{
             ParametricPlot[{
                Piecewise[{
                      {Cos[θ1[ra]]tt+Zo,tt<=c1[ra]},
                      {v Cos[θ2[ra]](tt−c1[ra])+za[ra],c1[ra]<tt<c2[ra]},
                      {zb[ra]+(tt−c2[ra])Cos[θ3[ra]],tt>=c2[ra]}}],

                Piecewise[{
```

```
                {ra,tt<=c1[ra]},
                {v Sin[θ2[ra]](tt−c1[ra])+ra,c1[ra]<tt<c2[ra]},
                {rb[ra]+(tt−c2[ra])Sin[θ3[ra]],tt>=c2[ra]}}]},
        {ra,−rmax,rmax},
        PlotRange→All,
        PlotStyle →{Blue, Thickness[0.002]}]},
        {tt,0,−Zo+nτ+zi,density}];

RaysV := Table[{
     ParametricPlot[{
        Piecewise[{
            {Cos[θ1[ra]]tt+Zo,tt<=c1[ra]},
            {v Cos[θ2[ra]](tt−c1[ra])+za[ra],c1[ra]<tt<c2[ra]},
            {zb[ra]+(tt−c2[ra])Cos[θ3[ra]],tt>=c2[ra]}}],

        Piecewise[{
            {ra,tt<=c1[ra]},
            {v Sin[θ2[ra]](tt−c1[ra])+ra,c1[ra]<tt<c2[ra]},
            {rb[ra]+(tt−c2[ra])Sin[θ3[ra]],tt>=c2[ra]}}]},

     {tt,0,−Zo+ n τ + zi},
     PlotRange→All,PlotStyle →{Lighter[Purple], Thickness[0.002]}]},
     {ra,−rmax,rmax,density}];


Show[SurfaceV,LensV,WavesV,RaysV,
   PlotRange→All,ImageSize→{640,Automatic}]
```

12.4.3 Case 3: Real infinity object—virtual finite image

```
"

(** Definition of set variables **)

za[ra_] := −(29−Sqrt[29^2−ra^2])

f[ra_] := n τ + ni zi − za[ra]

z[ra_] := za[ra] − τ − zi
```

$$\varrho z[ra\_] := \frac{1}{n(1+za'[ra]^2)^{3/2}}\sqrt{\frac{n^2+(-1+n^2)\ za'[ra]^2}{n^2\ (1+za'[ra]^2)}}$$

$$+\frac{\mathrm{za}'[\mathrm{ra}]^2}{\mathrm{n}\ \left(1+\mathrm{za}'[\mathrm{ra}]^2\right)^{3/2}}\left(\sqrt{1+\mathrm{za}'[\mathrm{ra}]^2}+\mathrm{n}\ \sqrt{\frac{\mathrm{n}^2+(-1+\mathrm{n}^2)\ \mathrm{za}'[\mathrm{ra}]^2}{\mathrm{n}^2\ \left(1+\mathrm{za}'[\mathrm{ra}]^2\right)}}\right)$$

$$\varrho\mathrm{r}[\mathrm{ra\_}]{:=}\frac{\mathrm{za}'[\mathrm{ra}]}{\mathrm{n}\ \left(1+\mathrm{za}'[\mathrm{ra}]^2\right)^{3/2}}\left(\sqrt{1+\mathrm{za}'[\mathrm{ra}]^2}\right.$$

$$\left.-\mathrm{n}\ \sqrt{\frac{\mathrm{n}^2+(-1+\mathrm{n}^2)\ \mathrm{za}'[\mathrm{ra}]^2}{\mathrm{n}^2\ \left(1+\mathrm{za}'[\mathrm{ra}]^2\right)}}-\mathrm{n}\ \mathrm{za}'[\mathrm{ra}]^2\ \sqrt{\frac{\mathrm{n}^2+(-1+\mathrm{n}^2)\ \mathrm{za}'[\mathrm{ra}]^2}{\mathrm{n}^2\ \left(1+\mathrm{za}'[\mathrm{ra}]^2\right)}}\right);$$

$$\vartheta[\mathrm{ra\_}] := \frac{1}{-1+\mathrm{n}^2}\left(\mathrm{n\ f}[\mathrm{ra}]+\mathrm{ni}^2\ (\mathrm{ra}\ \varrho\mathrm{r}[\mathrm{ra}]+z[\mathrm{ra}]\ \varrho\mathrm{z}[\mathrm{ra}]) - \mathrm{Sign}[\mathrm{n}]\mathrm{Sign}[\mathrm{zi}]\right.$$

$$\sqrt{}\left(\mathrm{ni}^2\ (\mathrm{f}[\mathrm{ra}]^2+\mathrm{n}^2\ (\mathrm{ra}^2+z[\mathrm{ra}]^2)-\mathrm{ni}^2\ (z[\mathrm{ra}]\ \varrho\mathrm{r}[\mathrm{ra}]-\mathrm{ra}\ \varrho\mathrm{z}[\mathrm{ra}])^2\right.$$

$$\left.\left.+\ 2\ \mathrm{n\ f}[\mathrm{ra}]\ (\mathrm{ra}\ \varrho\mathrm{r}[\mathrm{ra}]+z[\mathrm{ra}]\ \varrho\mathrm{z}[\mathrm{ra}]))\right)\right)$$

```
zb[ra_] := za[ra] + ϑ[ra] ϱz[ra];

rb[ra_] := ra + ϑ[ra] ϱr[ra];


(** Homotopy for Eikonal **)

θ1[ra_] := ArcTan[Limit[(ra)/(−zo+za[ra]),zo→−Infinity]]

θ2[ra_] := ArcTan[(rb[ra]−ra)/(zb[ra]−za[ra] )];

θ3[ra_] := ArcTan[rb[ra]/(zb[ra]−τ−zi)];

c1[ra_] := (−Zo+za[ra]);

c2[ra_] := (Sec[θ2[ra]](zb[ra]−za[ ra])/v)+c1[ra];


(** Setting variable's **)

c = 1;
v = c/n;

no = 1;
ni = no;
n = 1.5;

rmax = 28;
density = 4;
```

```
τ = 29;
(* zo --> -Infinity *)
Zo = −55;
zi = −55;



(** Plot Section **)

SurfaceV := ParametricPlot[{{za[ra],ra},{zb[ra],rb[ra]}},{ra,−rmax−1,rmax+1},
        PlotStyle →{Black,Red}, AspectRatio→Automatic, AxesLabel →{r,z}]

LensV := Graphics[{EdgeForm[Directive[{Black,Thin}]],
         Directive[Gray,Opacity[0.3]],
         Polygon[Join[#[[1,1]],#[[2,1]],Reverse[#[[4,1]]],Reverse[#[[3,1]]]]&
         [Cases[SurfaceV,_Line,∞]]]}];

 WavesV := Table[{
          ParametricPlot[{
            Piecewise[{
                  {Cos[θ1[ra]]tt+Zo,tt<=c1[ra]},
                  {v Cos[θ2[ra]](tt−c1[ra])+za[ra],c1[ra]<tt<c2[ra]},
                  {zb[ra]+(tt−c2[ra])Cos[θ3[ra]],tt>=c2[ra]}}],

            Piecewise[{
                  {Sin[θ1[ra]]tt+ra[ra],tt<=c1[ra]},
                  {v Sin[θ2[ra]](tt−c1[ra])+ra,c1[ra]<tt<c2[ra]},
                  {rb[ra]+(tt−c2[ra])Sin[θ3[ra]],tt>=c2[ra]}}]},
          {ra,−rmax,rmax},
          PlotRange→All,
          PlotStyle →{Blue, Thickness[0.002]}]},
          {tt,0,−Zo+nτ−zi,density}];

  RaysV := Table[{
        ParametricPlot[{
          Piecewise[{
               {Cos[θ1[ra]]tt+Zo,tt<=c1[ra]},
               {v Cos[θ2[ra]](tt−c1[ra])+za[ra],c1[ra]<tt<c2[ra]},
               {zb[ra]+(tt−c2[ra])Cos[θ3[ra]],tt>=c2[ra]}}],

          Piecewise[{
               {Sin[θ1[ra]]tt+ra[ra],tt<=c1[ra]},
               {v Sin[θ2[ra]](tt−c1[ra])+ra,c1[ra]<tt<c2[ra]},
               {rb[ra]+(tt−c2[ra])Sin[θ3[ra]],tt>=c2[ra]}}]},

        {tt,0,−Zo+ n τ − zi},
        PlotRange→All,PlotStyle →{Lighter[Purple], Thickness[0.002]}]},
        {ra,−rmax,rmax,density}];


 Show[SurfaceV,LensV,WavesV,RaysV,
     PlotRange→All,ImageSize→{640,Automatic}]
```

12.4.4 Case 4: Real finite object—virtual finite image

```
" ClearAll["Global`*"]
(** Definition of set variables **)
```

$$za[ra\_] := -(29-Sqrt[29\hat{}2-ra\hat{}2])$$

$$f[ra\_] := -no\ zo + n\ \tau + ni\ zi + Sign[zo]\ \sqrt{ra^2+(zo-za[ra])^2}$$

$$z[ra\_] := za[ra] - \tau - zi$$

$$\varrho z[ra\_] := \frac{no(za'[ra]\ (ra+(-zo+za[ra])\ za'[ra]))}{n\ \sqrt{ra^2+(zo-za[ra])^2}\ (1+(za'[ra])^2)} + \frac{1}{\sqrt{1+(za'[ra])^2}}\sqrt{\left(1-\frac{no^2(ra+(-zo+za[ra])\ za'[ra])^2}{n^2\ (ra^2+(zo-za[ra])^2)\ (1+(za'[ra])^2)}\right)}$$

$$\varrho r[ra\_]:=\frac{no(ra+(-zo+za[ra])\ za'[ra])}{n\ \sqrt{ra^2+(zo-za[ra])^2}\ (1+(za'[ra])^2)} - \frac{za'[ra]}{\sqrt{1+(za'[ra])^2}}\sqrt{\left(1-\frac{no^2(ra+(-zo+za[ra])\ za'[ra])^2)}{n^2\ (ra^2+(zo-za[ra])^2)\ (1+(za'[ra])^2)}\right)}$$

$$\vartheta[ra\_] := \frac{1}{n^2-1}\Bigg(f[ra]\ n + ni^2\ (ra\ \varrho r[ra] + z[ra]\ \varrho z[ra]) + Sign[n]Sign[zi] \sqrt{\Big(ni^2\ (f[ra]^2 + n^2\ (ra^2 + z[ra]^2) - ni^2\ (z[ra]\ \varrho r[ra] - ra\ \varrho z[ra])^2 + 2\ f[ra]\ n\ (ra\ \varrho r[ra] + z[ra]\ \varrho z[ra]))\Big)}\Bigg)$$

$$zb[ra\_] := za[ra] + \vartheta[ra]\ \varrho z[ra];$$

$$rb[ra\_] := ra + \vartheta[ra]\ \varrho r[ra];$$

```
(** Homotopy for Eikonal **)

θ1[ra_] := ArcTan[(ra)/(−zo+za[ra])];

θ2[ra_] := ArcTan[(rb[ra]−ra)/(zb[ra]−za[ra] )];

θ3[ra_] := ArcTan[rb[ra]/(zb[ra]−τ−zi)];

c1[ra_] := Sec[θ1[ra]](−zo+za[ra]);

c2[ra_] := (Sec[θ2[ra]](zb[ra]−za[ ra])/v)+c1[ra];


(** Setting variable's **)

c = 1;
v = c/n;

no = 1;
ni = no;
n = 1.5;

rmax = 28;
density = 4;
τ = 29;
zo = −55;
zi = −55;


(** Plot Section **)

SurfaceV := ParametricPlot[{{za[ra],ra},{zb[ra],rb[ra]}},{ra,−rmax−1,rmax+1},
        PlotStyle →{Black,Red}, AspectRatio→Automatic, AxesLabel →{r,z}]


LensV := Graphics[{EdgeForm[Directive[{Black,Thin}]],
         Directive[Gray,Opacity[0.3]],
         Polygon[Join[#[[1,1]],#[[2,1]],Reverse[#[[4,1]]],Reverse[#[[3,1]]]]&
         [Cases[SurfaceV,_Line,∞]]]}];

 WavesV := Table[{
          ParametricPlot[{
             Piecewise[{
                   {Cos[θ1[ra]]tt+zo,tt<=c1[ra]},
                   {v Cos[θ2[ra]](tt−c1[ra])+za[ra],c1[ra]<tt<c2[ra]},
                   {zb[ra]+(tt−c2[ra])Cos[θ3[ra]],tt>=c2[ra]}}],

             Piecewise[{
                   {Sin[θ1[ra]]tt,tt<=c1[ra]},
                   {v Sin[θ2[ra]](tt−c1[ra])+ra,c1[ra]<tt<c2[ra]},
                   {rb[ra]+(tt−c2[ra])Sin[θ3[ra]],tt>=c2[ra]}}]},
          { ra,−rmax,rmax},
```

```
            PlotRange→All,
            PlotStyle →{Blue, Thickness[0.002]}]},
            {tt,0,−zo+nτ−zi+0.25,density}];

RaysV := Table[{
        ParametricPlot[{
          Piecewise[{
                {Cos[θ1[ra]]tt+zo,tt<=c1[ra]},
                {v Cos[θ2[ra]](tt−c1[ra])+za[ra],c1[ra]<tt<c2[ra]},
                {zb[ra]+(tt−c2[ra])Cos[θ3[ra]],tt>=c2[ra]}}],

          Piecewise[{
                {Sin[θ1[ra]]tt,tt<=c1[ra]},
                {v Sin[θ2[ra]](tt−c1[ra])+ra,c1[ra]<tt<c2[ra]},
                {rb[ra]+(tt−c2[ra])Sin[θ3[ra]],tt>=c2[ra]}}]},

        {tt,0,−zo+ n τ − zi+0.25},
        PlotRange→All,PlotStyle →{Lighter[Purple], Thickness[0.002]}]},
        { ra,−rmax,rmax,density}];

Show[SurfaceV,LensV,WavesV,RaysV,
    PlotRange→All,ImageSize→{640,Automatic}]
```

12.4.5 Case 5: Real finite object—real infinite image

```
" ClearAll["Global`*"]
 (** Definition of set variables **)

 za[ra_] := −(29−Sqrt[29^2−ra^2])
```

$$\varrho z[ra\_] := \frac{no(za'[ra]\ (ra+(-zo+za[ra])\ za'[ra]))}{n\ \sqrt{ra^2+(zo-za[ra])^2}\ (1+(za'[ra])^2)} + \frac{1}{\sqrt{1+(za'[ra])^2}}\sqrt{\left(1-\frac{no^2(ra+(-zo+za[ra])\ za'[ra])^2}{n^2\ (ra^2+(zo-za[ra])^2)\ (1+(za'[ra])^2)}\right)}$$

$$\varrho r[ra\_] := \frac{no(ra+(-zo+za[ra])\ za'[ra])}{n\ \sqrt{ra^2+(zo-za[ra])^2}\ (1+(za'[ra])^2)}$$

$$-\frac{za'[ra]}{\sqrt{1+(za'[ra])^2}}\sqrt{\left(1-\frac{no^2(ra+(-zo+za[ra])\ za'[ra])^2)}{n^2\ (ra^2+(zo-za[ra])^2)\ (1+(za'[ra])^2)}\right)}$$

```
ϑ[ra_] := ((−1+n)τ −zo+za[ra] − Sign[n]Sign[Zi]Sqrt[ra^2 + (za[ra]−zo)^2]
)/(n−ϱz[ra]);
zb[ra_] := za[ra] + ϑ[ra] ϱz[ra];

rb[ra_] := ra + ϑ[ra] ϱr[ra];


(** Homotopy for Eikonal **)

θ1[ra_] := ArcTan[(ra)/(−zo+za[ra])];

θ2[ra_] := ArcTan[(rb[ra]−ra)/(zb[ra]−za[ra] )];

θ3[ra_] := ArcTan[Limit[rb[ra]/(zb[ra]−τ−zi),zi→Infinity]];

c1[ra_] := Sec[θ1[ra]](−zo+za[ra]);

c2[ra_] := (Sec[θ2[ra]](zb[ra]−za[ra])/v)+c1[ra];

(** Setting variable's **)

c = 1;
v = c/n;

no = 1;
ni = no;
n = 1.5;

rmax = 28;
density = 4;
τ = 29;
zo = −55;
(* zi −> Infinity*)
Zi = 30;


(** Plot Section **)

SurfaceV := ParametricPlot[{{za[ra],ra},{zb[ra],rb[ra]}},{ra,−rmax−1,rmax+1},
        PlotStyle →{Black,Red}, AspectRatio→Automatic, AxesLabel →{r,z}]

LensV := Graphics[{EdgeForm[Directive[{Black,Thin}]],
        Directive[Gray,Opacity[0.3]],
        Polygon[Join[#[[1,1]],#[[2,1]],Reverse[#[[4,1]]],Reverse[#[[3,1]]]]&
        [Cases[SurfaceV,_Line,∞]]]}];
```

```
WavesV := Table[{
            ParametricPlot[{
               Piecewise[{
                     {Cos[θ1[ra]]tt+zo,tt<=c1[ra]},
                     {v Cos[θ2[ra]](tt−c1[ra])+za[ra],c1[ra]<tt<c2[ra]},
                     {zb[ra]+(tt−c2[ra])Cos[θ3[ra]],tt>=c2[ra]}}],

               Piecewise[{
                     {Sin[θ1[ra]]tt,tt<=c1[ra]},
                     {v Sin[θ2[ra]](tt−c1[ra])+ra,c1[ra]<tt<c2[ra]},
                     {rb[ra]+(tt−c2[ra])Sin[θ3[ra]],tt>=c2[ra]}}]},
            {ra,−rmax,rmax},
            PlotRange→All,
            PlotStyle →{Blue, Thickness[0.002]}]},
            {tt,0,−zo + n τ + Zi + 0.25,density}];

 RaysV := Table[{
         ParametricPlot[{
            Piecewise[{
                  {Cos[θ1[ra]]tt+zo,tt<=c1[ra]},
                  {v Cos[θ2[ra]](tt−c1[ra])+za[ra],c1[ra]<tt<c2[ra]},
                  {zb[ra]+(tt−c2[ra])Cos[θ3[ra]],tt>=c2[ra]}}],

            Piecewise[{
                  {Sin[θ1[ra]]tt,tt<=c1[ra]},
                  {v Sin[θ2[ra]](tt−c1[ra])+ra,c1[ra]<tt<c2[ra]},
                  {rb[ra]+(tt−c2[ra])Sin[θ3[ra]],tt>=c2[ra]}}]},

         {tt,0,−zo + n τ + Zi + 0.25},
         PlotRange→All,PlotStyle →{Lighter[Purple], Thickness[0.002]}]},
         {ra,−rmax,rmax,density}];


Show[SurfaceV,LensV,WavesV,RaysV,
   PlotRange→All,ImageSize→{640,Automatic}]
```

12.4.6 Case 6: Virtual finite object—real infinite image

```
" ClearAll["Global`*"]
(** Definition of set variables **)
```

$$\varrho r[ra\_]:=\frac{no(ra+(-zo+za[ra])\ za'[ra])}{n\ \sqrt{ra^2+(zo-za[ra])^2}\ (1+(za'[ra])^2)} - \frac{za'[ra]}{\sqrt{1+(za'[ra])^2}}\sqrt{\left(1-\frac{no^2(ra+(-zo+za[ra])\ za'[ra])^2)}{n^2\ (ra^2+(zo-za[ra])^2)\ (1+(za'[ra])^2)}\right)}$$

```
ϑ[ra_] := ((−1+n)τ −zo+za[ra] + Sign[n]Sign[zo]Sqrt[ra^2 + (za[ra]−zo)^2] )
/(n−ϱz[ra]);

zb[ra_] := za[ra] + ϑ[ra] ϱz[ra];

rb[ra_] := ra + ϑ[ra] ϱr[ra];


(** Homotopy for Eikonal **)

θ1[ra_] := ArcTan[(ra)/(−zo+za[ra])];

θ2[ra_] := ArcTan[(rb[ra]−ra)/(zb[ra]−za[ra] )];

θ3[ra_] := ArcTan[Limit[rb[ra]/(zb[ra]−τ−zi),zi→Infinity]];

c1[ra_] := (n zo + n τ) + Sec[θ1[ra]](−zo + za[ra]);

c2[ra_] := (Sec[θ2[ra]](zb[ra]−za[ra])/v)+c1[ra];

(** Setting variable's **)

c = 1;
v = c/n;

no = 1;
ni = no;
n = 1.5;

rmax = 28;
density = 4;
τ =291;
zo = 70;
(* zi −> Infinity*)
Zi = 30;


(** Plot Section **)

SurfaceV := ParametricPlot[{{za[ra],ra},{zb[ra],rb[ra]}},{ra,−rmax−1,rmax+1},
         PlotStyle →{Black,Red}, AspectRatio→Automatic, AxesLabel →{r,z}]

OvalSurfaceV := ParametricPlot[{{za[ra],ra},{zb[ra],rb[ra]}},
          {ra,−rmax−500,rmax+500},
          PlotStyle →{Directive[{Black,Dashed}],Directive[{Red,Dashed}]}]

LensV := Graphics[{EdgeForm[Directive[{Black,Thin}]],
         Directive[Gray,Opacity[0.3]],
         Polygon[Join[#[[1,1]],#[[2,1]],Reverse[#[[4,1]]],Reverse[#[[3,1]]]]&
```

```
            [Cases[SurfaceV,_Line,∞]]]}];

WavesV := Table[{
            ParametricPlot[{
              Piecewise[{
                    {Cos[θ1[ra]](tt−(n zo+n τ))+zo,tt<=c1[ra]},
                    {v Cos[θ2[ra]](tt−c1[ra])+za[ra],c1[ra]<tt<c2[ra]},
                    {zb[ra]+(tt−c2[ra])Cos[θ3[ra]],tt>=c2[ra]}}],

              Piecewise[{
                    {Sin[θ1[ra]](tt−(n zo+n τ)),tt<=c1[ra]},
                    {v Sin[θ2[ra]](tt−c1[ra])+ra,c1[ra]<tt<c2[ra]},
                    {rb[ra]+(tt−c2[ra])Sin[θ3[ra]],tt>=c2[ra]}}]},
            {ra,−rmax,rmax},
            PlotRange→All,
            PlotStyle →{Blue, Thickness[0.002]}]},
            {tt,0, zo + n τ + n Zi ,density}];

RaysV := Table[{
        ParametricPlot[{
          Piecewise[{
                {Cos[θ1[ra]](tt−(n zo+n τ))+zo,tt<=c1[ra]},
                {v Cos[θ2[ra]](tt−c1[ra])+za[ra],c1[ra]<tt<c2[ra]},
                {zb[ra]+(tt−c2[ra])Cos[θ3[ra]],tt>=c2[ra]}}],

          Piecewise[{
                {Sin[θ1[ra]](tt−(n zo+n τ)),tt<=c1[ra]},
                {v Sin[θ2[ra]](tt−c1[ra])+ra,c1[ra]<tt<c2[ra]},
                {rb[ra]+(tt−c2[ra])Sin[θ3[ra]],tt>=c2[ra]}}]},

        {tt,0, zo + n τ + n Zi + 0.25},
        PlotRange→All,PlotStyle →{Lighter[Purple], Thickness[0.002]}]},
        {ra,−rmax,rmax,density}];


Show[SurfaceV,LensV,WavesV,RaysV,
    PlotRange→All,ImageSize→{640,Automatic}]
```

12.4.7 Case 7: Virtual finite object—real finite image

```
" ClearAll["Global'*"]
(** Definition of set variables **)

za[ra_] := −(29−Sqrt[29^2−ra^2])
```

$$f[ra\_] := -no\ zo + n\ \tau + ni\ zi + \mathrm{Sign}[zo]\sqrt{ra^2+(zo-za[ra])^2}$$

$$z[ra\_] := za[ra] - \tau - zi$$

$$\varrho z[ra\_] := \frac{no(za'[ra]\ (ra+(-zo+za[ra])\ za'[ra]))}{n\ \sqrt{ra^2+(zo-za[ra])^2}\ (1+(za'[ra])^2)} + \frac{1}{\sqrt{1+(za'[ra])^2}}\sqrt{\left(1-\frac{no^2(ra+(-zo+za[ra])\ za'[ra])^2}{n^2\ (ra^2+(zo-za[ra])^2)\ (1+(za'[ra])^2)}\right)}$$

$$\varrho r[ra\_] := \frac{no(ra+(-zo+za[ra])\ za'[ra])}{n\ \sqrt{ra^2+(zo-za[ra])^2}\ (1+(za'[ra])^2)} - \frac{za'[ra]}{\sqrt{1+(za'[ra])^2}}\sqrt{\left(1-\frac{no^2(ra+(-zo+za[ra])\ za'[ra])^2)}{n^2\ (ra^2+(zo-za[ra])^2)\ (1+(za'[ra])^2)}\right)}$$

$$\vartheta[ra\_] := \frac{1}{n^2-1}\left(f[ra]\ n + ni^2\ (ra\ \varrho r[ra] + z[ra]\ \varrho z[ra]) - \mathrm{Sign}[n]\mathrm{Sign}[zi]\sqrt{\left(ni^2\ (f[ra]^2 + n^2\ (ra^2 + z[ra]^2) - ni^2\ (z[ra]\ \varrho r[ra] - ra\ \varrho z[ra])^2 + 2\ f[ra]\ n\ (ra\ \varrho r[ra] + z[ra]\ \varrho z[ra]))\right)}\right)$$

```
zb[ra_] := za[ra] + ϑ[ra] ϱz[ra];

rb[ra_] := ra + ϑ[ra] ϱr[ra];


(** Homotopy for Eikonal **)

θ1[ra_] := ArcTan[(ra)/(−zo+za[ra])];

θ2[ra_] := ArcTan[(rb[ra]−ra)/(zb[ra]−za[ra] )];

θ3[ra_] := ArcTan[rb[ra]/(zb[ra]−τ−zi)];

c1[ra_] := (zo+zi) + Sec[θ1[ra]](−zo+za[ra]);

c2[ra_] := (Sec[θ2[ra]](zb[ra]−za[ ra])/v)+c1[ra];
```

```
(** Setting variable's **)

c = 1;
v = c/n;

no = 1;
ni = no;
n = 1.5;

rmax = 24;
density = 4;
τ = 29;
zo = 60;
zi = 50;

(** Plot Section **)

SurfaceV := ParametricPlot[{{za[ra],ra},{zb[ra],rb[ra]}},{ra,−rmax−1,rmax+1},
        PlotStyle →{Black,Red}, AspectRatio→Automatic, AxesLabel →{r,z}]

LensV := Graphics[{EdgeForm[Directive[{Black,Thin}]],
        Directive[Gray,Opacity[0.3]],
        Polygon[Join[#[[1,1]],#[[2,1]],Reverse[#[[4,1]]],Reverse[#[[3,1]]]]&
        [Cases[SurfaceV,_Line,∞]]]}];

 WavesV := Table[{
          ParametricPlot[{
             Piecewise[{
                    {Cos[θ1[ra]](tt−(zo+zi))+zo,tt<=c1[ra]},
                    {v Cos[θ2[ra]](tt−c1[ra])+za[ra],c1[ra]<tt<c2[ra]},
                    {zb[ra]+(tt−c2[ra])Cos[θ3[ra]],tt>=c2[ra]}}],

             Piecewise[{
                    {Sin[θ1[ra]](tt−(zo+zi)),tt<=c1[ra]},
                    {v Sin[θ2[ra]](tt−c1[ra])+ra,c1[ra]<tt<c2[ra]},
                    {rb[ra]+(tt−c2[ra])Sin[θ3[ra]],tt>=c2[ra]}}]},
          {ra,−rmax,rmax},
          PlotRange→All,
          PlotStyle →{Blue, Thickness[0.002]}]},
          {tt,0,zo+n τ+zi+0.25,density}];

 RaysV := Table[{
        ParametricPlot[{
          Piecewise[{
               {Cos[θ1[ra]](tt−(zo+zi))+zo,tt<=c1[ra]},
               {v Cos[θ2[ra]](tt−c1[ra])+za[ra],c1[ra]<tt<c2[ra]},
               {zb[ra]+(tt−c2[ra])Cos[θ3[ra]],tt>=c2[ra]}}],

          Piecewise[{
               {Sin[θ1[ra]](tt−(zo+zi)),tt<=c1[ra]},
               {v Sin[θ2[ra]](tt−c1[ra])+ra,c1[ra]<tt<c2[ra]},
               {rb[ra]+(tt−c2[ra])Sin[θ3[ra]],tt>=c2[ra]}}]},

        {tt,0,zo+ n τ+zi+0.25},
        PlotRange→All,PlotStyle →{Lighter[Purple], Thickness[0.002]}]},
        {ra,−rmax,rmax,density}];


 Show[OvalSurfaceV,SurfaceV,LensV,WavesV,RaysV,
    PlotRange→All,ImageSize→{640,Automatic}]
```

12.4.8 Case 8: Virtual finite object—virtual finite image

```
" ClearAll["Global'*"]
(** Definition of set variables **)
```

$$za[ra\_] := -(29-Sqrt[29^2-ra^2])$$

$$f[ra\_] := -no\ zo + n\ \tau + ni\ zi + Sign[zo]\ \sqrt{ra^2+(zo-za[ra])^2}$$

$$z[ra\_] := za[ra] - \tau - zi$$

$$\varrho z[ra\_] := \frac{no(za'[ra]\ (ra+(-zo+za[ra])\ za'[ra]))}{n\ \sqrt{ra^2+(zo-za[ra])^2}\ (1+(za'[ra])^2)} + \frac{1}{\sqrt{1+(za'[ra])^2}}\sqrt{\left(1-\frac{no^2(ra+(-zo+za[ra])\ za'[ra])^2}{n^2\ (ra^2+(zo-za[ra])^2)\ (1+(za'[ra])^2)}\right)}$$

$$\varrho r[ra\_] := \frac{no(ra+(-zo+za[ra])\ za'[ra])}{n\ \sqrt{ra^2+(zo-za[ra])^2}\ (1+(za'[ra])^2)} - \frac{za'[ra]}{\sqrt{1+(za'[ra])^2}}\sqrt{\left(1-\frac{no^2(ra+(-zo+za[ra])\ za'[ra])^2)}{n^2\ (ra^2+(zo-za[ra])^2)\ (1+(za'[ra])^2)}\right)}$$

$$\vartheta[ra\_] := \frac{1}{n^2-1}\Big(f[ra]\ n + ni^2\ (ra\ \varrho r[ra] + z[ra]\ \varrho z[ra]) + Sign[n]Sign[zi]$$

$$\sqrt{\Big(ni^2\ (f[ra]^2 + n^2\ (ra^2 + z[ra]^2) - ni^2\ (z[ra]\ \varrho r[ra] - ra\ \varrho z[ra])^2}$$

```
        + 2 f[ra] n (ra ϱr[ra] + z[ra] ϱz[ra]))))

zb[ra_] := za[ra] + ϑ[ra] ϱz[ra];

rb[ra_] := ra + ϑ[ra] ϱr[ra];


(** Homotopy for Eikonal **)


θ1[ra_] := ArcTan[(ra)/(−zo+za[ra])];

θ2[ra_] := ArcTan[(rb[ra]−ra)/(zb[ra]−za[ra] )];

θ3[ra_] := ArcTan[rb[ra]/(zb[ra]−τ−zi)];

c1[ra_] := (zo + nτ) + Sec[θ1[ra]](−zo+za[ra]);

c2[ra_] := (Sec[θ2[ra]](zb[ra]−za[ ra])/v)+c1[ra];


(** Setting variable's **)

c = 1;
v = c/n;

no = 1;
ni = no;
n = 1.5;

rmax = 28;
density = 4;
τ = 29;
zo = 50;
zi = −25;


(** Plot Section **)

SurfaceV := ParametricPlot[{{za[ra],ra},{zb[ra],rb[ra]}},{ra,−rmax−1,rmax+1},
        PlotStyle →{Black,Red}, AspectRatio→Automatic, AxesLabel →{r,z}]

OvalSurfaceV := ParametricPlot[{{za[ra],ra},{zb[ra],rb[ra]}},
        {ra,−rmax−500,rmax+500},
        PlotStyle →{Directive[{Black,Dashed}],Directive[{Red,Dashed}]}]

LensV := Graphics[{EdgeForm[Directive[{Black,Thin}]],
        Directive[Gray,Opacity[0.3]],
        Polygon[Join[#[[1,1]],#[[2,1]],Reverse[#[[4,1]]],Reverse[#[[3,1]]]]&
        [Cases[SurfaceV,_Line,∞]]]}];
```

```
WavesV := Table[{
            ParametricPlot[{
               Piecewise[{
                     {Cos[θ1[ra]](tt−(zo + nτ))+zo,tt<=c1[ra]},
                     {v Cos[θ2[ra]](tt−c1[ra])+za[ra],c1[ra]<tt<c2[ra]},
                     {zb[ra]+(tt−c2[ra])Cos[θ3[ra]],tt>=c2[ra]}}],

               Piecewise[{
                     {Sin[θ1[ra]](tt−(zo + nτ)),tt<=c1[ra]},
                     {v Sin[θ2[ra]](tt−c1[ra])+ra,c1[ra]<tt<c2[ra]},
                     {rb[ra]+(tt−c2[ra])Sin[θ3[ra]],tt>=c2[ra]}}]},
            {ra,−rmax,rmax},
            PlotRange→All,
            PlotStyle →{Blue, Thickness[0.002]}]},
            {tt,0,zo+n τ − n zi+0.25,density}];

 RaysV := Table[{
         ParametricPlot[{
            Piecewise[{
                  {Cos[θ1[ra]](tt−(zo + nτ))+zo,tt<=c1[ra]},
                  {v Cos[θ2[ra]](tt−c1[ra])+za[ra],c1[ra]<tt<c2[ra]},
                  {zb[ra]+(tt−c2[ra])Cos[θ3[ra]],tt>=c2[ra]}}],

            Piecewise[{
                  {Sin[θ1[ra]](tt−(zo + nτ)),tt<=c1[ra]},
                  {v Sin[θ2[ra]](tt−c1[ra])+ra,c1[ra]<tt<c2[ra]},
                  {rb[ra]+(tt−c2[ra])Sin[θ3[ra]],tt>=c2[ra]}}]},

         {tt,0,zo+ n τ − n zi+0.25},
         PlotRange→All,PlotStyle →{Lighter[Purple], Thickness[0.002]}]},
         {ra,−rmax,rmax,density}];


Show[OvalSurfaceV,SurfaceV,LensV,WavesV,RaysV,
    PlotRange→All,ImageSize→{640,Automatic}]
```

12.4.9 Case 9: Real infinite object—real infinite image

"

```
(** Definition of set variables **)

za[ra_] := -(29-Sqrt[29^2-ra^2])
```

$$\varrho z[ra\_] := \frac{1}{n(1+za'[ra]^2)^{3/2}}\sqrt{\frac{n^2+(-1+n^2)\ za'[ra]^2}{n^2\ (1+za'[ra]^2)}} + \frac{za'[ra]^2}{n\ (1+za'[ra]^2)^{3/2}}\left(\sqrt{1+za'[ra]^2} + n\sqrt{\frac{n^2+(-1+n^2)\ za'[ra]^2}{n^2\ (1+za'[ra]^2)}}\right)$$

$$\varrho r[ra\_] := \frac{za'[ra]}{n\ (1+za'[ra]^2)^{3/2}}\left(\sqrt{1+za'[ra]^2} - n\sqrt{\frac{n^2+(-1+n^2)\ za'[ra]^2}{n^2\ (1+za'[ra]^2)}} - n\ za'[ra]^2\sqrt{\frac{n^2+(-1+n^2)\ za'[ra]^2}{n^2\ (1+za'[ra]^2)}}\right);$$

```
ϑ[ra_] := ((-1+n)τ -zo+za[ra]+Sign[n]Sign[zo]Sqrt[ra^2+(za[ra]-zo)^2])/(n
-ϱz[ra]);
zb[ra_] := za[ra] + ϑ[ra] ϱz[ra];

rb[ra_] := ra + ϑ[ra] ϱr[ra];


(** Homotopy for Eikonal **)

θ1[ra_] := ArcTan[Limit[(ra)/(-zo+za[ra]),zo→-Infinity]];

θ2[ra_] := ArcTan[(rb[ra]-ra)/(zb[ra]-za[ra])];

θ3[ra_] := ArcTan[Limit[rb[ra]/(zb[ra]-τ-zi),zi→Infinity]];

c1[ra_] := Sec[θ1[ra]](-zo+za[ra]);

c2[ra_] := (Sec[θ2[ra]](zb[ra]-za[ra])/v)+c1[ra];


(** Setting variable's **)

c = 1;
v = c/n;

no = 1;
ni = no;
n = 1.5;

rmax = 28;
density = 4;
τ = 29;
```

```
(*zo --> -Infinity*)
Zo = -55;
(* zi --> Infinity*)
Zi = 55;


(** Plot Section **)

SurfaceV := ParametricPlot[{{za[ra],ra},{zb[ra],rb[ra]}},{ra,-rmax-1,rmax+1},
          PlotStyle →{Black,Red}, AspectRatio→Automatic, AxesLabel →{r,z}]


LensV := Graphics[{EdgeForm[Directive[{Black,Thin}]],
          Directive[Gray,Opacity[0.3]],
          Polygon[Join[#[[1,1]],#[[2,1]],Reverse[#[[4,1]]],Reverse[#[[3,1]]]]&
          [Cases[SurfaceV,_Line,∞]]]}];

 WavesV := Table[{
           ParametricPlot[{
               Piecewise[{
                   {-tt,tt<=c1[ra]},
                   {v (tt-c1[ra])+za[ra],c1[ra]<tt<c2[ra]},
                   {zb[ra]+(tt-c2[ra]),tt>=c2[ra]}}],

               Piecewise[{
                   {ra,tt<=c1[ra]},
                   {v ra,c1[ra]<tt<c2[ra]},
                   {rb[ra],tt>=c2[ra]}}]},

           {ra,-rmax,rmax},
           PlotRange→All,
           PlotStyle →{Blue, Thickness[0.002]}]},
           {tt,0,-Zo+nτ + Zi+0.25,density}];

 RaysV := Table[{
         ParametricPlot[{
           Piecewise[{
               {-tt,tt<=c1[ra]},
               {v (tt-c1[ra])+za[ra],c1[ra]<tt<c2[ra]},
               {zb[ra]+(tt-c2[ra]),tt>=c2[ra]}}],

           Piecewise[{
               {ra,tt<=c1[ra]},
               {v ra,c1[ra]<tt<c2[ra]},
               {rb[ra],tt>=c2[ra]}}]},

         {tt,0,-Zo+ n τ + zi+0.25},
         PlotRange→All,PlotStyle →{Lighter[Purple], Thickness[0.002]}]},
         {ra,-rmax,rmax,density}];


 Show[SurfaceV,LensV,WavesV,RaysV,
     PlotRange→All,ImageSize→{640,Automatic}]
```

www.ingramcontent.com/pod-product-compliance
Lightning Source LLC
Chambersburg PA
CBHW080515220326
41599CB00032B/6084
9780750364256